Ford Mustang
Automotive Repair Manual

by Mike Stubblefield
and John H Haynes
Member of the Guild of Motoring Writers

Models covered:
Ford Mustang models
2005 through 2007

*Does not include information specific to
Shelby GT500 Cobra models*

ABCDE
FGHIJ
KLMNO
PQRST

Haynes Publishing Group
Sparkford Nr Yeovil
Somerset BA22 7JJ England

Haynes North America, Inc
861 Lawrence Drive
Newbury Park
California 91320 USA

Acknowledgements

Technical writers who contributed to this project include John Wegmann, Rob Maddox and Joe L. Hamilton. Wiring diagrams provided exclusively for Haynes North America, Inc by Valley Forge Technical Information Services.

A book in the Haynes Automotive Repair Manual Series

Printed in the U.S.A.

ISBN-13: 978-1-56392-661-7
ISBN-10: 1-56392-661-X

Library of Congress Control Number 2007920605

Contents

Haynes mechanic with a 2005 Mustang GT

About this manual

Its purpose

The purpose of this manual is to help you get the best value from your vehicle. It can do so in several ways. It can help you decide what work must be done, even if you choose to have it done by a dealer service department or a repair shop; it provides information and procedures for routine maintenance and servicing; and it offers diagnostic and repair procedures to follow when trouble occurs.

We hope you use the manual to tackle the work yourself. For many simpler jobs, doing it yourself may be quicker than arranging an appointment to get the vehicle into a shop and making the trips to leave it and pick it up. More importantly, a lot of money can be saved by avoiding the expense the shop must pass on to you to cover its labor and overhead costs. An added benefit is the sense of satisfaction and accomplishment that you feel after doing the job yourself.

Using the manual

The manual is divided into Chapters. Each Chapter is divided into numbered Sections, which are headed in bold type between horizontal lines. Each Section consists of consecutively numbered paragraphs.

At the beginning of each numbered Section you will be referred to any illustrations which apply to the procedures in that Section. The reference numbers used in illustration captions pinpoint the pertinent Section and the Step within that Section. That is, illustration 3.2 means the illustration refers to Section 3 and Step (or paragraph) 2 within that Section.

Procedures, once described in the text, are not normally repeated. When it's necessary to refer to another Chapter, the reference will be given as Chapter and Section number. Cross references given without use of the word "Chapter" apply to Sections and/or paragraphs in the same Chapter. For example, "see Section 8" means in the same Chapter.

References to the left or right side of the vehicle assume you are sitting in the driver's seat, facing forward.

Even though we have prepared this manual with extreme care, neither the publisher nor the author can accept responsibility for any errors in, or omissions from, the information given.

NOTE

A **Note** provides information necessary to properly complete a procedure or information which will make the procedure easier to understand.

CAUTION

A **Caution** provides a special procedure or special steps which must be taken while completing the procedure where the Caution is found. Not heeding a Caution can result in damage to the assembly being worked on.

WARNING

A **Warning** provides a special procedure or special steps which must be taken while completing the procedure where the Warning is found. Not heeding a Warning can result in personal injury.

Introduction

The Ford Mustang is available in two-door sport coupe or convertible body styles and have a conventional front engine/rear-wheel drive layout.

The available engines on the models covered by this manual are: a Single Over-Head Cam two-valves per cylinder (SOHC-2V) 4.0L V6, and a Dual Over-Head Cam three-valves per cylinder (DOHC-3V) 4.6L V8. All models use an electronically controlled Sequential Multiport Fuel Injection (SFI) system.

Power from the engine is transferred through either a five- or six-speed manual or four speed automatic transmission and a driveshaft to the differential which is mounted in the solid rear axle assembly. Axles inside the assembly carry power from the differential to the rear wheels.

Suspension is independent in the front, utilizing MacPherson struts and control arms. The rear suspension features coil springs and shock absorbers.

The steering gear is a power assisted rack-and-pinion type that is mounted to the front of the engine crossmember with rubber insulators.

The brakes are disc at the front and the rear with vacuum assist as standard equipment. An Anti-lock Brake System (ABS) is available as an option on base models, and is standard equipment on GT models.

Vehicle identification numbers

Modifications are a continuing and unpublicized process in vehicle manufacturing. Since spare parts lists and manuals are compiled on a numerical basis, the individual vehicle numbers are necessary to correctly identify the component required.

Vehicle Identification Number (VIN)

This very important identification number is stamped on a plate attached to the dashboard inside the windshield on the driver's side of the vehicle (see illustration). The VIN also appears on the Vehicle Certificate of Title and Registration. It contains information such as where and when the vehicle was manufactured, the model year and the body style.

VIN engine and model year codes

Two particularly important pieces of information found in the VIN are the engine code and the model year code. Counting from the left, the engine code letter designation is the 8th digit and the model year code letter designation is the 10th digit.

On the models covered by this manual the engine codes are:

N .... 4.0L SOHC V6
H .... 4.6L DOHC V8

On the models covered by this manual the model year codes are:

5 2005
6 2006
7 2007

Vehicle Certification Label

The Vehicle Certification Label is attached to the driver's side door pillar (see illustration). Information on this label includes the name of the manufacturer, the month and year of production, as well as information on the options with which it is equipped. This label is especially useful for matching the color and type of paint for repair work.

Engine identification number

Labels containing the engine code, engine number and build date can be found on the valve cover. The engine number is also stamped onto a machined pad on the external surface of the engine block.

Automatic transmission identification number

The automatic transmission ID number is affixed to a label on the right side of the case (see illustration).

Manual transmission identification number

The manual transmission ID number is stamped on a tag which is bolted to the driver's side of the bellhousing.

Vehicle Emissions Control Information label

This label is found in the engine compartment. See Chapter 6 for more information on this label.

The VIN number is visible through the windshield on the driver's side

The vehicle certification label is affixed the to the driver's side door pillar

The automatic transmission identification tag is located on the right side of the transmission

Recall information

Vehicle recalls are carried out by the manufacturer in the rare event of a possible safety-related defect. The vehicle's registered owner is contacted at the address on file at the Department of Motor Vehicles and given the details of the recall. Remedial work is carried out free of charge at a dealer service department.

If you are the new owner of a used vehicle which was subject to a recall and you want to be sure that the work has been carried out, it's best to contact a dealer service department and ask about your individual vehicle - you'll need to furnish them your Vehicle Identification Number (VIN).

The table below is based on information provided by the National Highway Traffic Safety Administration (NHTSA), the body which oversees vehicle recalls in the United States. The recall database is updated constantly. For the latest information on vehicle recalls, check the NHTSA website at www.nhtsa.gov, or call the NHTSA hotline at 1-888-327-4236.

Recall date	Recall campaign number	Model(s) affected	Concern
Aug 7, 2006	06E071000	2005/2006 Mustang GT	Certain aftermarket Roush branded front struts as a stand alone component, part no. 401297, and Roush branded front struts as a suspension kit, part no. 401296, sold for use on 2005 and 2006 Ford Mustang GT vehicles. The front stabilizer bar attachment bracket on the front strut can separate from the strut housing due to inconsistent welding. This could result in a clunking noise and or poor handling in extreme or emergency maneuvers. It could also damage the tire sidewall, resulting in deflation and loss of control of the vehicle
Oct 16, 2006	06E089000	2005/2006/2007 Mustang	Certain aftermarket Eibach branded front struts only, part no. 35101.8001, and Eibach branded front struts as suspension kits, part nos. 35101.840, 35101.680, 35101.780, 35100.680, 35100.780, 4.10135.680, and 4.10035.780, sold for use on 2005 through 2007 Ford Mustang vehicles. The front stabilizer bar attachment bracket on the front strut can separate from the strut housing due to inconsistent welding. This could result in a clunking noise and or poor handling in extreme or emergency maneuvers. It could also damage the tire sidewall, resulting in deflation and loss of control of the vehicle

Buying parts

Replacement parts are available from many sources, which generally fall into one of two categories - authorized dealer parts departments and independent retail auto parts stores. Our advice concerning these parts is as follows:

Retail auto parts stores: Good auto parts stores will stock frequently needed components which wear out relatively fast, such as clutch components, exhaust systems, brake parts, tune-up parts, etc. These stores often supply new or reconditioned parts on an exchange basis, which can save a considerable amount of money. Discount auto parts stores are often very good places to buy materials and parts needed for general vehicle maintenance such as oil, grease, filters, spark plugs, belts, touch-up paint, bulbs, etc. They also usually sell tools and general accessories, have convenient hours, charge lower prices and can often be found not far from home.

Authorized dealer parts department: This is the best source for parts which are unique to the vehicle and not generally available elsewhere (such as major engine parts, transmission parts, trim pieces, etc.).

Warranty information: If the vehicle is still covered under warranty, be sure that any replacement parts purchased - regardless of the source - do not invalidate the warranty!

To be sure of obtaining the correct parts, have engine and chassis numbers available and, if possible, take the old parts along for positive identification.

Maintenance techniques, tools and working facilities

Maintenance techniques

There are a number of techniques involved in maintenance and repair that will be referred to throughout this manual. Application of these techniques will enable the home mechanic to be more efficient, better organized and capable of performing the various tasks properly, which will ensure that the repair job is thorough and complete.

Fasteners

Fasteners are nuts, bolts, studs and screws used to hold two or more parts together. There are a few things to keep in mind when working with fasteners. Almost all of them use a locking device of some type, either a lockwasher, locknut, locking tab or thread adhesive. All threaded fasteners should be clean and straight, with undamaged threads and undamaged corners on the hex head where the wrench fits. Develop the habit of replacing all damaged nuts and bolts with new ones. Special locknuts with nylon or fiber inserts can only be used once. If they are removed, they lose their locking ability and must be replaced with new ones.

Rusted nuts and bolts should be treated with a penetrating fluid to ease removal and prevent breakage. Some mechanics use turpentine in a spout-type oil can, which works quite well. After applying the rust penetrant, let it work for a few minutes before trying to loosen the nut or bolt. Badly rusted fasteners may have to be chiseled or sawed off or removed with a special nut breaker, available at tool stores.

If a bolt or stud breaks off in an assembly, it can be drilled and removed with a special tool commonly available for this purpose. Most automotive machine shops can perform this task, as well as other repair procedures, such as the repair of threaded holes that have been stripped out.

Flat washers and lockwashers, when removed from an assembly, should always be replaced exactly as removed. Replace any damaged washers with new ones. Never use a lockwasher on any soft metal surface (such as aluminum), thin sheet metal or plastic.

Fastener sizes

For a number of reasons, automobile manufacturers are making wider and wider use of metric fasteners. Therefore, it is important to be able to tell the difference between standard (sometimes called U.S. or SAE) and metric hardware, since they cannot be interchanged.

All bolts, whether standard or metric, are sized according to diameter, thread pitch and length. For example, a standard 1/2 - 13 x 1 bolt is 1/2 inch in diameter, has 13 threads per inch and is 1 inch long. An M12 - 1.75 x 25 metric bolt is 12 mm in diameter, has a thread pitch of 1.75 mm (the distance between threads) and is 25 mm long. The two bolts are nearly identical, and easily confused, but they are not interchangeable.

In addition to the differences in diameter, thread pitch and length, metric and standard bolts can also be distinguished by examining the bolt heads. To begin with, the distance across the flats on a standard bolt head is measured in inches, while the same dimension on a metric bolt is sized in millimeters (the same is true for nuts). As a result, a standard wrench should not be used on a metric bolt and a metric wrench should not be used on a standard bolt. Also, most standard bolts have slashes radiating out from the center of the head to denote the grade or strength of the bolt, which is an indication of the amount of torque that can be applied to it. The greater the number of slashes, the greater the strength of the bolt. Grades 0 through 5 are commonly used on automobiles. Metric bolts have a property class (grade) number, rather than a slash, molded into their heads to indicate bolt strength. In this case, the higher the number, the stronger the bolt. Property class numbers 8.8, 9.8 and 10.9 are commonly used on automobiles.

Strength markings can also be used to distinguish standard hex nuts from metric hex nuts. Many standard nuts have dots stamped into one side, while metric nuts are marked with a number. The greater the number of

dots, or the higher the number, the greater the strength of the nut.

Metric studs are also marked on their ends according to property class (grade). Larger studs are numbered (the same as metric bolts), while smaller studs carry a geometric code to denote grade.

It should be noted that many fasteners, especially Grades 0 through 2, have no distinguishing marks on them. When such is the case, the only way to determine whether it is standard or metric is to measure the thread pitch or compare it to a known fastener of the same size.

Standard fasteners are often referred to as SAE, as opposed to metric. However, it should be noted that SAE technically refers to a non-metric fine thread fastener only. Coarse thread non-metric fasteners are referred to as USS sizes.

Since fasteners of the same size (both standard and metric) may have different

strength ratings, be sure to reinstall any bolts, studs or nuts removed from your vehicle in their original locations. Also, when replacing a fastener with a new one, make sure that the new one has a strength rating equal to or greater than the original.

Tightening sequences and procedures

Most threaded fasteners should be tightened to a specific torque value (torque is the twisting force applied to a threaded component such as a nut or bolt). Overtightening the fastener can weaken it and cause it to break, while undertightening can cause it to eventually come loose. Bolts, screws and studs, depending on the material they are made of and their thread diameters, have specific torque values, many of which are noted in the Specifications at the beginning of each Chapter. Be sure to follow the torque recommendations closely. For fasteners not assigned a

Grade 1 or 2   Grade 5   Grade 8

Bolt strength marking (standard/SAE/USS; bottom - metric)

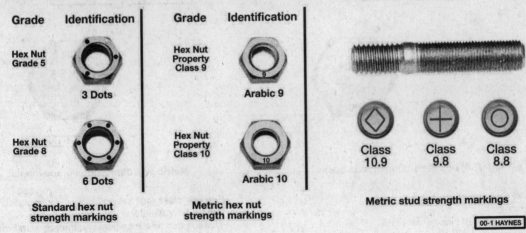

Grade	Identification
Hex Nut Grade 5	3 Dots
Hex Nut Grade 8	6 Dots

Standard hex nut strength markings

Grade	Identification
Hex Nut Property Class 9	Arabic 9
Hex Nut Property Class 10	Arabic 10

Metric hex nut strength markings

Class 10.9   Class 9.8   Class 8.8

Metric stud strength markings

00-1 HAYNES

specific torque, a general torque value chart is presented here as a guide. These torque values are for dry (unlubricated) fasteners threaded into steel or cast iron (not aluminum). As was previously mentioned, the size and grade of a fastener determine the amount of torque that can safely be applied to it. The figures listed here are approximate for Grade 2 and Grade 3 fasteners. Higher grades can tolerate higher torque values.

Fasteners laid out in a pattern, such as cylinder head bolts, oil pan bolts, differential cover bolts, etc., must be loosened or tightened in sequence to avoid warping the component. This sequence will normally be shown in the appropriate Chapter. If a specific pattern is not given, the following procedures can be used to prevent warping.

Initially, the bolts or nuts should be assembled finger-tight only. Next, they should be tightened one full turn each, in a crisscross or diagonal pattern. After each one has been tightened one full turn, return to the first one and tighten them all one-half turn, following the same pattern. Finally, tighten each of them one-quarter turn at a time until each fastener has been tightened to the proper torque. To loosen and remove the fasteners, the procedure would be reversed.

Component disassembly

Component disassembly should be done with care and purpose to help ensure that

Metric thread sizes	Ft-lbs	Nm
M-6	6 to 9	9 to 12
M-8	14 to 21	19 to 28
M-10	28 to 40	38 to 54
M-12	50 to 71	68 to 96
M-14	80 to 140	109 to 154

Pipe thread sizes		
1/8	5 to 8	7 to 10
1/4	12 to 18	17 to 24
3/8	22 to 33	30 to 44
1/2	25 to 35	34 to 47

U.S. thread sizes		
1/4 - 20	6 to 9	9 to 12
5/16 - 18	12 to 18	17 to 24
5/16 - 24	14 to 20	19 to 27
3/8 - 16	22 to 32	30 to 43
3/8 - 24	27 to 38	37 to 51
7/16 - 14	40 to 55	55 to 74
7/16 - 20	40 to 60	55 to 81
1/2 - 13	55 to 80	75 to 108

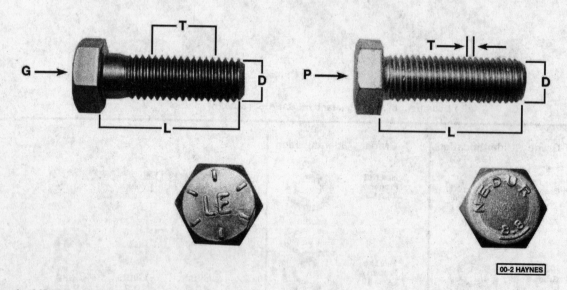

Standard (SAE and USS) bolt dimensions/grade marks

G   Grade marks (bolt strength)
L   Length (in inches)
T   Thread pitch (number of threads per inch)
D   Nominal diameter (in inches)

Metric bolt dimensions/grade marks

P   Property class (bolt strength)
L   Length (in millimeters)
T   Thread pitch (distance between threads in millimeters)
D   Diameter

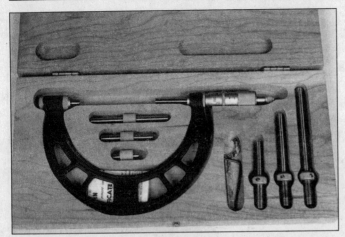

Micrometer set

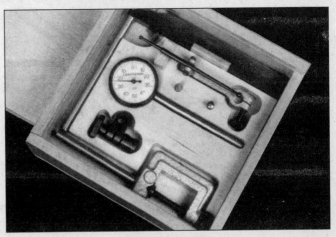

Dial indicator set

the parts go back together properly. Always keep track of the sequence in which parts are removed. Make note of special characteristics or marks on parts that can be installed more than one way, such as a grooved thrust washer on a shaft. It is a good idea to lay the disassembled parts out on a clean surface in the order that they were removed. It may also be helpful to make sketches or take instant photos of components before removal.

When removing fasteners from a component, keep track of their locations. Sometimes threading a bolt back in a part, or putting the washers and nut back on a stud, can prevent mix-ups later. If nuts and bolts cannot be returned to their original locations, they should be kept in a compartmented box or a series of small boxes. A cupcake or muffin tin is ideal for this purpose, since each cavity can hold the bolts and nuts from a particular area (i.e. oil pan bolts, valve cover bolts, engine mount bolts, etc.). A pan of this type is especially helpful when working on assemblies with very small parts, such as the carburetor, alternator, valve train or interior dash and trim pieces. The cavities can be marked with paint or tape to identify the contents.

Whenever wiring looms, harnesses or connectors are separated, it is a good idea to identify the two halves with numbered pieces of masking tape so they can be easily reconnected.

Gasket sealing surfaces

Throughout any vehicle, gaskets are used to seal the mating surfaces between two parts and keep lubricants, fluids, vacuum or pressure contained in an assembly.

Many times these gaskets are coated with a liquid or paste-type gasket sealing compound before assembly. Age, heat and pressure can sometimes cause the two parts to stick together so tightly that they are very difficult to separate. Often, the assembly can be loosened by striking it with a soft-face hammer near the mating surfaces. A regular hammer can be used if a block of wood is placed between the hammer and the part. Do

not hammer on cast parts or parts that could be easily damaged. With any particularly stubborn part, always recheck to make sure that every fastener has been removed.

Avoid using a screwdriver or bar to pry apart an assembly, as they can easily mar the gasket sealing surfaces of the parts, which must remain smooth. If prying is absolutely necessary, use an old broom handle, but keep in mind that extra clean up will be necessary if the wood splinters.

After the parts are separated, the old gasket must be carefully scraped off and the gasket surfaces cleaned. Stubborn gasket material can be soaked with rust penetrant or treated with a special chemical to soften it so it can be easily scraped off. **Caution:** *Never use gasket removal solutions or caustic chemicals on plastic or other composite components.* A scraper can be fashioned from a piece of copper tubing by flattening and sharpening one end. Copper is recommended because it is usually softer than the surfaces to be scraped, which reduces the chance of gouging the part. Some gaskets can be removed with a wire brush, but regardless of the method used, the mating surfaces must be left clean and smooth. If for some reason the gasket surface is gouged, then a gasket sealer thick enough to fill scratches will have to be used during reassembly of the components. For most applications, a non-drying (or semi-drying) gasket sealer should be used.

Hose removal tips

Warning: *If the vehicle is equipped with air conditioning, do not disconnect any of the A/C hoses without first having the system depressurized by a dealer service department or a service station.*

Hose removal precautions closely parallel gasket removal precautions. Avoid scratching or gouging the surface that the hose mates against or the connection may leak. This is especially true for radiator hoses. Because of various chemical reactions, the rubber in hoses can bond itself to the metal spigot that the hose fits over. To remove

a hose, first loosen the hose clamps that secure it to the spigot. Then, with slip-joint pliers, grab the hose at the clamp and rotate it around the spigot. Work it back and forth until it is completely free, then pull it off. Silicone or other lubricants will ease removal if they can be applied between the hose and the outside of the spigot. Apply the same lubricant to the inside of the hose and the outside of the spigot to simplify installation.

As a last resort (and if the hose is to be replaced with a new one anyway), the rubber can be slit with a knife and the hose peeled from the spigot. If this must be done, be careful that the metal connection is not damaged.

If a hose clamp is broken or damaged, do not reuse it. Wire-type clamps usually weaken with age, so it is a good idea to replace them with screw-type clamps whenever a hose is removed.

Tools

A selection of good tools is a basic requirement for anyone who plans to maintain and repair his or her own vehicle. For the owner who has few tools, the initial investment might seem high, but when compared to the spiraling costs of professional auto maintenance and repair, it is a wise one.

To help the owner decide which tools are needed to perform the tasks detailed in this manual, the following tool lists are offered: *Maintenance and minor repair, Repair/overhaul* and *Special.*

The newcomer to practical mechanics should start off with the *maintenance and minor repair* tool kit, which is adequate for the simpler jobs performed on a vehicle. Then, as confidence and experience grow, the owner can tackle more difficult tasks, buying additional tools as they are needed. Eventually the basic kit will be expanded into the *repair and overhaul* tool set. Over a period of time, the experienced do-it-yourselfer will assemble a tool set complete enough for most repair and overhaul procedures and will add tools from the special category when it is felt that the expense is justified by the frequency of use.

Dial caliper

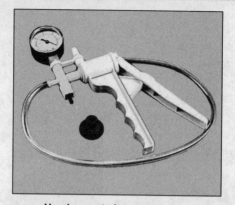

Hand-operated vacuum pump

Timing light

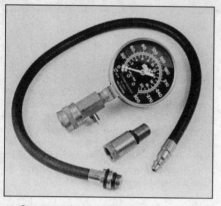

Compression gauge with spark plug
hole adapter

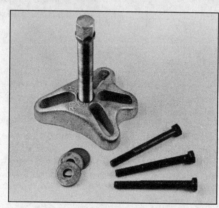

Damper/steering wheel puller

General purpose puller

Hydraulic lifter removal tool

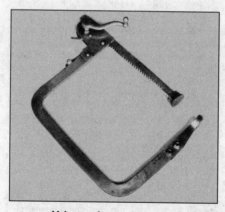

Valve spring compressor

Valve spring compressor

Ridge reamer

Piston ring groove cleaning tool

Ring removal/installation tool

Ring compressor

Cylinder hone

Brake hold-down spring tool

Torque angle gauge

Clutch plate alignment tool

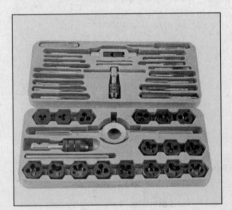

Tap and die set

Maintenance and minor repair tool kit

The tools in this list should be considered the minimum required for performance of routine maintenance, servicing and minor repair work. We recommend the purchase of combination wrenches (box-end and open-end combined in one wrench). While more expensive than open end wrenches, they offer the advantages of both types of wrench.

Combination wrench set (1/4-inch to
1 inch or 6 mm to 19 mm)
Adjustable wrench, 8 inch
Spark plug wrench with rubber insert
Spark plug gap adjusting tool
Feeler gauge set
Brake bleeder wrench
Standard screwdriver (5/16-inch x
6 inch)
Phillips screwdriver (No. 2 x 6 inch)
Combination pliers - 6 inch
Hacksaw and assortment of blades
Tire pressure gauge
Grease gun
Oil can
Fine emery cloth
Wire brush
Battery post and cable cleaning tool
Oil filter wrench
Funnel (medium size)
Safety goggles
Jackstands (2)
Drain pan

Note: If basic tune-ups are going to be part of routine maintenance, it will be necessary to purchase a good quality stroboscopic timing light and combination tachometer/dwell meter. Although they are included in the list of special tools, it is mentioned here because they are absolutely necessary for tuning most vehicles properly.

Repair and overhaul tool set

These tools are essential for anyone who plans to perform major repairs and are in addition to those in the maintenance and minor repair tool kit. Included is a comprehensive set of sockets which, though expensive, are invaluable because of their versatility, especially when various extensions and drives are available. We recommend the 1/2-inch drive over the 3/8-inch drive. Although the larger drive is bulky and more expensive, it has the capacity of accepting a very wide range of large sockets. Ideally, however, the mechanic should have a 3/8-inch drive set and a 1/2-inch drive set.

Socket set(s)
Reversible ratchet
Extension - 10 inch
Universal joint
Torque wrench (same size drive as
sockets)
Ball peen hammer - 8 ounce
Soft-face hammer (plastic/rubber)
Standard screwdriver (1/4-inch x 6 inch)

Standard screwdriver (stubby -
5/16-inch)
Phillips screwdriver (No. 3 x 8 inch)
Phillips screwdriver (stubby - No. 2)
Pliers - vise grip
Pliers - lineman's
Pliers - needle nose
Pliers - snap-ring (internal and external)
Cold chisel - 1/2-inch
Scribe
Scraper (made from flattened copper
tubing)
Centerpunch
Pin punches (1/16, 1/8, 3/16-inch)
Steel rule/straightedge - 12 inch
Allen wrench set (1/8 to 3/8-inch or
4 mm to 10 mm)
A selection of files
Wire brush (large)
Jackstands (second set)
Jack (scissor or hydraulic type)

Note: Another tool which is often useful is an electric drill with a chuck capacity of 3/8-inch and a set of good quality drill bits.

Special tools

The tools in this list include those which are not used regularly, are expensive to buy, or which need to be used in accordance with their manufacturer's instructions. Unless these tools will be used frequently, it is not very economical to purchase many of them. A consideration would be to split the cost and use between yourself and a friend or friends. In addition, most of these tools can be obtained from a tool rental shop on a temporary basis.

This list primarily contains only those tools and instruments widely available to the public, and not those special tools produced by the vehicle manufacturer for distribution to dealer service departments. Occasionally, references to the manufacturer's special tools are included in the text of this manual. Generally, an alternative method of doing the job without the special tool is offered. However, sometimes there is no alternative to their use. Where this is the case, and the tool cannot be purchased or borrowed, the work should be turned over to the dealer service department or an automotive repair shop.

> *Valve spring compressor*
> *Piston ring groove cleaning tool*
> *Piston ring compressor*
> *Piston ring installation tool*
> *Cylinder compression gauge*
> *Cylinder ridge reamer*
> *Cylinder surfacing hone*
> *Cylinder bore gauge*
> *Micrometers and/or dial calipers*
> *Hydraulic lifter removal tool*
> *Balljoint separator*
> *Universal-type puller*
> *Impact screwdriver*
> *Dial indicator set*
> *Stroboscopic timing light (inductive*
>   *pick-up)*
> *Hand operated vacuum/pressure pump*
> *Tachometer/dwell meter*
> *Universal electrical multimeter*
> *Cable hoist*
> *Brake spring removal and installation*
>   *tools*
> *Floor jack*

Buying tools

For the do-it-yourselfer who is just starting to get involved in vehicle maintenance and repair, there are a number of options available when purchasing tools. If maintenance and minor repair is the extent of the work to be done, the purchase of individual tools is satisfactory. If, on the other hand, extensive work is planned, it would be a good idea to purchase a modest tool set from one of the large retail chain stores. A set can usually be bought at a substantial savings over the individual tool prices, and they often come with a tool box. As additional tools are needed, add-on sets, individual tools and a larger tool box can be purchased to expand the tool selection. Building a tool set gradually allows the cost of the tools to be spread over a longer period of time and gives the mechanic the freedom to choose only those tools that will actually be used.

Tool stores will often be the only source of some of the special tools that are needed,

but regardless of where tools are bought, try to avoid cheap ones, especially when buying screwdrivers and sockets, because they won't last very long. The expense involved in replacing cheap tools will eventually be greater than the initial cost of quality tools.

Care and maintenance of tools

Good tools are expensive, so it makes sense to treat them with respect. Keep them clean and in usable condition and store them properly when not in use. Always wipe off any dirt, grease or metal chips before putting them away. Never leave tools lying around in the work area. Upon completion of a job, always check closely under the hood for tools that may have been left there so they won't get lost during a test drive.

Some tools, such as screwdrivers, pliers, wrenches and sockets, can be hung on a panel mounted on the garage or workshop wall, while others should be kept in a tool box or tray. Measuring instruments, gauges, meters, etc. must be carefully stored where they cannot be damaged by weather or impact from other tools.

When tools are used with care and stored properly, they will last a very long time. Even with the best of care, though, tools will wear out if used frequently. When a tool is damaged or worn out, replace it. Subsequent jobs will be safer and more enjoyable if you do.

How to repair damaged threads

Sometimes, the internal threads of a nut or bolt hole can become stripped, usually from overtightening. Stripping threads is an all-too-common occurrence, especially when working with aluminum parts, because aluminum is so soft that it easily strips out.

Usually, external or internal threads are only partially stripped. After they've been cleaned up with a tap or die, they'll still work. Sometimes, however, threads are badly damaged. When this happens, you've got three choices:

1) *Drill and tap the hole to the next suitable oversize and install a larger diameter bolt, screw or stud.*
2) *Drill and tap the hole to accept a threaded plug, then drill and tap the plug to the original screw size. You can also buy a plug already threaded to the original size. Then you simply drill a hole to the specified size, then run the threaded plug into the hole with a bolt and jam nut. Once the plug is fully seated, remove the jam nut and bolt.*
3) *The third method uses a patented thread repair kit like Heli-Coil or Slimsert. These*

easy-to-use kits are designed to repair damaged threads in straight-through holes and blind holes. Both are available as kits which can handle a variety of sizes and thread patterns. Drill the hole, then tap it with the special included tap. Install the Heli-Coil and the hole is back to its original diameter and thread pitch.

Regardless of which method you use, be sure to proceed calmly and carefully. A little impatience or carelessness during one of these relatively simple procedures can ruin your whole day's work and cost you a bundle if you wreck an expensive part.

Working facilities

Not to be overlooked when discussing tools is the workshop. If anything more than routine maintenance is to be carried out, some sort of suitable work area is essential.

It is understood, and appreciated, that many home mechanics do not have a good workshop or garage available, and end up removing an engine or doing major repairs outside. It is recommended, however, that the overhaul or repair be completed under the cover of a roof.

A clean, flat workbench or table of comfortable working height is an absolute necessity. The workbench should be equipped with a vise that has a jaw opening of at least four inches.

As mentioned previously, some clean, dry storage space is also required for tools, as well as the lubricants, fluids, cleaning solvents, etc. which soon become necessary.

Sometimes waste oil and fluids, drained from the engine or cooling system during normal maintenance or repairs, present a disposal problem. To avoid pouring them on the ground or into a sewage system, pour the used fluids into large containers, seal them with caps and take them to an authorized disposal site or recycling center. Plastic jugs, such as old antifreeze containers, are ideal for this purpose.

Always keep a supply of old newspapers and clean rags available. Old towels are excellent for mopping up spills. Many mechanics use rolls of paper towels for most work because they are readily available and disposable. To help keep the area under the vehicle clean, a large cardboard box can be cut open and flattened to protect the garage or shop floor.

Whenever working over a painted surface, such as when leaning over a fender to service something under the hood, always cover it with an old blanket or bedspread to protect the finish. Vinyl covered pads, made especially for this purpose, are available at auto parts stores.

Jacking and towing

Jacking

Warning: *The jack supplied with the vehicle should only be used for changing a tire or placing jackstands under the frame. Never work under the vehicle or start the engine while this jack is being used as the only means of support.*

The vehicle should be on level ground. Place the shift lever in Park, if you have an automatic, or Reverse if you have a manual transmission. Block the wheel diagonally opposite the wheel being changed. Set the parking brake.

Remove the spare tire and jack from stowage. Remove the wheel cover and trim ring (if so equipped) with the tapered end of the lug nut wrench by inserting and twisting the handle and then prying against the back of the wheel cover. Loosen the wheel lug nuts about 1/4-to-1/2 turn each.

Place the scissors-type jack under the side of the vehicle and adjust the jack height until it fits in the notch in the vertical rocker panel flange nearest the wheel to be changed. There is a front and rear jacking point on each side of the vehicle **(see illustration)**.

Turn the jack handle clockwise until the tire clears the ground. Remove the lug nuts and pull the wheel off. Replace it with the spare.

Install the lug nuts with the beveled edges facing in. Tighten them snugly. Don't attempt to tighten them completely until the vehicle is lowered or it could slip off the jack. Turn the jack handle counterclockwise to lower the vehicle. Remove the jack and tighten the lug nuts in a diagonal pattern.

Install the cover (and trim ring, if used) and be sure it's snapped into place all the way around.

Stow the tire, jack and wrench. Unblock the wheels.

Towing

As a general rule, the vehicle should be towed with the rear wheels off the ground. If they can't be raised, either place them on a dolly or disconnect the driveshaft from the differential. When a vehicle is towed with the rear wheels raised, the steering wheel must be clamped in the straight ahead position with a special device designed for use during towing. The ignition key must be in the OFF position, since the steering lock mechanism isn't strong enough to hold the front wheels straight while towing.

Vehicles equipped with an automatic transmission can be towed from the front only with all four wheels on the ground, provided that speeds don't exceed 35 mph and the distance is not over 50 miles. Before towing, check the transmission fluid level (see Chapter 1). If the level is below the HOT line on the dipstick, add fluid or use a towing dolly. Release the parking brake, put the transmission in Neutral and place the ignition key in the OFF position. There's no distance limitation when towing with either the rear wheels off the ground or the driveshaft disconnected, but don't exceed 50 mph.

Equipment specifically designed for towing should be used. It should be attached to the main structural members of the vehicle, not the bumpers or brackets.

Safety is a major consideration when towing and all applicable state and local laws must be obeyed. A safety chain system must be used at all times. Remember that power steering and power brakes will not work with the engine off.

Place the jack so it engages the notch in the rocker panel nearest the wheel to be raised

Booster battery (jump) starting

Observe the following precautions when using a booster battery to start a vehicle:

a) *Before connecting the booster battery, make sure the ignition switch is in the Off position.*
b) *Turn off the lights, heater and other electrical loads.*
c) *Your eyes should be shielded. Safety goggles are a good idea.*
d) *Make sure the booster battery is the same voltage as the dead one in the vehicle.*
e) *The two vehicles MUST NOT TOUCH each other!*
f) *Make sure the transmission is in Neutral (manual) or Park (automatic).*
g) *If the booster battery is not a maintenance-free type, remove the vent caps and lay a cloth over the vent holes.*

Connect one jumper lead between the positive (+) terminals of the two batteries **(see illustration)**. Connect the other jumper lead first to the negative (-) terminal of the booster battery, then to a good engine ground on the vehicle to be started.Attach the lead at least 18 inches from the battery, if possible. Make sure the that the jumper leads will not contact the fan, drivebelt or other moving parts of the engine.

Start the engine using the booster battery, then, with the engine running at idle speed, disconnect the jumper cables in the reverse order of connection.

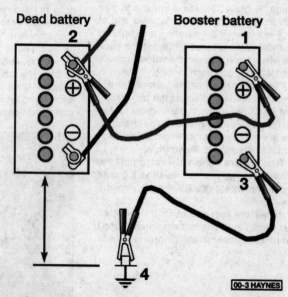

Make the booster battery cable connections in the numerical order shown (note that the negative cable of the booster battery is NOT attached to the negative terminal of the dead battery)

Automotive chemicals and lubricants

A number of automotive chemicals and lubricants are available for use during vehicle maintenance and repair. They include a wide variety of products ranging from cleaning solvents and degreasers to lubricants and protective sprays for rubber, plastic and vinyl.

Cleaners

Carburetor cleaner and choke cleaner is a strong solvent for gum, varnish and carbon. Most carburetor cleaners leave a dry-type lubricant film which will not harden or gum up. Because of this film it is not recommended for use on electrical components.

Brake system cleaner is used to remove brake dust, grease and brake fluid from the brake system, where clean surfaces are absolutely necessary. It leaves no residue and often eliminates brake squeal caused by contaminants.

Electrical cleaner removes oxidation, corrosion and carbon deposits from electrical contacts, restoring full current flow. It can also be used to clean spark plugs, carburetor jets, voltage regulators and other parts where an oil-free surface is desired.

Demoisturants remove water and moisture from electrical components such as alternators, voltage regulators, electrical connectors and fuse blocks. They are non-conductive and non-corrosive.

Degreasers are heavy-duty solvents used to remove grease from the outside of the engine and from chassis components. They can be sprayed or brushed on and, depending on the type, are rinsed off either with water or solvent.

Lubricants

Motor oil is the lubricant formulated for use in engines. It normally contains a wide variety of additives to prevent corrosion and reduce foaming and wear. Motor oil comes in various weights (viscosity ratings) from 0 to 50. The recommended weight of the oil depends on the season, temperature and the demands on the engine. Light oil is used in cold climates and under light load conditions. Heavy oil is used in hot climates and where high loads are encountered. Multi-viscosity oils are designed to have characteristics of both light and heavy oils and are available in a number of weights from 5W-20 to 20W-50.

Gear oil is designed to be used in differentials, manual transmissions and other areas where high-temperature lubrication is required.

Chassis and wheel bearing grease is a heavy grease used where increased loads and friction are encountered, such as for wheel bearings, balljoints, tie-rod ends and universal joints.

High-temperature wheel bearing grease is designed to withstand the extreme temperatures encountered by wheel bearings in disc brake equipped vehicles. It usually contains molybdenum disulfide (moly), which is a dry-type lubricant.

White grease is a heavy grease for metal-to-metal applications where water is a problem. White grease stays soft under both low and high temperatures (usually from -100 to +190-degrees F), and will not wash off or dilute in the presence of water.

Assembly lube is a special extreme pressure lubricant, usually containing moly, used to lubricate high-load parts (such as main and rod bearings and cam lobes) for initial start-up of a new engine. The assembly lube lubricates the parts without being squeezed out or washed away until the engine oiling system begins to function.

Silicone lubricants are used to protect rubber, plastic, vinyl and nylon parts.

Graphite lubricants are used where oils cannot be used due to contamination problems, such as in locks. The dry graphite will lubricate metal parts while remaining uncontaminated by dirt, water, oil or acids. It is electrically conductive and will not foul electrical contacts in locks such as the ignition switch.

Moly penetrants loosen and lubricate frozen, rusted and corroded fasteners and prevent future rusting or freezing.

Heat-sink grease is a special electrically non-conductive grease that is used for mounting electronic ignition modules where it is essential that heat is transferred away from the module.

Sealants

RTV sealant is one of the most widely used gasket compounds. Made from silicone, RTV is air curing, it seals, bonds, waterproofs, fills surface irregularities, remains flexible, doesn't shrink, is relatively easy to remove, and is used as a supplementary sealer with almost all low and medium temperature gaskets.

Anaerobic sealant is much like RTV in that it can be used either to seal gaskets or to form gaskets by itself. It remains flexible, is solvent resistant and fills surface imperfections. The difference between an anaerobic sealant and an RTV-type sealant is in the curing. RTV cures when exposed to air, while an anaerobic sealant cures only in the absence of air. This means that an anaerobic sealant cures only after the assembly of parts, sealing them together.

Thread and pipe sealant is used for sealing hydraulic and pneumatic fittings and vacuum lines. It is usually made from a Teflon compound, and comes in a spray, a paint-on liquid and as a wrap-around tape.

Chemicals

Anti-seize compound prevents seizing, galling, cold welding, rust and corrosion in fasteners. High-temperature ant-seize, usually made with copper and graphite lubricants, is used for exhaust system and exhaust manifold bolts.

Anaerobic locking compounds are used to keep fasteners from vibrating or working loose and cure only after installation, in the absence of air. Medium strength locking compound is used for small nuts, bolts and screws that may be removed later. High-strength locking compound is for large nuts, bolts and studs which aren't removed on a regular basis.

Oil additives range from viscosity index improvers to chemical treatments that claim to reduce internal engine friction. It should be noted that most oil manufacturers caution against using additives with their oils.

Gas additives perform several functions, depending on their chemical makeup. They usually contain solvents that help dissolve gum and varnish that build up on carburetor, fuel injection and intake parts. They also serve to break down carbon deposits that form on the inside surfaces of the combustion chambers. Some additives contain upper cylinder lubricants for valves and piston rings, and others contain chemicals to remove condensation from the gas tank.

Miscellaneous

Brake fluid is specially formulated hydraulic fluid that can withstand the heat and pressure encountered in brake systems. Care must be taken so this fluid does not come in contact with painted surfaces or plastics. An opened container should always be resealed to prevent contamination by water or dirt.

Weatherstrip adhesive is used to bond weatherstripping around doors, windows and trunk lids. It is sometimes used to attach trim pieces.

Undercoating is a petroleum-based, tar-like substance that is designed to protect metal surfaces on the underside of the vehicle from corrosion. It also acts as a sound-deadening agent by insulating the bottom of the vehicle.

Waxes and polishes are used to help protect painted and plated surfaces from the weather. Different types of paint may require the use of different types of wax and polish. Some polishes utilize a chemical or abrasive cleaner to help remove the top layer of oxidized (dull) paint on older vehicles. In recent years many non-wax polishes that contain a wide variety of chemicals such as polymers and silicones have been introduced. These non-wax polishes are usually easier to apply and last longer than conventional waxes and polishes.

Conversion factors

Length (distance)

Inches (in)	X	25.4	= Millimeters (mm)	X 0.0394	= Inches (in)
Feet (ft)	X	0.305	= Meters (m)	X 3.281	= Feet (ft)
Miles	X	1.609	= Kilometers (km)	X 0.621	= Miles

Volume (capacity)

Cubic inches (cu in; in^3)	X	16.387	= Cubic centimeters (cc; cm^3)	X 0.061	= Cubic inches (cu in; in^3)
Imperial pints (Imp pt)	X	0.568	= Liters (l)	X 1.76	= Imperial pints (Imp pt)
Imperial quarts (Imp qt)	X	1.137	= Liters (l)	X 0.88	= Imperial quarts (Imp qt)
Imperial quarts (Imp qt)	X	1.201	= US quarts (US qt)	X 0.833	= Imperial quarts (Imp qt)
US quarts (US qt)	X	0.946	= Liters (l)	X 1.057	= US quarts (US qt)
Imperial gallons (Imp gal)	X	4.546	= Liters (l)	X 0.22	= Imperial gallons (Imp gal)
Imperial gallons (Imp gal)	X	1.201	= US gallons (US gal)	X 0.833	= Imperial gallons (Imp gal)
US gallons (US gal)	X	3.785	= Liters (l)	X 0.264	= US gallons (US gal)

Mass (weight)

Ounces (oz)	X	28.35	= Grams (g)	X 0.035	= Ounces (oz)
Pounds (lb)	X	0.454	= Kilograms (kg)	X 2.205	= Pounds (lb)

Force

Ounces-force (ozf; oz)	X	0.278	= Newtons (N)	X 3.6	= Ounces-force (ozf; oz)
Pounds-force (lbf; lb)	X	4.448	= Newtons (N)	X 0.225	= Pounds-force (lbf; lb)
Newtons (N)	X	0.1	= Kilograms-force (kgf; kg)	X 9.81	= Newtons (N)

Pressure

Pounds-force per square inch (psi; lbf/in^2; lb/in^2)	X	0.070	= Kilograms-force per square centimeter (kgf/cm^2; kg/cm^2)	X 14.223	= Pounds-force per square inch (psi; lbf/in^2; lb/in^2)
Pounds-force per square inch (psi; lbf/in^2; lb/in^2)	X	0.068	= Atmospheres (atm)	X 14.696	= Pounds-force per square inch (psi; lbf/in^2; lb/in^2)
Pounds-force per square inch (psi; lbf/in^2; lb/in^2)	X	0.069	= Bars	X 14.5	= Pounds-force per square inch (psi; lbf/in^2; lb/in^2)
Pounds-force per square inch (psi; lbf/in^2; lb/in^2)	X	6.895	= Kilopascals (kPa)	X 0.145	= Pounds-force per square inch (psi; lbf/in^2; lb/in^2)
Kilopascals (kPa)	X	0.01	= Kilograms-force per square centimeter (kgf/cm^2; kg/cm^2)	X 98.1	= Kilopascals (kPa)

Torque (moment of force)

Pounds-force inches (lbf in; lb in)	X	1.152	= Kilograms-force centimeter (kgf cm; kg cm)	X 0.868	= Pounds-force inches (lbf in; lb in)
Pounds-force inches (lbf in; lb in)	X	0.113	= Newton meters (Nm)	X 8.85	= Pounds-force inches (lbf in; lb in)
Pounds-force inches (lbf in; lb in)	X	0.083	= Pounds-force feet (lbf ft; lb ft)	X 12	= Pounds-force inches (lbf in; lb in)
Pounds-force feet (lbf ft; lb ft)	X	0.138	= Kilograms-force meters (kgf m; kg m)	X 7.233	= Pounds-force feet (lbf ft; lb ft)
Pounds-force feet (lbf ft; lb ft)	X	1.356	= Newton meters (Nm)	X 0.738	= Pounds-force feet (lbf ft; lb ft)
Newton meters (Nm)	X	0.102	= Kilograms-force meters (kgf m; kg m)	X 9.804	= Newton meters (Nm)

Vacuum

Inches mercury (in. Hg)	X	3.377	= Kilopascals (kPa)	X 0.2961	= Inches mercury
Inches mercury (in. Hg)	X	25.4	= Millimeters mercury (mm Hg)	X 0.0394	= Inches mercury

Power

Horsepower (hp)	X	745.7	= Watts (W)	X 0.0013	= Horsepower (hp)

Velocity (speed)

Miles per hour (miles/hr; mph)	X	1.609	= Kilometers per hour (km/hr; kph)	X 0.621	= Miles per hour (miles/hr; mph)

Fuel consumption*

Miles per gallon, Imperial (mpg)	X	0.354	= Kilometers per liter (km/l)	X 2.825	= Miles per gallon, Imperial (mpg)
Miles per gallon, US (mpg)	X	0.425	= Kilometers per liter (km/l)	X 2.352	= Miles per gallon, US (mpg)

Temperature

Degrees Fahrenheit = (°C x 1.8) + 32   Degrees Celsius (Degrees Centigrade; °C) = (°F - 32) x 0.56

*It is common practice to convert from miles per gallon (mpg) to liters/100 kilometers (l/100km), where mpg (Imperial) x l/100 km = 282 and mpg (US) x l/100 km = 235

DECIMALS to MILLIMETERS

Decimal	mm	Decimal	mm
0.001	0.0254	0.500	12.7000
0.002	0.0508	0.510	12.9540
0.003	0.0762	0.520	13.2080
0.004	0.1016	0.530	13.4620
0.005	0.1270	0.540	13.7160
0.006	0.1524	0.550	13.9700
0.007	0.1778	0.560	14.2240
0.008	0.2032	0.570	14.4780
0.009	0.2286	0.580	14.7320
		0.590	14.9860
0.010	0.2540		
0.020	0.5080		
0.030	0.7620		
0.040	1.0160	0.600	15.2400
0.050	1.2700	0.610	15.4940
0.060	1.5240	0.620	15.7480
0.070	1.7780	0.630	16.0020
0.080	2.0320	0.640	16.2560
0.090	2.2860	0.650	16.5100
		0.660	16.7640
0.100	2.5400	0.670	17.0180
0.110	2.7940	0.680	17.2720
0.120	3.0480	0.690	17.5260
0.130	3.3020		
0.140	3.5560		
0.150	3.8100		
0.160	4.0640	0.700	17.7800
0.170	4.3180	0.710	18.0340
0.180	4.5720	0.720	18.2880
0.190	4.8260	0.730	18.5420
		0.740	18.7960
0.200	5.0800	0.750	19.0500
0.210	5.3340	0.760	19.3040
0.220	5.5880	0.770	19.5580
0.230	5.8420	0.780	19.8120
0.240	6.0960	0.790	20.0660
0.250	6.3500		
0.260	6.6040		
0.270	6.8580	0.800	20.3200
0.280	7.1120	0.810	20.5740
0.290	7.3660	0.820	20.8280
		0.830	21.0820
0.300	7.6200	0.840	21.3360
0.310	7.8740	0.850	21.5900
0.320	8.1280	0.860	21.8440
0.330	8.3820	0.870	22.0980
0.340	8.6360	0.880	22.3520
0.350	8.8900	0.890	22.6060
0.360	9.1440		
0.370	9.3980		
0.380	9.6520		
0.390	9.9060	0.900	22.8600
0.400	10.1600	0.910	23.1140
0.410	10.4140	0.920	23.3680
0.420	10.6680	0.930	23.6220
0.430	10.9220	0.940	23.8760
0.440	11.1760	0.950	24.1300
0.450	11.4300	0.960	24.3840
0.460	11.6840	0.970	24.6380
0.470	11.9380	0.980	24.8920
0.480	12.1920	0.990	25.1460
0.490	12.4460	1.000	25.4000

FRACTIONS to DECIMALS to MILLIMETERS

Fraction	Decimal	mm	Fraction	Decimal	mm
1/64	0.0156	0.3969	33/64	0.5156	13.0969
1/32	0.0312	0.7938	17/32	0.5312	13.4938
3/64	0.0469	1.1906	35/64	0.5469	13.8906
1/16	0.0625	1.5875	9/16	0.5625	14.2875
5/64	0.0781	1.9844	37/64	0.5781	14.6844
3/32	0.0938	2.3812	19/32	0.5938	15.0812
7/64	0.1094	2.7781	39/64	0.6094	15.4781
1/8	0.1250	3.1750	5/8	0.6250	15.8750
9/64	0.1406	3.5719	41/64	0.6406	16.2719
5/32	0.1562	3.9688	21/32	0.6562	16.6688
11/64	0.1719	4.3656	43/64	0.6719	17.0656
3/16	0.1875	4.7625	11/16	0.6875	17.4625
13/64	0.2031	5.1594	45/64	0.7031	17.8594
7/32	0.2188	5.5562	23/32	0.7188	18.2562
15/64	0.2344	5.9531	47/64	0.7344	18.6531
1/4	0.2500	6.3500	3/4	0.7500	19.0500
17/64	0.2656	6.7469	49/64	0.7656	19.4469
9/32	0.2812	7.1438	25/32	0.7812	19.8438
19/64	0.2969	7.5406	51/64	0.7969	20.2406
5/16	0.3125	7.9375	13/16	0.8125	20.6375
21/64	0.3281	8.3344	53/64	0.8281	21.0344
11/32	0.3438	8.7312	27/32	0.8438	21.4312
23/64	0.3594	9.1281	55/64	0.8594	21.8281
3/8	0.3750	9.5250	7/8	0.8750	22.2250
25/64	0.3906	9.9219	57/64	0.8906	22.6219
13/32	0.4062	10.3188	29/32	0.9062	23.0188
27/64	0.4219	10.7156	59/64	0.9219	23.4156
7/16	0.4375	11.1125	15/16	0.9375	23.8125
29/64	0.4531	11.5094	61/64	0.9531	24.2094
15/32	0.4688	11.9062	31/32	0.9688	24.6062
31/64	0.4844	12.3031	63/64	0.9844	25.0031
1/2	0.5000	12.7000	1	1.0000	25.4000

Safety first!

Regardless of how enthusiastic you may be about getting on with the job at hand, take the time to ensure that your safety is not jeopardized. A moment's lack of attention can result in an accident, as can failure to observe certain simple safety precautions. The possibility of an accident will always exist, and the following points should not be considered a comprehensive list of all dangers. Rather, they are intended to make you aware of the risks and to encourage a safety conscious approach to all work you carry out on your vehicle.

Essential DOs and DON'Ts

DON'T rely on a jack when working under the vehicle. Always use approved jackstands to support the weight of the vehicle and place them under the recommended lift or support points.

DON'T attempt to loosen extremely tight fasteners (i.e. wheel lug nuts) while the vehicle is on a jack - it may fall.

DON'T start the engine without first making sure that the transmission is in Neutral (or Park where applicable) and the parking brake is set.

DON'T remove the radiator cap from a hot cooling system - let it cool or cover it with a cloth and release the pressure gradually.

DON'T attempt to drain the engine oil until you are sure it has cooled to the point that it will not burn you.

DON'T touch any part of the engine or exhaust system until it has cooled sufficiently to avoid burns.

DON'T siphon toxic liquids such as gasoline, antifreeze and brake fluid by mouth, or allow them to remain on your skin.

DON'T inhale brake lining dust - it is potentially hazardous (see *Asbestos* below).

DON'T allow spilled oil or grease to remain on the floor - wipe it up before someone slips on it.

DON'T use loose fitting wrenches or other tools which may slip and cause injury.

DON'T push on wrenches when loosening or tightening nuts or bolts. Always try to pull the wrench toward you. If the situation calls for pushing the wrench away, push with an open hand to avoid scraped knuckles if the wrench should slip.

DON'T attempt to lift a heavy component alone - get someone to help you.

DON'T *rush or take unsafe shortcuts to finish a job.*

DON'T allow children or animals in or around the vehicle while you are working on it.

DO wear eye protection when using power tools such as a drill, sander, bench grinder, etc. and when working under a vehicle.

DO keep loose clothing and long hair well out of the way of moving parts.

DO make sure that any hoist used has a safe working load rating adequate for the job.

DO get someone to check on you periodically when working alone on a vehicle.

DO carry out work in a logical sequence and make sure that everything is correctly assembled and tightened.

DO keep chemicals and fluids tightly capped and out of the reach of children and pets.

DO remember that your vehicle's safety affects that of yourself and others. If in doubt on any point, get professional advice.

Asbestos

Certain friction, insulating, sealing, and other products - such as brake linings, brake bands, clutch linings, torque converters, gaskets, etc. - may contain asbestos. Extreme care must be taken to avoid inhalation of dust from such products, since it is hazardous to health. If in doubt, assume that they do contain asbestos.

Fire

Remember at all times that gasoline is highly flammable. Never smoke or have any kind of open flame around when working on a vehicle. But the risk does not end there. A spark caused by an electrical short circuit, by two metal surfaces contacting each other, or even by static electricity built up in your body under certain conditions, can ignite gasoline vapors, which in a confined space are highly explosive. Do not, under any circumstances, use gasoline for cleaning parts. Use an approved safety solvent.

Always disconnect the battery ground (-) cable at the battery before working on any part of the fuel system or electrical system. Never risk spilling fuel on a hot engine or exhaust component. It is strongly recommended that a fire extinguisher suitable for use on fuel and electrical fires be kept handy in the garage or workshop at all times. Never try to extinguish a fuel or electrical fire with water.

Fumes

Certain fumes are highly toxic and can quickly cause unconsciousness and even death if inhaled to any extent. Gasoline vapor falls into this category, as do the vapors from some cleaning solvents. Any draining or pouring of such volatile fluids should be done in a well ventilated area.

When using cleaning fluids and solvents, read the instructions on the container carefully. Never use materials from unmarked containers.

Never run the engine in an enclosed space, such as a garage. Exhaust fumes contain carbon monoxide, which is extremely poisonous. If you need to run the engine, always do so in the open air, or at least have the rear of the vehicle outside the work area.

If you are fortunate enough to have the use of an inspection pit, never drain or pour gasoline and never run the engine while the vehicle is over the pit. The fumes, being heavier than air, will concentrate in the pit with possibly lethal results.

The battery

Never create a spark or allow a bare light bulb near a battery. They normally give off a certain amount of hydrogen gas, which is highly explosive.

Always disconnect the battery ground (-) cable at the battery before working on the fuel or electrical systems.

If possible, loosen the filler caps or cover when charging the battery from an external source (this does not apply to sealed or maintenance-free batteries). Do not charge at an excessive rate or the battery may burst.

Take care when adding water to a non maintenance-free battery and when carrying a battery. The electrolyte, even when diluted, is very corrosive and should not be allowed to contact clothing or skin.

Always wear eye protection when cleaning the battery to prevent the caustic deposits from entering your eyes.

Household current

When using an electric power tool, inspection light, etc., which operates on household current, always make sure that the tool is correctly connected to its plug and that, where necessary, it is properly grounded. Do not use such items in damp conditions and, again, do not create a spark or apply excessive heat in the vicinity of fuel or fuel vapor.

Secondary ignition system voltage

A severe electric shock can result from touching certain parts of the ignition system (such as the spark plug wires) when the engine is running or being cranked, particularly if components are damp or the insulation is defective. In the case of an electronic ignition system, the secondary system voltage is much higher and could prove fatal.

Troubleshooting

Contents

Engine

1 Engine will not rotate when attempting to start

1   Battery terminal connections loose or corroded. Check the cable terminals at the battery; tighten cable clamp and/or clean off corrosion as necessary (see Chapter 1).
2   Battery discharged or faulty. If the cable ends are clean and tight on the battery posts, turn the key to the On position and switch on the headlights or windshield wipers. If they won't run, the battery is discharged.
3   Automatic transmission not engaged in park (P) or Neutral (N).
4   Broken, loose or disconnected wires in the starting circuit. Inspect all wires and connectors at the battery, starter solenoid and ignition switch (on steering column).
5   Starter motor pinion jammed in flywheel/driveplate ring gear. Remove the starter (Chapter 5) and inspect the pinion and ring gear (Chapter 2).
6   Starter solenoid faulty (Chapter 5).
7   Starter motor faulty (Chapter 5).
8   Ignition switch faulty (Chapter 12).
9   Engine seized. Try to turn the crankshaft with a large socket and breaker bar on the pulley bolt.
10   Starter relay faulty (Chapter 4).
11   Transmission Range (TR) sensor out of adjustment or defective (Chapter 6).

2 Engine rotates but will not start

1   Fuel tank empty.
2   Battery discharged (engine rotates slowly).
3   Battery terminal connections loose or corroded.
4   Fuel not reaching fuel injectors. Check for clogged fuel filter or lines and defective fuel pump. Also make sure the tank vent lines aren't clogged (Chapter 4).
5   Low cylinder compression. Check as described in Chapter 2.
6   Water in fuel. Drain tank and fill with new fuel.
7   Defective ignition coil(s) (Chapter 5).
8   Dirty or clogged fuel injector(s) (Chapter 4).
9   Wet or damaged ignition components (Chapters 1 and 5).
10   Worn, faulty or incorrectly gapped spark plugs (Chapter 1).
11   Broken, loose or disconnected wires in the starting circuit (see previous Section).
12   Broken, loose or disconnected wires at the ignition coil or faulty coil (Chapter 5).
13   Timing chain failure or wear affecting valve timing (Chapter 2).
14   Fuel injection or engine control systems failure (Chapters 4 and 6).
15   Defective MAF sensor (Chapter 6)

3 Starter motor operates without turning engine

1   Starter pinion sticking. Remove the starter (Chapter 5) and inspect.
2   Starter pinion or flywheel/driveplate teeth worn or broken. Remove the inspection cover and inspect.

4 Engine hard to start when cold

1   Battery discharged or low. Check as described in Chapter 1.
2   Fuel not reaching the fuel injectors. Check the fuel filter, lines and fuel pump (Chapters 1 and 4).
3   Defective spark plugs (Chapter 1).
4   Defective engine coolant temperature sensor (Chapter 6).
5   Fuel injection or engine control systems malfunction (Chapters 4 and 6).

5 Engine hard to start when hot

1   Air filter dirty (Chapter 1).
2   Fuel not reaching the fuel injection (see Section 4). Check for a vapor lock situation, brought about by clogged fuel tank vent lines.
3   Bad engine ground connection.
4   Fuel injection or engine control systems malfunction (Chapters 4 and 6).

6 Starter motor noisy or engages roughly

1   Pinion or driveplate teeth worn or broken. Remove the inspection cover on the left side of the engine and inspect.
2   Starter motor mounting bolts loose or missing.

7 Engine starts but stops immediately

1   Loose or damaged wire harness connections at coil or alternator.
2   Intake manifold vacuum leaks. Make sure all mounting bolts/nuts are tight and all vacuum hoses connected to the manifold are attached properly and in good condition.
3   Insufficient fuel pressure (see Chapter 4).
4   Fuel injection or engine control systems malfunction (Chapters 4 and 6).

8 Engine 'lopes' while idling or idles erratically

1   Vacuum leaks. Check mounting bolts at the intake manifold for tightness. Make sure that all vacuum hoses are connected and in good condition. Use a stethoscope or a length of fuel hose held against your ear to listen for vacuum leaks while the engine is running. A hissing sound will be heard. A soapy water solution will also detect leaks. Check the intake manifold gasket surfaces.
2   Leaking EGR valve or plugged PCV valve (see Chapters 1 and 6).
3   Air filter clogged (Chapter 1).
4   Fuel pump not delivering sufficient fuel (Chapter 4).
5   Leaking head gasket. Perform a cylinder compression check (Chapter 2).
6   Timing chain(s) worn (Chapter 2).
7   Camshaft lobes worn (Chapter 2).
8   Valves burned or otherwise leaking (Chapter 2).
9   Ignition system not operating properly (Chapters 1 and 5).
10   Fuel injection or engine control systems malfunction (Chapters 4 and 6).

9 Engine misses at idle speed

1   Spark plugs faulty or not gapped properly (Chapter 1).
2   Faulty spark plug wires (Chapter 1).
3   Short circuits in ignition, coil or spark plug wires.
4   Sticking or faulty emissions systems (see Chapter 6).
5   Clogged fuel filter and/or foreign matter in fuel. Remove the fuel filter (Chapter 1) and inspect.
6   Vacuum leaks at intake manifold or hose connections. Check as described in Section 12.
7   Low or uneven cylinder compression. Check as described in Chapter 2.
8   Fuel injection or engine control systems malfunction (Chapters 4 and 6).

10 Excessively high idle speed

1   Sticking throttle linkage (Chapter 4).
2   Vacuum leaks at intake manifold or hose connections. Check as described in Section 8.
3   Fuel injection or engine control systems malfunction (Chapters 4 and 6).

11 Battery will not hold a charge

1   Alternator drivebelt defective or not adjusted properly (Chapter 1).
2   Battery cables loose or corroded (Chapter 1).
3   Alternator not charging properly (Chapter 5).
4   Loose, broken or faulty wires in the charging circuit (Chapter 5).
5   Short circuit causing a continuous drain on the battery.
6   Battery defective internally.

12 Alternator light stays on

1 Fault in alternator or charging circuit (Chapter 5).
2 Alternator drivebelt defective or not properly adjusted (Chapter 1).

13 Alternator light fails to come on when key is turned on

1 Faulty bulb (Chapter 12).
2 Defective alternator (Chapter 5).
3 Fault in the printed circuit, dash wiring or bulb holder (Chapter 12).

14 Engine misses throughout driving speed range

1 Fuel filter clogged and/or impurities in the fuel system. Check fuel filter (Chapter 1) or clean system (Chapter 4).
2 Faulty or incorrectly gapped spark plugs (Chapter 1).
3 Defective spark plug wires (Chapter 1).
4 Emissions system components faulty (Chapter 6).
5 Low or uneven cylinder compression pressures. Check as described in Chapter 2.
6 Weak or faulty ignition coil(s) (Chapter 5).
7 Weak or faulty ignition system (Chapter 5).
8 Vacuum leaks at intake manifold or vacuum hoses (see Section 8).
9 Dirty or clogged fuel injector(s) (Chapter 4).
10 Leaky EGR valve (Chapter 6).
11 Fuel injection or engine control systems malfunction (Chapters 4 and 6).

15 Hesitation or stumble during acceleration

1 Ignition system not operating properly (Chapter 5).
2 Dirty or clogged fuel injector(s) (Chapter 4).
3 Low fuel pressure. Check for proper operation of the fuel pump and for restrictions in the fuel filter and lines (Chapter 4).
4 Fuel injection or engine control systems malfunction (Chapters 4 and 6).

16 Engine stalls

1 Fuel filter clogged and/or water and impurities in the fuel system (Chapter 1).
2 Emissions system components faulty (Chapter 6).
3 Faulty or incorrectly gapped spark plugs (Chapter 1). Also check the spark plug wires (Chapter 1).
4 Vacuum leak at the intake manifold or vacuum hoses. Check as described in Section 12.
5 Fuel injection or engine control systems malfunction (Chapters 4 and 6).

17 Engine lacks power

1 Faulty or incorrectly gapped spark plugs (Chapter 1).
2 Air filter dirty (Chapter 1).
3 Faulty ignition coil(s) (Chapter 5).
4 Brakes binding (Chapters 1 and 10).
5 Automatic transmission fluid level incorrect, causing slippage (Chapter 1).
6 Fuel filter clogged and/or impurities in the fuel system (Chapters 1 and 4).
7 EGR system not functioning properly (Chapter 6).
8 Use of sub-standard fuel. Fill tank with proper octane fuel.
9 Low or uneven cylinder compression pressures. Check as described in Chapter 2.
10 Vacuum leak at intake manifold or vacuum hoses (check as described in Section 12).
11 Dirty or clogged fuel injector(s) (Chapters 1 and 4).
12 Fuel injection or engine control systems malfunction (Chapters 4 and 6).
13 Restricted exhaust system (Chapter 4).

18 Engine backfires

1 EGR system not functioning properly (Chapter 6).
2 Vacuum leak (refer to Section 12).
3 Damaged valve springs or sticking valves (Chapter 2).
4 Vacuum leak at the intake manifold or vacuum hoses (see Section12).

19 Engine surges while holding accelerator steady

1 Vacuum leak at the intake manifold or vacuum hoses (see Section 12).
2 Restricted air filter (Chapter 1).
3 Fuel pump or pressure regulator defective (Chapter 4).
4 Fuel injection or engine control systems malfunction (Chapters 4 and 6).

20 Pinging or knocking engine sounds when engine is under load

1 Incorrect grade of fuel. Fill tank with fuel of the proper octane rating.
2 Carbon build-up in combustion chambers. Remove cylinder head(s) and clean combustion chambers (Chapter 2).
3 Incorrect spark plugs (Chapter 1).
4 Fuel injection or engine control systems malfunction (Chapters 4 and 6).
5 Restricted exhaust system (Chapter 4).

21 Engine diesels (continues to run) after being turned off

1 Incorrect spark plug heat range (Chapter 1).
2 Vacuum leak at the intake manifold or vacuum hoses (see Section 8).
3 Carbon build-up in combustion chambers. Remove the cylinder head(s) and clean the combustion chambers (Chapter 2).
4 Valves sticking (Chapter 2).
5 EGR system not operating properly (Chapter 6).
6 Fuel injection or engine control systems malfunction (Chapters 4 and 6).
7 Check for causes of overheating (Section 27).

22 Low oil pressure

1 Improper grade of oil.
2 Oil pump worn or damaged (Chapter 2).
3 Engine overheating (refer to Section 27).
4 Clogged oil filter (Chapter 1).
5 Clogged oil strainer (Chapter 2).
6 Oil pressure gauge not working properly (Chapter 2).

23 Excessive oil consumption

1 Loose oil drain plug.
2 Loose bolts or damaged oil pan gasket (Chapter 2).
3 Loose bolts or damaged front cover gasket (Chapter 2).
4 Front or rear crankshaft oil seal leaking (Chapter 2).
5 Loose bolts or damaged valve cover gasket (Chapter 2).
6 Loose oil filter (Chapter 1).
7 Loose or damaged oil pressure switch (Chapter 2).
8 Pistons and cylinders excessively worn (Chapter 2).
9 Piston rings not installed correctly on pistons (Chapter 2).
10 Worn or damaged piston rings (Chapter 2).
11 Intake and/or exhaust valve oil seals worn or damaged (Chapter 2).
12 Worn valve stems or guides.
13 Worn or damaged valves/guides (Chapter 2).
14 Faulty or incorrect PCV valve allowing too much crankcase airflow.

24 Excessive fuel consumption

1 Dirty or clogged air filter element (Chapter 1).

2   Low tire pressure or incorrect tire size (Chapter 10).
3   Inspect for binding brakes.
4   Fuel leakage. Check all connections, lines and components in the fuel system (Chapter 4).
5   Dirty or clogged fuel injectors (Chapter 4).
6   Fuel injection or engine control systems malfunction (Chapters 4 and 6).
7   Thermostat stuck open or not installed.
8   Improperly operating transmission.

25   Fuel odor

1   Fuel leakage. Check all connections, lines and components in the fuel system (Chapter 4).
2   Fuel tank overfilled. Fill only to automatic shut-off.
3   Charcoal canister filter in Evaporative Emissions Control system clogged (Chapter 1).
4   Vapor leaks from Evaporative Emissions Control system lines (Chapter 6).

26   Miscellaneous engine noises

1   A strong dull noise that becomes more rapid as the engine accelerates indicates worn or damaged crankshaft bearings or an unevenly worn crankshaft. To pinpoint the trouble spot, remove the spark plug wire from one plug at a time (V6 engine) or disconnect the electrical connector from one coil at a time (V8 engine) and crank the engine over. If the noise stops, the cylinder with the removed plug wire or disconnected coil indicates the problem area. Replace the bearing and/or service or replace the crankshaft (Chapter 2).
2   A similar (yet slightly higher pitched) noise to the crankshaft knocking described in the previous paragraph, that becomes more rapid as the engine accelerates, indicates worn or damaged connecting rod bearings (Chapter 2). The procedure for locating the problem cylinder is the same as described in Paragraph 1.
3   An overlapping metallic noise that increases in intensity as the engine speed increases, yet diminishes as the engine warms up indicates abnormal piston and cylinder wear (Chapter 2). To locate the problem cylinder, use the procedure described in Paragraph 1.
4   A rapid clicking noise that becomes faster as the engine accelerates indicates a worn piston pin or piston pin hole. This sound will happen each time the piston hits the highest and lowest points in the stroke (Chapter 2). The procedure for locating the problem piston is described in Paragraph 1.
5   A metallic clicking noise coming from the water pump indicates worn or damaged water pump bearings or pump. Replace the water pump with a new one (Chapter 3).

6   A rapid tapping sound or clicking sound that becomes faster as the engine speed increases indicates "valve tapping." This can be identified by holding one end of a section of hose to your ear and placing the other end at different spots along the valve cover. The point where the sound is loudest indicates the problem valve. If the pushrod and rocker arm components are in good shape, you likely have a collapsed valve lifter. Changing the engine oil and adding a high viscosity oil treatment will sometimes cure a stuck lifter problem. If the problem persists, the lifters, pushrods and rocker arms must be removed for inspection (see Chapter 2).
7   A steady metallic rattling or rapping sound coming from the area of the timing chain cover indicates a worn, damaged or out-of-adjustment timing chain. Service or replace the chain and related components (Chapter 2).

Cooling system

27   Overheating

1   Insufficient coolant in system (Chapter 1).
2   Drivebelt defective or not adjusted properly (Chapter 1).
3   Radiator core blocked or dirty and restricted (Chapter 3).
4   Thermostat faulty (Chapter 3).
5   Cooling fan not functioning properly (Chapter 3).
6   Expansion tank cap not maintaining proper pressure. Have cap pressure tested by gas station or repair shop.
7   Defective water pump (Chapter 3).
8   Improper grade of engine oil.
9   Inaccurate temperature gauge (Chapter 12).

28   Overcooling

1   Thermostat faulty (Chapter 3).
2   Inaccurate temperature gauge (Chapter 12).

29   External coolant leakage

1   Deteriorated or damaged hoses. Loose clamps at hose connections (Chapter 1).
2   Water pump seals defective. If this is the case, water will drip from the weep hole in the water pump body (Chapter 3).
3   Leakage from radiator core or header tank. This will require the radiator to be professionally repaired (see Chapter 3 for removal procedures).
4   Leakage from the expansion tank or cap.
5   Engine drain plugs or water jacket freeze plugs leaking (see Chapters 1 and 2).

6   Leak from coolant temperature switch (Chapter 3).
7   Leak from damaged gaskets or small cracks (Chapter 2).

30   Internal coolant leakage

Note: *Internal coolant leaks can usually be detected by examining the oil. Check the dipstick and the underside of the engine oil filler cap for water deposits and an oil consistency like that of a milkshake.*
1   Leaking cylinder head gasket. Have the system pressure tested or remove the cylinder head (Chapter 2) and inspect.
2   Cracked cylinder bore or cylinder head. Dismantle engine and inspect (Chapter 2).

31   Abnormal coolant loss

1   Overfilled cooling system (Chapter 1).
2   Coolant boiling away due to overheating (see causes in Section 27).
3   Internal or external leakage (see Sections 29 and 30).
4   Faulty expansion tank cap. Have the cap pressure tested.
5   Cooling system being pressurized by engine compression. This could be due to a cracked head or block or leaking head gasket(s). Have the system tested for the presence of combustion gas in the coolant at a shop. (Combustion leak detectors are also available at some auto parts stores.)

32   Poor coolant circulation

1   Inoperative water pump. A quick test is to pinch the top radiator hose closed with your hand while the engine is idling, then release it. You should feel a surge of coolant if the pump is working properly (Chapter 3).
2   Restriction in cooling system. Drain, flush and refill the system (Chapter 1). If necessary, remove the radiator (Chapter 3) and have it reverse flushed or professionally cleaned.
3   Loose water pump drivebelt (Chapter 1).
4   Thermostat sticking (Chapter 3).
5   Insufficient coolant (Chapter 1).

33   Corrosion

1   Excessive impurities in the water. Soft, clean water is recommended. Distilled or rainwater is satisfactory.
2   Insufficient antifreeze solution (refer to Chapter 1 for the proper ratio of water to antifreeze).
3   Infrequent flushing and draining of system. Regular flushing of the cooling system should be carried out at the specified intervals as described in (Chapter 1).

Clutch

Note: *All clutch service information is located in Chapter 8, unless otherwise noted.*

34 Fails to release (pedal pressed to the floor - shift lever does not move freely in and out of Reverse)

1   Clutch plate warped, distorted or otherwise damaged.
2   Diaphragm spring fatigued. Remove clutch cover/pressure plate assembly and inspect.
3   Insufficient pedal stroke. Check and adjust as necessary.
4   Lack of grease on pilot bushing.

35 Clutch slips (engine speed increases with no increase in vehicle speed)

1   Worn or oil soaked clutch plate.
2   Clutch plate not broken in. It may take 30 or 40 normal starts for a new clutch to seat.
3   Air in clutch hydraulic release system. Bleed system and check for leaks.
4   Defective clutch master or release cylinder.

36 Grabbing (chattering) as clutch is engaged

1   Oil on clutch plate. Remove and inspect. Repair any leaks.
2   Worn or loose engine or transmission mounts. They may move slightly when clutch is released. Inspect mounts and bolts.
3   Worn splines on transmission input shaft. Remove clutch components and inspect.
4   Warped pressure plate or flywheel. Remove clutch components and inspect.
5   Diaphragm spring fatigued. Remove clutch cover/pressure plate assembly and inspect.
6   Clutch linings hardened or warped.
7   Clutch lining rivets loose.

37 Squeal or rumble with clutch engaged (pedal released)

1   Improper pedal adjustment. Adjust pedal free play.
2   Release bearing binding on transmission shaft. Remove clutch components and check bearing. Remove any burrs or nicks, clean and relubricate before reinstallation.
3   Clutch plate cracked.
4   Fatigued clutch plate torsion springs. Replace clutch plate.

38 Squeal or rumble with clutch disengaged (pedal depressed)

1   Worn or damaged release bearing.
2   Worn or broken pressure plate diaphragm fingers.
3   Defective pilot bearing.

39 Clutch pedal stays on floor when disengaged

Defective release system.

Manual transmission

Note: *All manual transmission service information is located in Chapter 7A, unless otherwise noted.*

40 Noisy in Neutral with engine running

1   Input shaft bearing worn.
2   Damaged main drive gear bearing.
3   Insufficient transmission oil (Chapter 1).
4   Transmission oil in poor condition. Drain and fill with proper grade oil. Check old oil for water and debris (Chapter 1).

41 Noisy in all gears

1   Any of the above causes, and/or:
2   Worn or damaged output gear bearings or shaft.

42 Noisy in one particular gear

1   Worn, damaged or chipped gear teeth.
2   Worn or damaged synchronizer.

43 Slips out of gear

1   Shift linkage binding.
2   Broken or loose input gear bearing retainer.
3   Worn linkage.
4   Damaged or worn check balls, fork rod ball grooves or check springs.
5   Worn mainshaft or countershaft bearings.
6   Excessive gear end play.
7   Worn synchronizers.
8   Chipped or worn gear teeth.

44 Oil leaks

1   Excessive amount of lubricant in transmission (see Chapter 1 for correct checking procedures). Drain lubricant as required.

2   Oil seal damaged.
3   To pinpoint a leak, first remove all built-up dirt and grime from the transmission. Degreasing agents and/or steam cleaning will achieve this. With the underside clean, drive the vehicle at low speeds so the air flow will not blow the leak far from its source. Raise the vehicle and determine where the leak is located.

45 Difficulty engaging gears

1   Clutch not releasing completely.
2   Insufficient transmission oil (Chapter 1).
3   Transmission oil in poor condition. Drain and fill with proper grade oil. Check oil for water and debris (Chapter 1).
4   Damaged shift fork.
5   Worn or damaged synchronizer.

46 Noise occurs while shifting gears

1   Check for proper operation of the clutch (Chapter 8).
2   Faulty synchronizer assemblies.

Automatic transmission

Note: *Due to the complexity of the automatic transmission, it's difficult for the home mechanic to properly diagnose and service. For problems other than the following, the vehicle should be taken to a reputable mechanic.*

47 Fluid leakage

1   Automatic transmission fluid is a deep red color, and fluid leaks should not be confused with engine oil which can easily be blown by air flow to the transmission.
2   To pinpoint a leak, first remove all built-up dirt and grime from the transmission. Degreasing agents and/or steam cleaning will achieve this. With the underside clean, drive the vehicle at low speeds so the air flow will not blow the leak far from its source. Raise the vehicle and determine where the leak is located. Common areas of leakage are:

a) *Fluid pan: tighten mounting bolts and/or replace pan gasket as necessary (Chapter 1).*
b) *Rear extension: tighten bolts and/or replace oil seal as necessary.*
c) *Filler pipe: replace the rubber oil seal where pipe enters transmission case.*
d) *Transmission oil lines: tighten fittings where lines enter transmission case and/or replace lines.*
e) *Vent pipe: transmission overfilled and/or water in fluid (see checking procedures, Chapter 1).*
f) *Vehicle speed sensor: replace the O-ring where speed sensor enters transmission case.*

48 General shift mechanism problems

Chapter 7 deals with checking and adjusting the shift linkage on automatic transmissions. Common problems which may be caused by out of adjustment linkage are:

a) *Engine starting in gears other than P (park) or N (Neutral).*
b) *Indicator pointing to a gear other than the one actually engaged.*
c) *Vehicle moves with transmission in P (Park) position.*

49 Transmission will not downshift with the accelerator pedal pressed to the floor

Since these transmissions are electronically controlled, check for any diagnostic trouble codes stored in the PCM. The actual repair will most likely have to be performed by a qualified repair shop with the proper equipment.

50 Engine will start in gears other than Park or Neutral

Chapter 7 deals with adjusting the Neutral start switch installed on automatic transmissions.

51 Transmission slips, shifts rough, is noisy or has no drive in forward or Reverse gears

1   There are many probable causes for the above problems, but the home mechanic should concern himself only with one possibility: fluid level.
2   Before taking the vehicle to a shop, check the fluid level and condition as described in Chapter 1. Add fluid, if necessary, or change the fluid and filter if needed. If problems persist, have a professional diagnose the transmission.
3   Transmission fluid break down after 30,000 miles.

Driveshaft

Note: *Refer to Chapter 8, unless otherwise specified, for service information.*

52 Leaks at front of driveshaft

Defective transmission or transfer case seal. See Chapter 7 for replacement procedure. As this is done, check the splined yoke for burrs or roughness that could damage the new seal. Remove burrs with a fine file or whetstone.

53 Knock or clunk when transmission is under initial load (just after transmission is put into gear)

1   Loose or disconnected rear suspension components. Check all mounting bolts and bushings (Chapters 7 and 10).
2   Loose driveshaft bolts. Inspect all bolts and nuts and tighten them securely.
3   Worn or damaged universal joint bearings (Chapter 8).
4   Worn sleeve yoke and mainshaft spline.

54 Metallic grating sound consistent with vehicle speed

Pronounced wear in the universal joint bearings. Replace U-joints or driveshaft, as necessary.

55 Vibration

Note: *Before blaming the driveshaft, make sure the tires are perfectly balanced and perform the following test.*
1   Install a tachometer inside the vehicle to monitor engine speed as the vehicle is driven. Drive the vehicle and note the engine speed at which the vibration (roughness) is most pronounced. Now shift the transmission to a different gear and bring the engine speed to the same point.
2   If the vibration occurs at the same engine speed (rpm) regardless of which gear the transmission is in, the driveshaft is NOT at fault since the driveshaft speed varies.
3   If the vibration decreases or is eliminated when the transmission is in a different gear at the same engine speed, refer to the following probable causes:

a) *Bent or dented driveshaft. Inspect and replace as necessary.*
b) *Undercoating or built-up dirt, etc. on the driveshaft. Clean the shaft thoroughly.*
c) *Worn universal joint bearings. Replace the U-joints or driveshaft as necessary.*
d) *Driveshaft and/or companion flange out of balance. Check for missing weights on the shaft. Remove driveshaft and reinstall 180-degrees from original position, then recheck. Have the driveshaft balanced if problem persists.*
e) *Loose driveshaft mounting bolts/nuts.*
f) *Worn transmission rear bushing (Chapter 7).*

56 Scraping noise

Make sure there is nothing, such as an exhaust heat shield, rubbing on the driveshaft.

Axle(s) and differential

Note: *For differential servicing information, refer to Chapter 8, unless otherwise specified.*

57 Noise - same when in drive as when vehicle is coasting

1   Road noise. No corrective action available.
2   Tire noise. Inspect tires and check tire pressures (Chapter 1).
3   Front wheel bearings loose, worn or damaged (Chapter 1).
4   Insufficient differential oil (Chapter 1).
5   Defective differential.

58 Knocking sound when starting or shifting gears

Defective or incorrectly adjusted differential.

59 Noise when turning

Defective differential.

60 Vibration

See probable causes under Driveshaft. Proceed under the guidelines listed for the driveshaft. If the problem persists, check the rear wheel bearings by raising the rear of the vehicle and spinning the wheels by hand. Listen for evidence of rough (noisy) bearings. Remove and inspect (Chapter 8).

61 Oil leaks

1   Pinion oil seal damaged (Chapter 8).
2   Axleshaft oil seals damaged (Chapter 8).
3   Differential cover leaking. Tighten mounting bolts or replace the gasket as required.
4   Loose filler plug on differential (Chapter 1).
5   Clogged or damaged breather on differential.

Brakes

Note: *Before assuming a brake problem exists, make sure the tires are in good condition and inflated properly, the front end alignment is correct and the vehicle is not loaded with weight in an unequal manner. All service procedures for the brakes are included in Chapter 9, unless otherwise noted.*

62 Vehicle pulls to one side during braking

1   Defective, damaged or contaminated brake pad on one side. Inspect as described in Chapter 1. Refer to Chapter 9 if replacement is required.
2   Excessive wear of brake pad material or disc on one side. Inspect and repair as necessary.
3   Loose or disconnected front suspension components. Inspect and tighten all bolts securely (Chapters 1 and 10).
4   Defective front brake caliper assembly. Remove caliper and inspect for stuck piston or damage.
5   Scored or out of round disc.
6   Loose brake caliper mounting bolts.

63 Noise (high-pitched squeal or scraping sound)

1   Brake pads worn out. Replace pads with new ones immediately!
2   Glazed or contaminated pads.
3   Dirty or scored disc.
4   Bent support plate.

64 Excessive brake pedal travel

1   Partial brake system failure. Inspect entire system (Chapter 1) and correct as required.
2   Insufficient fluid in master cylinder. Check (Chapter 1) and add fluid - bleed system if necessary.
3   Air in system. Bleed system.
4   Defective master cylinder.

65 Brake pedal feels spongy when depressed

1   Air in brake lines. Bleed the brake system.
2   Deteriorated rubber brake hoses. Inspect all system hoses and lines. Replace parts as necessary.
3   Master cylinder mounting nuts loose. Inspect master cylinder bolts (nuts) and tighten them securely.
4   Master cylinder faulty.
5   Incorrect brake pad clearance.
6   Clogged reservoir cap vent hole.
7   Deformed rubber brake lines.
8   Soft or swollen caliper seals.
9   Poor quality brake fluid. Bleed entire system and fill with new approved fluid.

66 Excessive effort required to stop vehicle

1   Power brake booster not operating properly.
2   Excessively worn brake pads. Check

and replace if necessary.
3   One or more caliper pistons seized or sticking. Inspect and rebuild as required.
4   Brake pads contaminated with oil or grease. Inspect and replace as required.
5   Worn or damaged master cylinder or caliper assemblies. Check particularly for frozen pistons.

67 Pedal travels to the floor with little resistance

Little or no fluid in the master cylinder reservoir caused by leaking caliper piston(s) or loose, damaged or disconnected brake lines. Inspect entire system and repair as necessary.

68 Brake pedal pulsates during brake application

1   Wheel bearings damaged, worn or out of adjustment.
2   Caliper not sliding properly due to improper installation or obstructions. Remove and inspect.
3   Disc not within specifications. Check for excessive lateral runout and parallelism. Have the discs resurfaced or replace them with new ones. Also make sure that all discs are the same thickness.

69 Brakes drag (indicated by sluggish engine performance or wheels being very hot after driving)

1   Pushrod adjustment incorrect at the brake pedal or power booster.
2   Master cylinder piston seized in bore. Replace master cylinder.
3   Caliper piston seized in bore.
4   Parking brake assembly will not release.
5   Clogged or internally split brake lines.
6   Brake pedal height improperly adjusted.

70 Rear brakes lock up under light brake application

1   Tire pressures too high.
2   Tires excessively worn (Chapter 1).
3   Defective proportioning valve.

71 Rear brakes lock up under heavy brake application

1   Tire pressures too high.
2   Tires excessively worn (Chapter 1).
3   Front brake pads contaminated with oil, mud or water. Clean or replace the pads.
4   Front brake pads excessively worn.
5   Defective proportioning valve.

Suspension and steering

Note: *All service procedures for the suspension and steering systems are included in Chapter 10, unless otherwise noted.*

72 Vehicle pulls to one side

1   Tire pressures uneven (Chapter 1).
2   Defective tire (Chapter 1).
3   Excessive wear in suspension or steering components (Chapter 1).
4   Front end alignment incorrect.
5   Front brakes dragging. Inspect as described in Section 71.
6   Wheel bearings improperly adjusted (Chapter 1).
7   Wheel lug nuts loose.

73 Shimmy, shake or vibration

1   Tire or wheel out of balance or out of round.
2   Loose, worn or out of adjustment wheel bearings (Chapter 1).
3   Shock absorbers and/or suspension components worn or damaged (see Chapter 10).

74 Excessive pitching and/or rolling around corners or during braking

1   Defective shock absorbers. Replace as a set.
2   Sagging springs.
3   Worn or damaged stabilizer bar or bushings.

75 Wandering or general instability

1   Improper tire pressures.
2   Incorrect front end alignment.
3   Worn or damaged steering linkage or suspension components.
4   Improperly adjusted steering gear.
5   Out-of-balance wheels.
6   Loose wheel lug nuts.
7   Worn rear shock absorbers.

76 Excessively stiff steering

1   Lack of fluid in the power steering fluid reservoir, where appropriate (Chapter 1).
2   Incorrect tire pressures (Chapter 1).
3   Front end out of alignment.
4   Steering gear out of adjustment or lacking lubrication.
5   Worn or damaged steering gear.
6   Low tire pressures.
7   Worn or damaged balljoints.
8   Worn or damaged tie-rod ends.

77 Excessive play in steering

1   Worn wheel bearings (Chapter 1).
2   Excessive wear in suspension bushings (Chapter 1).
3   Steering gear worn.
4   Incorrect front end alignment.
5   Steering gear mounting bolts loose.
6   Worn or damaged tie-rod ends.

78 Lack of power assistance

1   Steering pump drivebelt faulty or tensioner defective (Chapter 1).
2   Fluid level low (Chapter 1).
3   Hoses or pipes restricting the flow. Inspect and replace parts as necessary.
4   Air in power steering system. Bleed system.
5   Defective power steering pump.

79 Steering wheel fails to return to straight-ahead position

1   Incorrect front end alignment.
2   Tire pressures low.
3   Worn or damaged balljoint.
4   Worn or damaged tie-rod end.
5   Lack of fluid in power steering pump.

80 Steering effort not the same in both directions

1   Leaks in steering gear.
2   Clogged fluid passage in steering gear.

81 Noisy power steering pump

1   Insufficient fluid in pump.
2   Clogged hoses or oil filter in pump.
3   Loose pulley.
4   Drivebelt faulty or tensioner defective (Chapter 1).
5   Defective pump.

82 Miscellaneous noises

1   Improper tire pressures.
2   Defective balljoint or tie-rod end.
3   Loose or worn steering gear or suspension components.
4   Defective shock absorber.
5   Defective wheel bearing.
6   Worn or damaged suspension bushings.
7   Loose wheel lug nuts.
8   Worn or damaged shock absorber mounting bushing.
9   Worn stabilizer bar bushings.
10   Incorrect rear axle endplay.
11   See also causes of noises at the rear axle and driveshaft.

83 Excessive tire wear (not specific to one area)

1   Incorrect tire pressures.
2   Tires out of balance.
3   Wheels damaged. Inspect and replace as necessary.
4   Suspension or steering components worn (Chapter 1).
5   Front end alignment incorrect.
6   Lack of proper tire rotation routine. See Routine Maintenance Schedule, Chapter 1.

84 Excessive tire wear on outside edge

1   Incorrect tire pressure.
2   Excessive speed in turns.
3   Front end alignment incorrect.

85 Excessive tire wear on inside edge

1   Incorrect tire pressure.
2   Front end alignment incorrect.

86 Tire tread worn in one place

1   Tires out of balance.
2   Damaged or buckled wheel. Inspect and replace if necessary.
3   Defective tire.

Chapter 1
Tune-up and routine maintenance

Contents

Specifications

Recommended lubricants and fluids

Note: *Listed here are manufacturer recommendations at the time this manual was written. Manufacturers occasionally upgrade their fluid and lubricant specifications, so check with your local auto parts store for current recommendations*

Engine oil	
Type	API "Certified for gasoline engines"
Viscosity	
V6 engine	SAE 5W-30
V8 engine	SAE 5W-20
Fuel	Unleaded gasoline, 87 minimum
Engine coolant *	Motorcraft Premium Gold Engine Coolant (yellow colored)
Brake fluid	DOT 3 brake fluid
Clutch fluid	DOT 3 brake fluid
Power steering fluid	MERCON® automatic transmission fluid
Automatic transmission fluid	MERCON® V automatic transmission fluid
Manual transmission fluid	MERCON® automatic transmission fluid
Differential lubricant	SAE 75W-140 synthetic rear axle lubricant**

  * **Caution:** *Do not mix coolants of different colors. Doing so might damage the cooling system and/or the engine. The manufacturer specifies a yellow colored coolant to be used in these systems.*

** *Trak-Lok axles add 4 oz. of friction modifier XL-3 when oil is changed.*

Capacities*

Engine oil (with filter change)
    V6 engine.. 5.0 qts
    V8 engine.. 6.0 qts
Fuel tank .... 16.0 gallons
Cooling system
    V6 engine.. 12.5 qts
    V8 engine.. 13.6 qts
Automatic transmission** .... 11.9 qts
Manual transmission
    V6 engine.. 2.75 qts
    V8 engine.. 3.2 qts
Rear axle differential
    7.5 inch ring gear .... 3.25 pts
    8.8 inch ring gear .... 4.25 pts

All capacities approximate. Add as necessary to bring to appropriate level.

*** This is a dry-fill specification. Follow the procedure in Section 27 when adding fluid to prevent overfilling.*

Ignition system

Spark plugs
    Type
        V6 engine .... Motorcraft AGSF-24FM
        V8 engine PZT1F
    Gap
        V6 engine .... 0.052 to 0.054 inch
        V8 engine .... 0.040 to 0.050 inch
Firing order
    V6 engine.. 1-4-2-5-3-6
    V8 engine.. 1-3-7-2-6-5-4-8

Brakes

Disc brake pad thickness (minimum) .... 1/8 inch

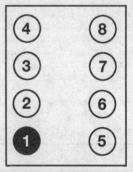

**Cylinder locations -
V8 engine**

**4.6L V8 ENGINE
1-3-7-2-6-5-4-8**

36061-1-specs.C HAYNES

**4.0L
V6 Engine
1-4-2-5-3-6**

36025-specs.b HAYNES

**Cylinder location and coil terminal
identification - V6 engine**

Torque specifications

	Ft-lbs (unless otherwise indicated)
Wheel lug nuts...	100
Spark plugs	
V6 engine..	156 in-lbs
V8 engine..	25
Oil pan drain plug	19
Automatic transmission	
Fill plug...	88 in-lbs
Drain plug...	19
Manual transmission drain and fill plug	
T5OD	156 in-lbs
TR3650	18 to 20
Drivebelt tensioner mounting bolt(s)	
V6 engine..	35
V8 engine..	18

Engine compartment layout (4.6L V8 model shown, V6 model similar)

1	Battery	6	Coolant expansion tank
2	Brake fluid reservoir	7	Engine oil filler cap
3	Engine oil dipstick	8	Upper radiator hose
4	Air filter housing	9	Windshield washer fluid reservoir
5	Power steering fluid reservoir	10	Underhood fuse/relay block

Typical underside components

1	Radiator drain plug	3	Brake caliper	5	Automatic transmission fluid
2	Engine oil filter	4	Engine oil drain plug		drain/fill plug

Typical rear underside components

1 Muffler	*2* Shock absorber	*3* Fuel tank	*4* Exhaust pipe

1   Maintenance schedule

The following maintenance intervals are based on the assumption that the vehicle owner will be doing the maintenance or service work, as opposed to having a dealer service department or other repair shop do the work. Although the time/mileage intervals are loosely based on factory recommendations, most have been shortened to ensure, for example, that such items as lubricants and fluids are checked/changed at intervals that promote maximum engine/driveline service life. Also, subject to the preference of the individual owner interested in keeping his or her vehicle in peak condition at all times, and with the vehicle's ultimate resale in mind, many of the maintenance procedures may be performed more often than recommended in the following schedule. We encourage such owner initiative.

When the vehicle is new it should be serviced initially by a factory authorized dealer service department to protect the factory warranty. In many cases the initial maintenance check is done at no cost to the owner (check with your dealer service department for more information).

Every 250 miles or weekly, whichever comes first

Check the engine oil level (Section 4)
Check the engine coolant level (Section 4)
Check the brake and clutch fluid level (Section 4)
Check the windshield washer fluid level (Section 4)
Check the power steering fluid level (Section 4)
Check the tires and tire pressures (Section 5)

Every 3000 miles or 3 months, whichever comes first

All items listed above, plus . . .
Change the engine oil and filter (Section 6)
Rotate the tires (Section 7)

Every 7500 miles or 6 months, whichever comes first

All items listed above, plus . . .
Inspect and replace, if necessary, the windshield wiper blades (Section 8)
Check and service the battery (Section 9)
Inspect the cooling system (Section 10)
Check the seatbelts (Section 11)

Every 15,000 miles or 12 months, whichever comes first

All items listed above, plus . . .
Check the manual transmission lubricant level (Section 4)
Check the rear axle (differential) lubricant level (Section 4)
Inspect and replace, if necessary, all underhood hoses (Section 12)
Inspect the brakes (Section 13)
Inspect the steering and suspension components (Section 14)

Check the fuel system (Section 15)
Replace the air filter (Section 16)*
Replace the interior ventilation filter (Section 17)

Every 30,000 miles or 24 months, whichever comes first

All items listed above, plus . . .
Inspect the exhaust system (Section 18)
Replace the fuel filter (Section 19)
Change the brake fluid (Section 20)
Check the engine drivebelt (Section 21)
Change the automatic transmission fluid (severe duty only) (Section 27)

Every 60,000 miles or 48 months, whichever comes first

All items listed above, plus . . .
Change the manual transmission lubricant (Section 22)
Change the rear axle (differential) lubricant (Section 22)
Inspect the spark plugs (Section 23)
Inspect the ignition coils (Section 24)
Inspect the spark plug wires - V6 engines (Section 28)

Every 100,000 miles

Replace the spark plugs (Section 23)
Service the cooling system (drain, flush and refill) (Section 25)
Replace the PCV valve (Section 26)

Every 150,000 miles

Change the automatic transmission fluid (Section 27)
Replace more often if the vehicle is driven in dusty areas

2   Introduction

This Chapter is designed to help the home mechanic maintain the Ford Mustang with the goals of maximum performance, economy, safety and reliability in mind.

Included is a master maintenance schedule, followed by procedures dealing specifically with each item on the schedule. Visual checks, adjustments, component replacement and other helpful items are included. Refer to the accompanying illustrations of the engine compartment and the underside of the vehicle for the locations of various components.

Servicing the vehicle, in accordance with the mileage/time maintenance schedule and the step-by-step procedures will result in a planned maintenance program that should produce a long and reliable service life. Keep in mind that it is a comprehensive plan, so maintaining some items but not others at the specified intervals will not produce the same results.

As you service the vehicle, you will discover that many of the procedures can - and should - be grouped together because of the nature of the particular procedure you're performing or because of the close proximity of two otherwise unrelated components to one another.

For example, if the vehicle is raised for chassis lubrication, you should inspect the exhaust, suspension, steering and fuel systems while you're under the vehicle. When you're rotating the tires, it makes good sense to check the brakes since the wheels are already removed. Finally, let's suppose you have to borrow or rent a torque wrench. Even if you only need it to tighten the spark plugs, you might as well check the torque of as many critical fasteners as time allows.

The first step in this maintenance program is to prepare yourself before the actual work begins. Read through all the procedures you're planning to do, then gather up all the parts and tools needed. If it looks like you might run into problems during a particular job, seek advice from a mechanic or an experienced do-it-yourselfer.

Owner's Manual and VECI label information

Your vehicle owner's manual was written for your year and model and contains very specific information on component locations, specifications, fuse ratings, part numbers, etc. The Owner's Manual is an important resource for the do-it-yourselfer to have; if one was not supplied with your vehicle, it can generally be ordered from a dealer parts department.

Among other important information, the Vehicle Emissions Control Information (VECI) label contains specifications and procedures for applicable tune-up adjustments and, in some instances, spark plugs. The information on this label is the exact maintenance data recommended by the manufacturer. This data often varies by intended operating altitude, local emissions regulations, month of manufacture, etc.

This Chapter contains procedural details, safety information and more ambitious maintenance intervals than you might find in manufacturer's literature. However, you may also find procedures or specifications in your Owner's Manual or VECI label that differ with what's printed here. In these cases, the Owner's Manual or VECI label can be considered correct, since it is specific to your particular vehicle.

3   Tune-up general information

The term tune-up is used in this manual to represent a combination of individual operations rather than one specific procedure.

If, from the time the vehicle is new, the routine maintenance schedule is followed closely and frequent checks are made of fluid levels and high wear items, as suggested throughout this manual, the engine will be kept in relatively good running condition and the need for additional work will be minimized.

More likely than not, however, there will be times when the engine is running poorly due to lack of regular maintenance. This is even more likely if a used vehicle, which has not received regular and frequent maintenance checks, is purchased. In such cases, an engine tune-up will be needed outside of the regular routine maintenance intervals.

The first step in any tune-up or diagnostic procedure to help correct a poor running engine is a cylinder compression check. A compression check (see Chapter 2C) will help determine the condition of internal engine components and should be used as a guide for tune-up and repair procedures. If, for instance, a compression check indicates serious internal engine wear, a conventional tune-up will not improve the performance of the engine and would be a waste of time and money. Because of its importance, the compression check should be done by someone with the right equipment and the knowledge to use it properly.

The following procedures are those most often needed to bring a generally poor running engine back into a proper state of tune.

Minor tune-up

Check all engine related fluids (Section 4)
Clean, inspect and test the battery (Section 9)
Check the cooling system (Section 10)
Check all underhood hoses (Section 12)
Check the air filter (Section 16)
Check the drivebelt (Section 21)
Inspect the spark plug and coils (Section 23)

Major tune-up

All items listed under Minor tune-up, plus . . .

Check the fuel system (Section 15)
Replace the air filter (Section 16)
Replace the fuel filter (Section 19)
Replace the spark plugs (Section 23)
Replace the PCV valve (Section 24)
Check the charging system (Chapter 5)

4   Fluid level checks (every 250 miles or weekly)

1   Fluids are an essential part of the lubrication, cooling, brake and windshield washer systems. Because the fluids gradually become depleted and/or contaminated during normal operation of the vehicle, they must be periodically replenished. See *Recommended lubricants and fluids* at the beginning of this Chapter before adding fluid to any of the following components. **Note:** *The vehicle must be on level ground when fluid levels are checked.*

Engine oil

Refer to illustrations 4.2, 4.4 and 4.6
2   The oil level is checked with a dipstick, which is located on the left (driver's) side of the engine **(see illustration)**. The dipstick extends through a metal tube down into the oil pan.
3   The oil level should be checked before

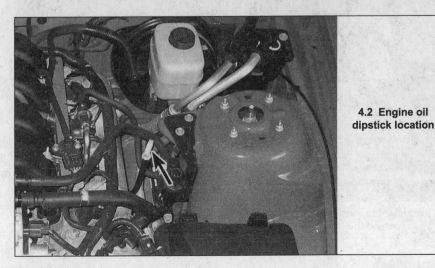

4.2 Engine oil dipstick location

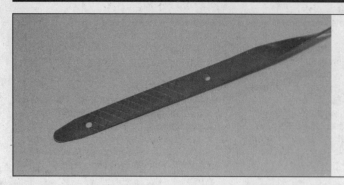

4.4 It takes one quart of oil to raise the level from the ADD mark (lower hole) to the FULL mark (upper hole). Don't allow the oil level to exceed the upper hole

4.6 Engine oil filler cap location

the vehicle has been driven, or about 5 minutes after the engine has been shut off. If the oil is checked immediately after driving the vehicle, some of the oil will remain in the upper part of the engine, resulting in an inaccurate reading on the dipstick.

4   Pull the dipstick out of the tube and wipe all the oil from the end with a clean rag or paper towel. Insert the clean dipstick all the way back into the tube and pull it out again. Note the oil at the end of the dipstick. At its highest point, the level should be above the MIN mark or lower hole, within the hatched marked section of the dipstick **(see illustration)**.

5   It takes one quart of oil to raise the level from the MIN mark (or lower hole) to the MAX mark (or upper hole) on the dipstick. Do not allow the level to drop below the MIN mark or oil starvation may cause engine damage. Conversely, overfilling the engine (adding oil above the MAX mark) may cause oil fouled spark plugs, oil leaks or oil seal failures.

6   To add oil, remove the filler cap from the valve cover **(see illustration)**. After adding oil, wait a few minutes to allow the level to stabilize, then pull out the dipstick and check the level again. Add more oil if required. Install the filler cap and tighten it by hand only.

7   Checking the oil level is an important preventive maintenance step. A consistently low oil level indicates oil leakage through damaged seals, defective gaskets or past worn rings or valve guides. If the oil looks milky in

color or has water droplets in it, the cylinder head gasket(s) may be blown or the head(s) or block may be cracked. The engine should be checked immediately. The condition of the oil should also be checked. Whenever you check the oil level, slide your thumb and index finger up the dipstick before wiping off the oil. If you see small dirt or metal particles clinging to the dipstick, the oil should be changed (see Section 6).

Engine coolant

Refer to illustrations 4.8 and 4.9
Warning: *Do not allow antifreeze to come in contact with your skin or painted surfaces of the vehicle. Flush contaminated areas immediately with plenty of water. Don't store new coolant or leave old coolant lying around where it's accessible to children or pets, they're attracted by its sweet smell. Ingestion of even a small amount of coolant can be fatal! Wipe up garage floor and drip pan spills immediately. Keep antifreeze containers covered and repair cooling system leaks as soon as they're noticed.*

8   All vehicles covered by this manual are equipped with a pressurized coolant recovery system. A plastic expansion tank located directly behind the radiator is connected by a hose to the engine **(see illustration)**. As the engine heats up during operation, the expanding coolant fills the tank.

9   The coolant level in the tank should be

checked regularly. The level in the tank varies with the temperature of the engine. When the engine is cold, the coolant level should be in the COLD FILL RANGE on the expansion tank **(see illustration)**. If it isn't, remove the cap from the tank and add a 50/50 mixture of ethylene glycol based antifreeze and water. **Warning:** *Do not remove the expansion tank cap when the engine is warm!*

10   Drive the vehicle, let the engine cool completely then recheck the coolant level. Don't use rust inhibitors or additives. If only a small amount of coolant is required to bring the system up to the proper level, water can be used. However, repeated additions of water will dilute the antifreeze and water solution. In order to maintain the proper ratio of antifreeze and water, always top up the coolant level with the correct mixture. An empty plastic milk jug or bleach bottle makes an excellent container for mixing coolant.

11   If the coolant level drops consistently, there may be a leak in the system. Inspect the radiator, hoses, filler cap, drain plugs and water pump (see Section 10). If no leaks are noted, have the expansion tank cap pressure tested by a service station.

12   If you have to remove the expansion tank

4.8 The coolant expansion tank is located behind the radiator

4.9 Keep the level in the COLD FILL RANGE - DO NOT remove the cap until the engine has cooled completely

4.15 The brake fluid level should be kept between the MIN and MAX marks on the translucent plastic reservoir

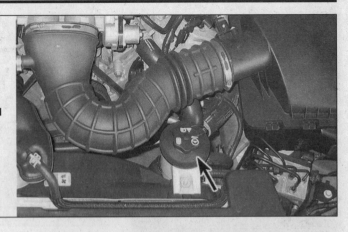

4.25 The power steering fluid reservoir is located on the left side of the radiator

4.22 The windshield washer fluid reservoir is located in the front corner of the engine compartment

cap wait until the engine has cooled completely, then wrap a thick cloth around the cap and unscrew it slowly, stopping if you hear a hissing noise. If coolant or steam escapes, let the engine cool down longer, then remove the cap.

13 Check the condition of the coolant as well. It should be relatively clear. If it's brown or rust colored, the system should be drained, flushed and refilled. Even if the coolant appears to be normal, the corrosion inhibitors wear out, so it must be replaced at the specified intervals.

Brake and clutch fluid

Refer to illustration 4.15

14 The brake master cylinder is mounted on the front of the power booster unit in the engine compartment. The hydraulic clutch master cylinder used on manual transaxle vehicles is located next to the brake master cylinder.

15 The brake master cylinder and the clutch master cylinder share a common reservoir. To check the fluid level of either system, simply look at the MAX and MIN marks on the brake fluid reservoir **(see illustration)**.

16 If the level is low, wipe the top of the reservoir with a clean rag to prevent contamination of the brake system before unscrewing the cap.

17 Add only the specified brake fluid to the reservoir (refer to *Recommended lubricants and fluids* at the front of this Chapter or to your owner's manual). Mixing different types of brake fluid can damage the system. Fill the brake master cylinder reservoir only to the MAX line. **Warning:** *Use caution when filling the reservoir - brake fluid can harm your eyes and damage painted surfaces. Do not use brake fluid that is more than one year old or has been left open. Brake fluid absorbs moisture from the air. Excess moisture can cause a dangerous loss of braking.*

18 While the reservoir cap is removed, inspect the master cylinder reservoir for contamination. If deposits, dirt particles or water droplets are present, the system should be

drained and refilled.

19 After filling the reservoir to the proper level, make sure the cap is tightened securely to prevent fluid leakage.

20 The fluid in the brake master cylinder will drop slightly as the brake pads at each wheel wear down during normal operation. If the master cylinder requires repeated replenishing to keep it at the proper level, this is an indication of leakage in the brake or clutch system, which should be corrected immediately. If the brake system shows an indication of leakage check all brake lines and connections, along with the calipers, wheel cylinders and booster (see Section 13 for more information). If the hydraulic clutch system shows an indication of leakage check all clutch lines and connections, along with the clutch release cylinder (see Chapter 8 for more information).

21 If, upon checking the brake or clutch master cylinder fluid level, you discover the reservoir empty or nearly empty, the systems should be bled (see Chapters 8 and 9).

Windshield washer fluid

Refer to illustration 4.22

22 Fluid for the windshield washer system is stored in a plastic reservoir located at the right front of the engine compartment **(see illustration)**.

23 In milder climates, plain water can be used in the reservoir, but it should be kept no more than 2/3 full to allow for expansion if the water freezes. In colder climates, use windshield washer system antifreeze, available at any auto parts store, to lower the freezing point of the fluid. Mix the antifreeze with water in accordance with the manufacturer's directions on the container. **Caution:** *Do not use cooling system antifreeze - it will damage the vehicle's paint.*

Power steering fluid

Refer to illustrations 4.25 and 4.28

24 Check the power steering fluid level periodically to avoid steering system problems, such as damage to the pump. **Caution:** *DO NOT hold the steering wheel against either stop (extreme left or right turn) for more than five seconds. If you do, the power steering pump could be damaged.*

4.28 With the engine cold, the fluid level should be between the MIN and MAX marks on the reservoir

25 The power steering reservoir, located at the left side of the radiator **(see illustration)**, has MIN and MAX fluid level marks on the side. The fluid level can be seen without removing the reservoir cap.

26 Park the vehicle on level ground and apply the parking brake.

27 Run the engine until it has reached normal operating temperature. With the engine at idle, turn the steering wheel back and forth about 10 times to get any air out of the steering system. Shut the engine off with the wheels in the straight-ahead position.

28 Note the fluid level on the side of the reservoir. It should be between the two marks **(see illustration)**.

29 Add small amounts of fluid until the level is correct. **Caution:** *Do not overfill the reservoir. If too much fluid is added, remove the excess with a clean syringe or suction pump.*

30 Check the power steering hoses and connections for leaks and wear.

Manual transmission lubricant level check

Note: *It isn't necessary to check this lubricant weekly; every 15000 miles or 12 months will be adequate*

31 The manual transmission has a filler plug which must be removed to check the lubricant

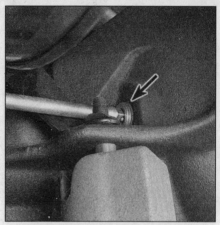

4.36 Use a 3/8-inch drive ratchet or breaker bar and an extension to remove the differential fill plug

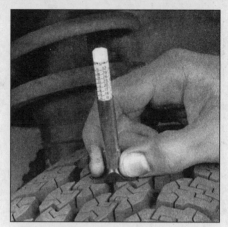

5.2 A tire tread depth indicator should be used to monitor tire wear - they are available at auto parts stores and service stations and cost very little

35 The differential has a check/fill plug which must be removed to check the lubricant level. If the vehicle is raised to gain access to the plug, be sure to support it safely on jackstands - DO NOT crawl under the vehicle when it's supported only by the jack!

36 Remove the check/fill plug from the differential (**see illustration**).

37 Use your little finger as a dipstick to make sure the lubricant level is within one inch of the bottom of the plug hole. If not, use a syringe to add the recommended lubricant. On some models a tag is located in the area of the plug which gives information regarding lubricant type, particularly on models equipped with a limited slip differential.

38 Install the plug and tighten it securely.

level. If the vehicle is raised to gain access to the plug, be sure to support it safely on jackstands - DO NOT crawl under a vehicle which is supported only by a jack! Be sure the vehicle is level or the check may be inaccurate.

32 Using an open-end wrench, unscrew the plug from the transmission and use a finger to reach inside the housing to determine the lubricant level. On V6 models, the level should even with the plug hole, and on V8 models, the level should be a 1/2-inch below the bottom of the plug hole.

33 If it isn't, add the recommended lubricant through the plug hole with a pump or squeeze bottle.

34 Install and tighten the plug and check for leaks after the first few miles of driving.

Differential lubricant level check

Refer to illustration 4.36

Note: *It isn't necessary to check this lubricant weekly; every 15000 miles or 12 months will be adequate*

5   Tire and tire pressure checks (every 250 miles or weekly)

Refer to illustrations 5.2, 5.3, 5.4a, 5.4b and 5.8

1 Periodic inspection of the tires may spare you the inconvenience of being stranded with a flat tire. It can also provide you with vital information regarding possible problems in the steering and suspension systems before major damage occurs.

2 The original tires on this vehicle are equipped with 1/2-inch wide bands that will appear when tread depth reaches 1/16-inch, at which point they can be considered worn out. Tread wear can be monitored with a

UNDERINFLATION

INCORRECT TOE-IN OR EXTREME CAMBER

CUPPING

Cupping may be caused by:

- Underinflation and/or mechanical irregularities such as out-of-balance condition of wheel and/or tire, and bent or damaged wheel.
- Loose or worn steering tie-rod or steering idler arm.
- Loose, damaged or worn front suspension parts.

OVERINFLATION

FEATHERING DUE TO MISALIGNMENT

5.3 This chart will help you determine the condition of your tires, the probable cause(s) of abnormal wear and the corrective action necessary

5.4a If a tire loses air on a steady basis, check the valve core first to make sure it's snug (special inexpensive wrenches are commonly available at auto parts stores)

5.4b If the valve core is tight, raise the corner of the vehicle with the low tire and spray a soapy water solution onto the tread as the tire is turned slowly - slow leaks will cause small bubbles to appear

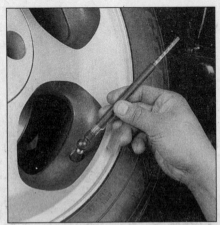

5.8 To extend the life of your tires, check the air pressure at least once a week with an accurate gauge (don't forget the spare!)

simple, inexpensive device known as a tread depth indicator **(see illustration)**.

3   Note any abnormal tread wear **(see illustration)**. Tread pattern irregularities such as cupping, flat spots and more wear on one side than the other are indications of front end alignment and/or balance problems. If any of these conditions are noted, take the vehicle to a tire shop or service station to correct the problem.

4   Look closely for cuts, punctures and embedded nails or tacks. Sometimes a tire will hold air pressure for a short time or leak down very slowly after a nail has embedded itself in the tread. If a slow leak persists, check the valve stem core to make sure it is tight **(see illustration)**. Examine the tread for an object that may have embedded itself in the tire or for a "plug" that may have begun to leak (radial tire punctures are repaired with a plug that is installed in a puncture). If a puncture is suspected, it can be easily verified by spraying a solution of soapy water onto the puncture area **(see illustration)**. The soapy solution will bubble if there is a leak. Unless the puncture is unusually large, a tire shop or service station can usually repair the tire.

5   Carefully inspect the inner sidewall of each tire for evidence of brake fluid leakage. If you see any, inspect the brakes immediately.

6   Correct air pressure adds miles to the life span of the tires, improves mileage and enhances overall ride quality. Tire pressure cannot be accurately estimated by looking at a tire, especially if it's a radial. A tire pressure gauge is essential. Keep an accurate gauge in the glove compartment. The pressure gauges attached to the nozzles of air hoses at gas stations are often inaccurate.

7   Always check tire pressure when the tires are cold. Cold, in this case, means the vehicle has not been driven over a mile in the three hours preceding a tire pressure check. A pressure rise of four to eight pounds is not uncommon once the tires are warm.

8   Unscrew the valve cap protruding from

the wheel or hubcap and push the gauge firmly onto the valve stem **(see illustration)**. Note the reading on the gauge and compare the figure to the recommended tire pressure shown on the tire placard on the driver's side door. Be sure to reinstall the valve cap to keep dirt and moisture out of the valve stem mechanism. Check all four tires and, if necessary, add enough air to bring them up to the recommended pressure.

9   Don't forget to keep the spare tire inflated to the specified pressure (refer to the pressure molded into the tire sidewall).

6   Engine oil and filter change (every 3000 miles or 3 months)

Refer to illustrations 6.2, 6.7, 6.12 and 6.15

1   Frequent oil changes are the most important preventive maintenance procedures that can be done by the home mechanic. As engine oil ages, it becomes diluted and contaminated, which leads to premature engine wear.

2   Make sure that you have all the necessary tools before you begin this procedure **(see illustration)**. You should also have plenty of rags or newspapers handy for mopping up oil spills.

3   Access to the oil drain plug and filter will be improved if the vehicle can be lifted on a hoist, driven onto ramps or supported by jackstands. **Warning:** *Do not work under a vehicle supported only by a jack - always use jackstands!*

4   If you haven't changed the oil on this vehicle before, get under it and locate the oil drain plug and the oil filter. The exhaust components will be warm as you work, so note how they are routed to avoid touching them when you are under the vehicle.

5   Start the engine and allow it to reach normal operating temperature - oil and sludge will flow out more easily when warm. If new oil, a

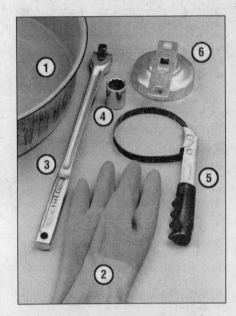

6.2 These tools are required when changing the engine oil and filter

1   *Drain pan - It should be fairly shallow in depth, but wide in order to prevent spills*
2   *Rubber gloves - When removing the drain plug and filter, it is inevitable that you will get oil on your hands (the gloves will prevent burns)*
3   *Breaker bar - Sometimes the oil drain plug is pretty tight and a long breaker bar is needed to loosen it*
4   *Socket - To be used with the breaker bar or a ratchet (must be the correct size to fit the drain plug)*
5   *Filter wrench - This is a metal band-type wrench, which requires clearance around the filter to be effective*
6   *Filter wrench - This type fits on the bottom of the filter and can be turned with a ratchet or beaker bar (different size wrenches are available for different types of filters)*

6.7 Use a proper size box-end wrench or socket to remove the oil drain plug and avoid rounding it off

6.12 Use an oil filter wrench to remove the filter

filter or tools are needed, use the vehicle to go get them and warm up the engine/oil at the same time. Park on a level surface and shut off the engine when it's warmed up. Remove the oil filler cap from the valve cover.

6   Raise the vehicle and support it on jackstands. Make sure it is safely supported!

7   Being careful not to touch the hot exhaust components, position a drain pan under the plug in the bottom of the engine, then remove the plug **(see illustration)**. It's a good idea to wear a rubber glove while unscrewing the plug the final few turns to avoid being scalded by hot oil.

8   It may be necessary to move the drain pan slightly as oil flow slows to a trickle. Inspect the old oil for the presence of metal particles.

9   After all the oil has drained, wipe off the drain plug with a clean rag. Any small metal particles clinging to the plug would immediately contaminate the new oil.

10   Clean the area around the drain plug opening, reinstall the plug and tighten it securely, but don't strip the threads.

11   Move the drain pan into position under the oil filter.

12   Loosen the oil filter by turning it counter-clockwise with a filter wrench **(see illustration)**. Any standard filter wrench will work.

13   Once the filter is loose, use your hands to unscrew it from the block. Just as the filter is detached from the block, immediately tilt the open end up to prevent the oil inside the filter from spilling out.

14   Using a clean rag, wipe off the mounting surface on the block. Also, make sure that none of the old gasket remains stuck to the mounting surface. It can be removed with a scraper if necessary.

15   Compare the old filter with the new one to make sure they are the same type. Smear some engine oil on the rubber gasket of the new filter and screw it into place **(see illustration)**. Overtightening the filter will damage the gasket, so don't use a filter wrench. Most filter manufacturers recommend tightening the filter by hand only. Normally they should be tightened 3/4-turn after the gasket contacts the block, but be sure to follow the directions on the filter or container.

16   Remove all tools and materials from under the vehicle, being careful not to spill the oil in the drain pan, then lower the vehicle.

17   Add new oil to the engine through the oil filler cap. Use a funnel to prevent oil from spilling onto the top of the engine. Pour four quarts of fresh oil into the engine. Wait a few minutes to allow the oil to drain into the pan, then check the level on the dipstick (see Section 4 if necessary). If the oil level is in the OK range, install the filler cap.

18   Start the engine and run it for about a minute. While the engine is running, look under the vehicle and check for leaks at the oil pan drain plug and around the oil filter. If either one is leaking, stop the engine and tighten the plug or filter slightly.

19   Wait a few minutes, then recheck the level on the dipstick. Add oil as necessary to bring the level into the OK range.

20   During the first few trips after an oil change, make it a point to check frequently for leaks and proper oil level.

21   The old oil drained from the engine can-

not be reused in its present state and should be disposed of. Check with your local auto parts store, disposal facility or environmental agency to see if they will accept the oil for recycling. After the oil has cooled it can be drained into a container (capped plastic jugs, topped bottles, milk cartons, etc.) for transport to one of these disposal sites. Don't dispose of the oil by pouring it on the ground or down a drain!

7   Tire rotation (every 3000 miles or 3 months)

Refer to illustrations 7.2a and 7.2b

1   The tires should be rotated at the specified intervals and whenever uneven wear is noticed. Since the vehicle will be raised and the tires removed anyway, check the brakes also (see Section 13).

2   Radial tires must be rotated in a specific

6.15 Lubricate the oil filter gasket with clean engine oil before installing the filter on the engine

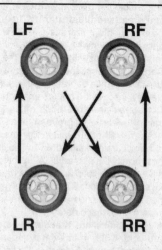

7.2a The recommended tire rotation pattern *for non-directional tires*

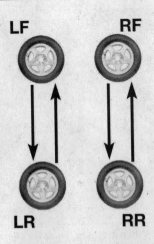

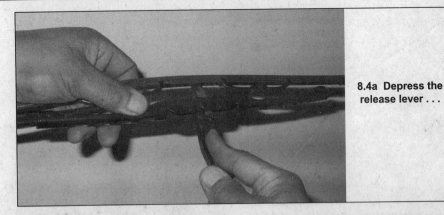

8.4a Depress the release lever . . .

1-AJ HAYNES

7.2b The recommended tire rotation pattern *for directional tires*

pattern **(see illustrations)**. If your vehicle has a compact spare tire, don't include it in the rotation pattern.

3   Refer to the information in *Jacking and towing* at the front of this manual for the proper procedure to follow when raising the vehicle and changing a tire. If the brakes must be checked, don't apply the parking brake as stated.

4   The vehicle must be raised on a hoist or supported on jackstands to get all four wheels off the ground. Make sure the vehicle is safely supported!

5   After the rotation procedure is finished, check and adjust the tire pressures as necessary and be sure to check the lug nut tightness.

8   Windshield wiper blade inspection and replacement (every 7500 miles or 6 months)

Refer to illustrations 8.4a and 8.4b

1   The windshield wiper and blade assembly should be inspected periodically for damage, loose components and cracked or worn blade elements.

2   Road film can build up on the wiper blades and affect their efficiency, so they should be washed regularly with a mild detergent solution.

3   If the wiper blade elements are cracked, worn or warped, or no longer clean adequately, they should be replaced with new ones.

4   Lift the arm assembly away from the glass for clearance, press on the release lever, then slide the wiper blade assembly out of the hook in the end of the arm **(see illustrations)**.

5   Attach the new wiper to the arm. Connection can be confirmed by an audible click.

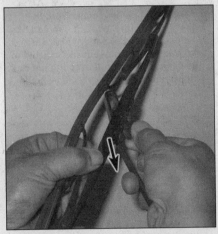

8.4b . . . and slide the wiper assembly down the wiper arm and out of the hook in the end of the arm

9   Battery check, maintenance and charging (every 7500 miles or 6 months)

Refer to illustrations 9.1, 9.6a, 9.6b, 9.7a and 9.7b

Warning: *Certain precautions must be followed when checking and servicing the battery. Hydrogen gas, which is highly flammable, is always present in the battery cells, so keep lighted tobacco and all other open flames and sparks away from the battery. The electrolyte inside the battery is actually diluted sulfuric acid, which will cause injury if splashed on your skin or in your eyes. It will also ruin clothes and painted surfaces. When removing the battery cables, always detach the negative cable first and hook it up last!*

1   A routine preventive maintenance program for the battery in your vehicle is the only way to ensure quick and reliable starts. But before performing any battery maintenance, make sure that you have the proper equipment necessary to work safely around the battery **(see illustration)**.

2   There are also several precautions that should be taken whenever battery maintenance is performed. Before servicing the battery, always turn the engine and all acces-

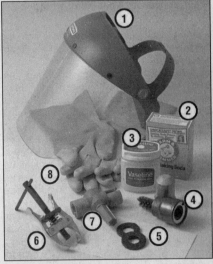

9.1 Tools and materials required for battery maintenance

1   *Face shield/safety goggles - When removing corrosion with a brush, the acidic particles can easily fly up into your eyes*

2   *Baking soda - A solution of baking soda and water can be used to neutralize corrosion*

3   *Petroleum jelly - A layer of this on the battery posts will help prevent corrosion*

4   *Battery post/cable cleaner - This wire brush cleaning tool will remove all traces of corrosion from the battery posts and cable clamps*

5   *Treated felt washers - Placing one of these on each post, directly under the cable clamps, will help prevent corrosion*

6   *Puller - Sometimes the cable clamps are very difficult to pull off the posts, even after the nut/bolt has been completely loosened. This tool pulls the clamp straight up and off the post without damage*

7   *Battery post/cable cleaner - Here is another cleaning tool which is a slightly different version of number 4 above, but it does the same thing*

8   *Rubber gloves - Another safety item to consider when servicing the battery; remember that's acid inside the battery*

9.6a Battery terminal corrosion usually appears as light, fluffy powder

9.6b Removing a cable from the battery post with a wrench - sometimes a pair of special battery pliers are required for this procedure if corrosion has caused deterioration of the nut hex (always remove the ground (-) cable first and hook it up last!)

9.7a When cleaning the cable clamps, all corrosion must be removed (the inside of the clamp is tapered to match the taper on the post, so don't remove too much material)

sories off and disconnect the cable from the negative terminal of the battery (see Chapter 5).

3   The battery produces hydrogen gas, which is both flammable and explosive. Never create a spark, smoke or light a match around the battery. Always charge the battery in a ventilated area.

4   Electrolyte contains poisonous and corrosive sulfuric acid. Do not allow it to get in your eyes, on your skin, or on your clothes. Never ingest it. Wear protective safety glasses when working near the battery. Keep children away from the battery.

5   Note the external condition of the battery. If the positive terminal and cable clamp on your vehicle's battery is equipped with a rubber protector, make sure that it's not torn or damaged. It should completely cover the terminal. Look for any corroded or loose connections, cracks in the case or cover or loose hold-down clamps. Also check the entire length of each cable for cracks and frayed conductors.

6   If corrosion, which looks like white, fluffy deposits **(see illustration)** is evident, particularly around the terminals, the battery should be removed for cleaning. Loosen the cable clamp bolts with a wrench, being careful to

9.7b Regardless of the type of tool used to clean the battery posts, a clean, shiny surface should be the result

remove the ground cable first, and slide them off the terminals **(see illustration)**. Then disconnect the hold-down clamp bolt and nut, remove the clamp and lift the battery from the engine compartment.

7   Clean the cable clamps thoroughly with a battery brush or a terminal cleaner and a solution of warm water and baking soda **(see illustration)**. Wash the terminals and the top of the battery case with the same solution but make sure that the solution doesn't get into the battery. When cleaning the cables, terminals and battery top, wear safety goggles and rubber gloves to prevent any solution from coming in contact with your eyes or hands. Wear old clothes too - even diluted, sulfuric acid splashed onto clothes will burn holes in them. If the terminals have been extensively corroded, clean them up with a terminal cleaner **(see illustration)**. Thoroughly wash all cleaned areas with plain water.

8   Make sure that the battery tray is in good condition and the hold-down clamp fasteners are tight. If the battery is removed from the tray, make sure no parts remain in the bottom of the tray when the battery is reinstalled. When reinstalling the hold-down clamp bolts, do not overtighten them.

9   Information on removing and installing the battery can be found in Chapter 5. Information on jump starting can be found at the front of this manual. For more detailed battery checking procedures, refer to the *Haynes Automotive Electrical Manual*.

Cleaning

10   Corrosion on the hold-down components, battery case and surrounding areas can be removed with a solution of water and baking soda. Thoroughly rinse all cleaned areas with plain water.

11   Any metal parts of the vehicle damaged by corrosion should be covered with a zinc-based primer, then painted.

Charging

Warning: *When batteries are being charged, hydrogen gas, which is very explosive and flammable, is produced. Do not smoke or allow open flames near a charging or a recently charged battery. Wear eye protection when near the battery during charging. Also, make sure the charger is unplugged before connecting or disconnecting the battery from the charger.*

12   Slow-rate charging is the best way to restore a battery that's discharged to the point where it will not start the engine. It's also a good way to maintain the battery charge in a vehicle that's only driven a few miles between starts. Maintaining the battery charge is particularly important in the winter when the battery must work harder to start the engine and electrical accessories that drain the battery are in greater use.

13   It's best to use a one or two-amp battery charger (sometimes called a "trickle" charger). They are the safest and put the least strain on the battery. They are also the least expensive. For a faster charge, you can use a higher amperage charger, but don't use one rated more than 1/10th the amp/hour rating of the battery. Rapid boost charges that claim to restore the power of the battery in one to two hours are hardest on the battery and can damage batteries not in good condition. This type of charging should only be used in emergency situations.

14   The average time necessary to charge a battery should be listed in the instructions that come with the charger. As a general rule, a trickle charger will charge a battery in 12 to 16 hours.

10   Cooling system check (every 7,500 miles or 6 months)

Refer to illustration 10.4

Warning: *Wait until the engine is completely*

Check for a chafed area that could fail prematurely.

Check for a soft area indicating the hose has deteriorated inside.

Overtightening the clamp on a hardened hose will damage the hose and cause a leak.

Check each hose for swelling and oil-soaked ends. Cracks and breaks can be located by squeezing the hose.

10.4 Hoses, like drivebelts, have a habit of failing at the worst possible time - to prevent the inconvenience of a blown radiator or heater hose, inspect them carefully as shown here

cool before performing this procedure.

1   Many major engine failures can be caused by a faulty cooling system.

2   The engine must be cold for the cooling system check, so perform the following procedure before the vehicle is driven for the day or after it has been shut off for at least three hours.

3   Remove the pressure cap from the expansion tank at the right side of the radiator. Clean the cap thoroughly, inside and out, with clean water. The presence of rust or corrosion in the expansion tank means the coolant should be changed (see Section 25). The coolant inside the expansion tank should be relatively clean and transparent. If it's rust colored, drain the system and refill it with new coolant.

4   Carefully check the radiator hoses and the smaller diameter heater hoses. Inspect

each coolant hose along its entire length, replacing any hose which is cracked, swollen or deteriorated **(see illustration)**. Cracks will show up better if the hose is squeezed. Pay close attention to hose clamps that secure the hoses to cooling system components. Hose clamps can pinch and puncture hoses, resulting in coolant leaks.

5   Make sure that all hose connections are tight. A leak in the cooling system will usually show up as white or rust colored deposits on the area adjoining the leak. If wire-type clamps are used on the hoses, it may be a good idea to replace them with screw-type clamps.

6   Clean the front of the radiator and air conditioning condenser with compressed air, if available, or a soft brush. Remove all bugs, leaves, etc. embedded in the radiator fins. Be extremely careful not to damage the cooling fins or cut your fingers on them.

7   If the coolant level has been dropping consistently and no leaks are detectable, have the expansion tank cap and cooling system pressure checked at a service station.

---

11   Seat belt check (every 7,500 miles or 6 months)

1   Check seat belts, buckles, latch plates and guide loops for obvious damage and signs of wear.

2   See if the seat belt reminder light comes on when the key is turned to the Run or Start position. A chime should also sound.

3   The seat belts are designed to lock up during a sudden stop or impact, yet allow free movement during normal driving. Make sure the retractors return the belt against your chest while driving and rewind the belt fully when the buckle is unlatched.

4   If any of the above checks reveal problems with the seat belt system, replace parts as necessary.

---

12   Underhood hose check and replacement (every 15,000 miles or 12 months)

Warning: *Replacement of air conditioning hoses must be left to a dealer service department or air conditioning shop that has the equipment to depressurize the system safely. Never remove air conditioning components or hoses until the system has been depressurized.*

General

1   High temperatures under the hood can cause deterioration of the rubber and plastic hoses used for engine, accessory and emission systems operation. Periodic inspection should be made for cracks, loose clamps, material hardening and leaks.

2   Information specific to the cooling system hoses can be found in Section 10.

3   Most (but not all) hoses are secured to

the fittings with clamps. Where clamps are used, check to be sure they haven't lost their tension, allowing the hose to leak. If clamps aren't used, make sure the hose has not expanded and/or hardened where it slips over the fitting, allowing it to leak.

PCV system hose

4   To reduce hydrocarbon emissions, crankcase blow-by gas is vented through the PCV valve (V6 models only) in the valve cover to the intake manifold via a rubber hose on most models. The blow-by gases mix with incoming air in the intake manifold before being burned in the combustion chambers.

5   Check the PCV hose for cracks, leaks and other damage. Disconnect it from the valve cover and the intake manifold and check the inside for obstructions. If it's clogged, clean it out with solvent.

Vacuum hoses

6   It's quite common for vacuum hoses, especially those in the emissions system, to be color coded or identified by colored stripes molded into them. Various systems require hoses with different wall thickness, collapse resistance and temperature resistance. When replacing hoses, be sure the new ones are made of the same material.

7   Often the only effective way to check a hose is to remove it completely from the vehicle. If more than one hose is removed, be sure to label the hoses and fittings to ensure correct installation.

8   When checking vacuum hoses, be sure to include any plastic T-fittings in the check. Inspect the fittings for cracks and the hose where it fits over each fitting for distortion, which could cause leakage.

9   A small piece of vacuum hose (1/4-inch inside diameter) can be used as a stethoscope to detect vacuum leaks. Hold one end of the hose to your ear and probe around vacuum hoses and fittings, listening for the "hissing" sound characteristic of a vacuum leak. **Warning:** *When probing with the vacuum hose stethoscope, be careful not to come into contact with moving engine components such as drivebelts, the cooling fan, etc.*

Fuel hose

Warning: *Gasoline is flammable, so take extra precautions when you work on any part of the fuel system. Don't smoke or allow open flames or bare light bulbs near the work area, and don't work in a garage where a gas-type appliance (such as a water heater or clothes dryer) is present. Since fuel is carcinogenic, wear fuel-resistant gloves when there's a possibility of being exposed to fuel, and, if you spill any fuel on your skin, rinse it off immediately with soap and water. Mop up any spills immediately and do not store fuel-soaked rags where they could ignite. The fuel system is under constant pressure, so, if any fuel lines are to be disconnected, the fuel pressure in the system must be relieved first (see*

Chapter 4 for more information). *When you perform any kind of work on the fuel system, wear safety glasses and have a Class B type fire extinguisher on hand.*

10   The fuel lines are usually under pressure, so if any fuel lines are to be disconnected be prepared to catch spilled fuel. **Warning:** *Your vehicle is equipped with fuel injection and you must relieve the fuel system pressure before servicing the fuel lines. Refer to Chapter 4 for the fuel system pressure relief procedure.*

11   Check all flexible fuel lines for deterioration and chafing. Check especially for cracks in areas where the hose bends and just before fittings, such as where a hose attaches to the fuel pump, fuel filter and fuel injection unit.

12   When replacing a hose, use only hose that is specifically designed for your fuel injection system.

13   Spring-type clamps are sometimes used on fuel return or vapor lines. These clamps often lose their tension over a period of time, and can be "sprung" during removal. Replace all spring-type clamps with screw clamps whenever a hose is replaced. Some fuel lines use spring-lock type couplings, which require a special tool to disconnect. See Chapter 4 for more information on this type of coupling.

Metal lines

14   Sections of metal line are often used for fuel line between the fuel pump and the fuel injection unit. Check carefully to make sure the line isn't bent, crimped or cracked.

15   If a section of metal fuel line must be replaced, use seamless steel tubing only, since copper and aluminum tubing do not have the strength necessary to withstand vibration caused by the engine.

16   Check the metal brake lines where they enter the master cylinder and brake proportioning unit (if used) for cracks in the lines and loose fittings. Any sign of brake fluid leakage calls for an immediate thorough inspection of the brake system.

13   Brake check (every 15,000 miles or 12 months)

Refer to illustrations 13.7a, 13.7b and 13.9

Warning: *The dust created by the brake system is harmful to your health. Never blow it out with compressed air and don't inhale any of it. An approved filtering mask should be worn when working on the brakes. Do not, under any circumstances, use petroleum-based solvents to clean brake parts. Use brake system cleaner only! Try to use non-asbestos replacement parts whenever possible.*

Note: *For detailed photographs of the brake system, refer to* Chapter 9.

1   In addition to the specified intervals, the brakes should be inspected every time the wheels are removed or whenever a defect is suspected.

2   Any of the following symptoms could indicate a potential brake system defect: The

13.7a With the wheel off, check the thickness of the inner pad through the inspection hole (front disc shown, rear disc caliper similar)

vehicle pulls to one side when the brake pedal is depressed; the brakes make squealing or dragging noises when applied; brake pedal travel is excessive; the pedal pulsates; or brake fluid leaks, usually onto the inside of the tire or wheel.

3   Loosen the wheel lug nuts.

4   Raise the vehicle and place it securely on jackstands.

5   Remove the wheels (see *Jacking and towing* at the front of this book, or your owner's manual, if necessary).

6   There are two pads (an outer and an inner) in each caliper. The pads are visible with the wheels removed.

7   Check the pad thickness by looking at each end of the caliper and through the inspection window in the caliper body **(see illustrations)**. If the lining material is less than the thickness listed in this Chapter's Specifications, replace the pads. **Note:** *Keep in mind that the lining material is riveted or bonded to a metal backing plate and the metal portion is not included in this measurement.*

8   If it is difficult to determine the exact thickness of the remaining pad material by the above method, or if you are at all concerned about the condition of the pads, remove the caliper(s), then remove the pads from the calipers for further inspection (refer to Chapter 9).

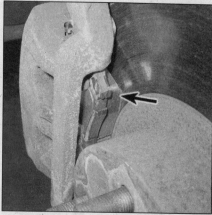

13.7b The outer pad is more easily checked at the edge of the caliper

9   Once the pads are removed from the calipers, clean them with brake cleaner and re-measure them with a ruler or a vernier caliper **(see illustration)**.

10   Measure the disc thickness with a micrometer to make sure that it still has service life remaining. If any disc is thinner than the specified minimum thickness, replace it (refer to Chapter 9). Even if the disc has service life remaining, check its condition. Look for scoring, gouging and burned spots. If these conditions exist, remove the disc and have it resurfaced (see Chapter 9).

11   Before installing the wheels, check all brake lines and hoses for damage, wear, deformation, cracks, corrosion, leakage, bends and twists, particularly in the vicinity of the rubber hoses at the calipers. Check the clamps for tightness and the connections for leakage. Make sure that all hoses and lines are clear of sharp edges, moving parts and the exhaust system. If any of the above conditions are noted, repair, reroute or replace the lines and/or fittings as necessary (see Chapter 9).

Brake booster check

12   Sit in the driver's seat and perform the following sequence of tests.

13   With the brake fully depressed, start the engine - the pedal should move down a little when the engine starts.

14   With the engine running, depress the

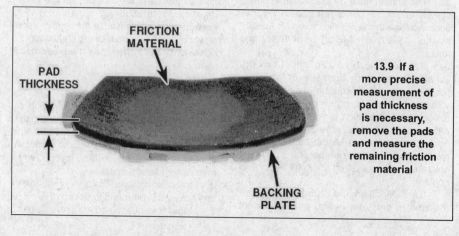

FRICTION
MATERIAL

PAD
THICKNESS

13.9 If a more precise measurement of pad thickness is necessary, remove the pads and measure the remaining friction material

BACKING
PLATE

brake pedal several times - the travel distance should not change.

15   Depress the brake, stop the engine and hold the pedal in for about 30 seconds - the pedal should neither sink nor rise.

16   Restart the engine, run it for about a minute and turn it off. Then firmly depress the brake several times - the pedal travel should decrease with each application.

17   If your brakes do not operate as described, the brake booster has failed. Refer to Chapter 9 for the replacement procedure.

Parking brake

18   One method of checking the parking brake is to park the vehicle on a steep hill with the parking brake set and the transmission in Neutral (be sure to stay in the vehicle for this check). If the parking brake cannot prevent the vehicle from rolling, it's in need of attention (see Chapter 9).

---

14   Steering and suspension check (every 15,000 miles or 12 months)

Note: *The steering linkage and suspension components should be checked periodically. Worn or damaged suspension and steering linkage components can result in excessive and abnormal tire wear, poor ride quality and vehicle handling and reduced fuel economy. For detailed illustrations of the steering and suspension components, refer to Chapter 10.*

Shock absorber check

Refer to illustration 14.6

1   Park the vehicle on level ground, turn the engine off and set the parking brake. Check the tire pressures.

2   Push down at one corner of the vehicle, then release it while noting the movement of

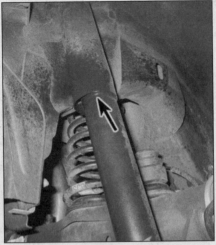

14.6 Check for signs of fluid leakage at this point on shock absorbers

the body. It should stop moving and come to rest in a level position within one or two bounces.

3   If the vehicle continues to move up-and-down or if it fails to return to its original position, a worn or weak shock absorber is probably the reason.

4   Repeat the above check at each of the three remaining corners of the vehicle.

5   Raise the vehicle and support it securely on jackstands.

6   Check the shock absorbers for evidence of fluid leakage (**see illustration**). A light film of fluid is no cause for concern. Make sure that any fluid noted is from the shocks and not from some other source. If leakage is noted, replace the shocks as a set.

7   Check the shocks to be sure that they are securely mounted and undamaged. Check the upper mounts for damage and wear. If damage or wear is noted, replace the shocks

as a set (front or rear).

8   If the shocks must be replaced, refer to Chapter 10 for the procedure.

Steering and suspension check

Refer to illustrations 14.9 and 14.11

9   Visually inspect the steering and suspension components (front and rear) for damage and distortion. Look for damaged seals, boots and bushings and leaks of any kind. Examine the bushings where the control arms meet the chassis (**see illustration**).

10   Clean the lower end of the steering knuckle. Have an assistant grasp the lower edge of the tire and move the wheel in-and-out while you look for movement at the steering knuckle-to-control arm balljoint. If there is any movement the suspension balljoint(s) must be replaced.

11   Grasp each front tire at the front and rear edges, push in at the front, pull out at the rear and feel for play in the steering system components. If any freeplay is noted, check the idler arm and the tie-rod ends for looseness (**see illustration**).

12   Additional steering and suspension system information and illustrations can be found in Chapter 10.

---

15   Fuel system check (every 15,000 miles or 12 months)

Warning: *Gasoline is flammable, so take extra precautions when you work on any part of the fuel system. Don't smoke or allow open flames or bare light bulbs near the work area, and don't work in a garage where a gas-type appliance (such as a water heater or clothes dryer) is present. Since fuel is carcinogenic, wear fuel-resistant gloves when there's a possibility of being exposed to fuel, and, if you*

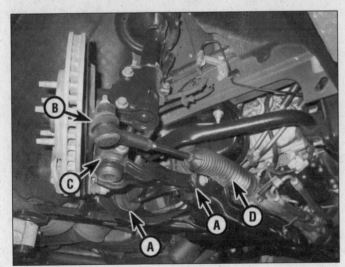

14.9 Examine the mounting points for the control arms (A), the tie-rod ends (B), the balljoints (C), and the steering gear boots (D)

14.11 With the steering wheel in the locked position and the vehicle raised, grasp the front tire as shown and try to move it back-and-forth - if any play is noted, check the steering gear mounts and tie-rod ends for looseness

16.2a Release the two clamps . . .

16.2b . . . then remove the air filter element

spill any fuel on your skin, rinse it off immediately with soap and water. Mop up any spills immediately and do not store fuel-soaked rags where they could ignite. When you perform any kind of work on the fuel system, wear safety glasses and have a Class B type fire extinguisher on hand. The fuel system is under constant pressure, so, before any lines are disconnected, the fuel system pressure must be relieved (see Chapter 4).

1   If you smell gasoline while driving or after the vehicle has been sitting in the sun, inspect the fuel system immediately.

2   Remove the fuel filler cap and inspect it for damage and corrosion. The gasket should have an unbroken sealing imprint. If the gasket is damaged or corroded, install a new cap.

3   Inspect the fuel feed line for cracks. Make sure that the connections between the fuel lines and the fuel injection system and between the fuel lines and the in-line fuel filter are tight. **Warning:** *Your vehicle is fuel injected, so you must relieve the fuel system pressure before servicing fuel system components. The fuel system pressure relief procedure is outlined in* Chapter 4.

4   Since some components of the fuel system - the fuel tank and part of the fuel feed line, for example - are underneath the vehicle, they can be inspected more easily with the vehicle raised on a hoist. If that's not possible, raise the vehicle and support it on jackstands.

5   With the vehicle raised and safely supported, inspect the gas tank and filler neck for punctures, cracks and other damage. The connection between the filler neck and the tank is particularly critical. Sometimes a rubber filler neck will leak because of loose clamps or deteriorated rubber. Inspect all fuel tank mounting brackets and straps to be sure that the tank is securely attached to the vehicle. **Warning:** *Do not, under any circumstances, try to repair a fuel tank (except rubber components). A welding torch or any open flame can easily cause fuel vapors inside the tank to explode.*

6   Carefully check all rubber hoses and

metal lines leading away from the fuel tank. Check for loose connections, deteriorated hoses, crimped lines and other damage. Repair or replace damaged sections as necessary (see Chapter 4).

16   Air filter check and replacement (every 15,000 miles or 12 months)

Refer to illustrations 16.2a and 16.2b

1   The air filter is located inside a housing at the left (driver's) side of the engine compartment.

2   To remove the air filter, release the tabs that secure the two halves of the air cleaner housing together, then lift the cover up and remove the air filter element **(see illustrations).**

3   Inspect the outer surface of the filter element. If it is dirty, replace it. If it is only moderately dusty, it can be reused by blowing it clean from the back to the front surface with compressed air. Because it is a pleated paper type filter, it cannot be washed or oiled. If it cannot be cleaned satisfactorily with compressed air, discard and replace it. While the cover is off, be careful not to drop anything down into the housing. **Caution:** *Never drive the vehicle with the air cleaner removed. Excessive engine wear could result and backfiring could even cause a fire under the hood.*

4   Wipe out the inside of the air cleaner housing.

5   Place the new filter into the air cleaner housing, making sure it seats properly.

6   Installation of the housing cover is the reverse of removal.

17   Interior ventilation filter replacement (every 15,000 miles or 12 months)

Refer to illustration 17.3

1   These models are equipped with an air

17.3 Release the tab and remove the filter

filtering element for the air conditioning system, located under the right side cowl cover.

2   Remove the right side cowl cover (see Chapter 11).

3   Release the tab and remove the filter **(see illustration).**

4   Installation is the reverse of the removal procedure.

18   Exhaust system check (every 30,000 miles or 24 months)

Refer to illustration 18.2

1   With the engine cold (at least three hours after the vehicle has been driven), check the complete exhaust system from the engine to the end of the tailpipe. Ideally, the inspection should be done with the vehicle on a hoist to permit unrestricted access. If a hoist isn't available, raise the vehicle and support it securely on jackstands.

2   Check the exhaust pipes and connections for evidence of leaks, severe corrosion and damage. Make sure that all brackets and hangers are in good condition and tight **(see illustration).**

18.2 Inspect the hangers for signs of deterioration

19.1 The fuel filter is mounted under a cover on the left side, in front of the gas tank

3   At the same time, inspect the underside of the body for holes, corrosion, open seams, etc. which may allow exhaust gases to enter the passenger compartment. Seal all body openings with silicone or body putty.

4   Rattles and other noises can often be traced to the exhaust system, especially the mounts and hangers. Try to move the pipes, muffler and catalytic converter. If the components can come in contact with the body or suspension parts, secure the exhaust system with new mounts.

5   Check the running condition of the engine by inspecting inside the end of the tail-pipe. The exhaust deposits here are an indication of engine state-of-tune. If the pipe is black and sooty or coated with white deposits, the engine may need a tune-up, including a thorough fuel system inspection.

19 Fuel filter replacement (every 30,000 miles or 24 months)

Refer to illustration 19.1

Warning: *Gasoline is extremely flammable, so take extra precautions when you work on any part of the fuel system. Don't smoke or allow open flames or bare light bulbs near the work area, and don't work in a garage where a gas-type appliance (such as a water heater or clothes dryer) is present. Since fuel is carcinogenic, wear fuel-resistant gloves when there's a possibility of being exposed to fuel, and, if you spill any fuel on your skin, rinse it off immediately with soap and water. Mop up any spills immediately and do not store fuel-soaked rags where they could ignite. When you perform any kind of work on the fuel system, wear safety glasses and have a Class B type fire extinguisher on hand.*

1   The fuel filter is mounted under the vehicle on the left side, in front of the gas tank **(see illustration).**

2   Relieve the fuel system pressure (see Chapter 4), then disconnect the cable from the negative terminal of the battery.

3   If necessary, raise the vehicle and support it securely on jackstands.

4   Inspect the fittings at both ends of the filter to see if they're clean. If more than a light coating of dust is present, clean the fittings before proceeding.

5   Release the clips holding the fuel lines to the filter (see Chapter 4).

6   Detach the fuel hoses, one at a time, from the filter. Be prepared for fuel spillage.

7   After the lines are detached, check the fittings for damage and distortion. If they were damaged in any way during removal, new ones must be used when the lines are reattached to the new filter (if new clips are packaged with the filter, be sure to use them in place of the originals).

8   Remove the fuel filter from the mounting bracket, while noting the direction the fuel filter is installed.

9   Install the new filter in the same direction. Carefully push each hose onto the filter until it's seated against the collar on the fitting. Make sure the clips are securely attached to the hose fittings - if they come off, the hoses could back off the filter and a fire could result!

10   Start the engine and check for fuel leaks.

20 Brake fluid change (every 30,000 miles or 24 months)

Warning: *Brake fluid can harm your eyes and damage painted surfaces, so use extreme caution when handling or pouring it. Do not use brake fluid that has been standing open or is more than one year old. Brake fluid absorbs moisture from the air. Excess moisture can cause a dangerous loss of braking effectiveness.*

1   At the specified intervals, the brake fluid should be drained and replaced. Since the brake fluid may drip or splash when pouring it, place plenty of rags around the master cylinder to protect any surrounding painted surfaces.

2   Before beginning work, purchase the specified brake fluid (see *Recommended lubricants and fluids* at the beginning of this Chapter).

3   Remove the cap from the master cylinder reservoir.

4   Using a hand suction pump or similar device, withdraw the fluid from the master cylinder reservoir.

5   Add new fluid to the master cylinder until it rises to the base of the filler neck.

6   Bleed the brake system as described in Chapter 9 at all four brakes until new and uncontaminated fluid is expelled from the bleeder screw. Be sure to maintain the fluid level in the master cylinder as you perform the bleeding process. If you allow the master cylinder to run dry, air will enter the system.

7   Refill the master cylinder with fluid and check the operation of the brakes. The pedal should feel solid when depressed, with no sponginess. **Warning:** *Do not operate the vehicle if you are in doubt about the effectiveness of the brake system.*

21 Drivebelt check and replacement (every 30,000 miles or 24 months)

1   A single serpentine drivebelt is located at the front of the engine and plays an important role in the overall operation of the engine and its components. Due to its function and material make up, the belt is prone to wear and should be periodically inspected. Although the belt should be inspected at the recommended intervals, replacement may not be necessary for more than 100,000 miles.

2   The vehicles covered by this manual are equipped with a single self-adjusting serpentine drivebelt, which is used to drive all of the accessory components such as the alternator, power steering pump, water pump and air conditioning compressor.

Check

Refer to illustrations 21.4 and 21.5

3   With the engine off, open the hood and locate the drivebelt at the front of the engine. Using your fingers (and a flashlight, if necessary), move along the belts checking for cracks and separation of the belt plies. Also

check for fraying and glazing, which gives the belt a shiny appearance. Both sides of each belt should be inspected, which means you will have to twist the belt to check the underside.

4   Check the ribs on the underside of the belt. They should all be the same depth, with none of the surface uneven (see illustration).

5   The tension of the belt is automatically adjusted by the belt tensioner and does not require any adjustments. Drivebelt wear can be checked visually by inspecting the wear indicator marks located on the side of the tensioner body. Locate the belt tensioner at the front of the engine, adjacent to the lower crankshaft pulley, then find the tensioner operating marks. If the indicator mark is outside the operating range, the belt should be replaced (see illustration).

Replacement

Refer to illustration 21.6

6   To replace the belt, rotate the tensioner counterclockwise on V6 models, or clockwise on V8 models to relieve the tension on the belt (see illustration).

7   Remove the belt from the auxiliary components.

8   Route the new belt over the various pulleys, again rotating the tensioner to allow the belt to be installed, then release the belt tensioner. Make sure the belt fits properly into the pulley grooves - it must be completely engaged. **Note:** *Most models have a drivebelt routing decal on the upper radiator panel to help during drivebelt installation.*

Tensioner replacement

9   Remove the drivebelt as described previously.

10   On V6 models, remove the bolt in the center of the tensioner, then detach the tensioner from the engine.

11   On V8 models, remove the three bolts securing the tensioner to the engine block.

12   Installation is the reverse of removal.

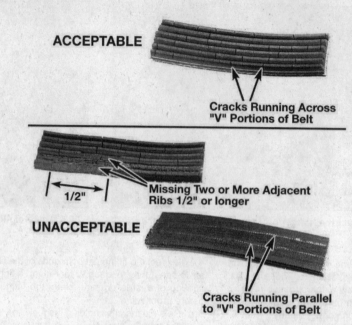

ACCEPTABLE

Cracks Running Across "V" Portions of Belt

1/2"

Missing Two or More Adjacent Ribs 1/2" or longer

UNACCEPTABLE

Cracks Running Parallel to "V" Portions of Belt

21.4 Small cracks in the underside of a V-ribbed belt are acceptable - lengthwise cracks, or missing pieces that cause the belt to make noise, are cause for replacement

Be sure to tighten the tensioner bolt(s) to the torque listed in this Chapter's Specifications.

22   Manual transmission and differential lubricant change (every 60,000 miles or 48 months)

Manual transmission lubricant

1   Raise the vehicle and support it securely on jackstands.

2   Move a drain pan, rags, newspapers and wrenches under the transmission.

3   Remove the fill plug from the side of the transmission case, then remove the transmission drain plug at the bottom of the case and allow the lubricant to drain into the pan.

4   After the lubricant has drained completely, reinstall the plug and tighten it

securely.

5   Using a hand pump, syringe or funnel, fill the transmission with the specified lubricant until it is level with the lower edge of the filler hole. Reinstall the fill plug and tighten it securely.

6   Lower the vehicle.

7   Drive the vehicle for a short distance, then check the drain and fill plugs for leakage.

Differential lubricant

Refer to illustrations 22.11a, 22.11b, 22.11c and 22.13

8   On these models there is no drain plug, so a hand suction pump will be required to remove the differential lubricant through the filler hole. If a suction pump isn't available it will be necessary to remove the cover plate.

9   Raise the vehicle and support it securely

21.5 Belt wear indicator marks are located on the side of the tensioner body - when the belt reaches the maximum wear mark it must be replaced

21.6 Rotate the tensioner arm to relieve belt tension (V8 engine shown)

22.11a Remove the bolts from the lower edge of the cover . . .

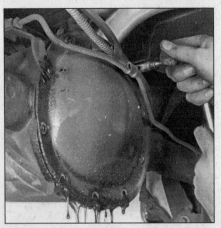

22.11b . . . then loosen the top bolts and let the lubricant drain

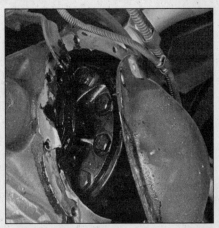

22.11c Once the lubricant has drained, remove the cover

22.13 Carefully scrape off the old sealant to ensure a leak-free seal

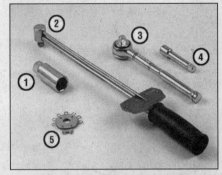

23.2 Tools required for changing spark plugs

1 **Spark plug socket** - This will have special padding inside to protect the spark plug porcelain insulator
2 **Torque wrench** - Although not mandatory, use of this tool is the best way to ensure that the plugs are tightened properly
3 **Ratchet** - Standard hand tool to fit the plug socket
4 **Extension** - Depending on model and accessories, you may need a longer extension and/or a universal joint to reach one or more of the plugs
5 **Spark plug gap gauge** - This gauge for checking the gap comes in a variety of styles. Make sure the gap for your engine is included

on jackstands. Move a drain pan, rags, newspapers and wrenches under the vehicle.

10   Remove the fill plug from the differential (see Section 4). If a suction pump is being used, insert the flexible hose. Work the hose down to the bottom of the differential housing and pump the lubricant out.

11   If the differential is being drained by removing the cover plate, remove the bolts on the lower half of the plate **(see illustration)**. Loosen the bolts on the upper half and use them to keep the cover loosely attached **(see illustration)**. Allow the oil to drain into the pan, then completely remove the cover **(see illustration)**.

12   Using a lint-free rag, clean the inside of the cover and the accessible areas of the differential housing. As this is done, check for chipped gears and metal particles in the lubricant, indicating that the differential should be more thoroughly inspected and/or repaired.

13   Thoroughly clean the mating surfaces of the differential housing and the cover plate. Use a gasket scraper or putty knife to remove all traces of the old sealant **(see illustration)**.

14   Apply a 1/8- to 3/16-inch diameter bead of RTV sealant to the cover flange, routing the bead around the inner side of the bolt holes. **Note:** *The cover must be installed within 15 minutes of sealant application.*

15   Place the cover on the differential housing and install the bolts. Tighten the bolts securely.

16   Use a hand pump, syringe or funnel to fill the differential housing with the specified lubricant until it's level with the bottom of the plug hole.

17   Install the filler plug and make sure it is secure.

23 Spark plug check and replacement (see the Maintenance schedule for service intervals)

Refer to illustrations 23.2, 23.5a, 23.5b, 23.6a, 23.6b, 23.6c, 23.8, 23.10a and 23.10b

1   The spark plugs are located in the cylinder heads.

2   In most cases, the tools necessary for spark plug replacement include a spark plug socket which fits onto a ratchet (spark plug sockets are padded inside to prevent damage to the porcelain insulators on the new plugs), various extensions and a gap gauge to check and adjust the gaps on the new plugs **(see illustration)**. A torque wrench should be used to tighten the new plugs.

3   The best approach when replacing the spark plugs is to purchase the new ones in

advance, adjust them to the proper gap and replace the plugs one at a time. When buying the new spark plugs, be sure to obtain the correct plug type for your particular engine. This information can be found in the Specifications Section at the beginning of this Chapter or in your owner's manual.

4   Allow the engine to cool completely before attempting to remove any of the plugs. These engines are equipped with aluminum cylinder heads, which can be damaged if the spark plugs are removed when the engine is hot. While you are waiting for the engine to cool, check the new plugs for defects and adjust the gaps. **Note:** *Spark plugs on 4.6L*

23.5a The manufacturer recommends using a wire-type gauge to check the spark plug gap - if the wire doesn't slide between the electrodes with a slight drag, adjustment is required

23.5b To change the gap, bend the side electrode only, and be very careful not to crack or chip the porcelain insulator surrounding the center electrode

Chapter's Specifications. The gauge should just slide between the electrodes with a slight amount of drag. If the gap is incorrect, use the adjuster on the gauge body to bend the curved side electrode slightly until the proper gap is obtained (**see illustration**). If the side electrode is not exactly over the center electrode, bend it with the adjuster until it is. Check for cracks in the porcelain insulator (if any are found, the plug should not be used). **Note:** *We recommend using a wire-type gauge when checking platinum- or iridium-type spark plugs. Other types of gauges may scrape the thin coating from the electrodes, thus dramatically shortening the life of the plugs.*

6   V8 engines are equipped with individual ignition coils which must be removed first to access the spark plugs (**see illustrations**). On V6 engines, remove the spark plug wire from one spark plug. Pull only on the boot at the end of the wire - do not pull on the wire. A plug wire removal tool should be used if available (**see illustration**).

7   If compressed air is available, use it to blow any dirt or foreign material away from the spark plug hole. The idea here is to eliminate the possibility of debris falling into the cylinder

V8 models are NOT adjustable. If the gap is out of specification, replace the spark plug.

5   The gap is checked by inserting the proper-thickness gauge between the elec-

trodes at the tip of the plug (**see illustration**). The gap between the electrodes should be the same as the one specified on the Vehicle Emissions Control Information label or in this

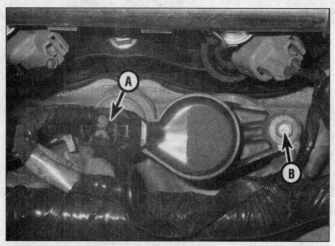

23.6a Disconnect the electrical connector (A), remove the mounting bolt (B) . . .

23.6b . . . and the individual coil(s) to access the spark plug(s)

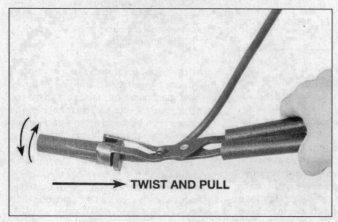

TWIST AND PULL

23.6c A tool like this one makes the job of removing the spark plug boot easier

23.8 Use a socket and extension to unscrew the spark plugs

as the spark plug is removed.

8   Place the spark plug socket over the plug and remove it from the engine by turning it in a counterclockwise direction **(see illustration).**

9   Compare the spark plug to those shown in the photos located on the inside back cover to get an indication of the general running condition of the engine.

10   Apply a small amount of anti-seize compound to the spark plug threads **(see illustration).** Install one of the new plugs into the hole until you can no longer turn it with your fingers, then tighten it with a torque wrench (if available) or the ratchet. It is a good idea to slip a short length of rubber hose over the end of the plug to use as a tool to thread it into place **(see illustration).** The hose will grip the plug well enough to turn it, but will start to slip if the plug begins to cross-thread in the hole - this will prevent damaged threads and the accompanying repair costs.

11   Before pushing the ignition coil onto the end of the plug, inspect the ignition coil following the procedures outlined in Section 24. On

V6 engines, inspect the plug wire following the procedures outlined in Section 28.

12   Repeat the procedure for the remaining spark plugs.

24   Ignition coil check (every 60,000 miles or 48 months)

1   Remove the ignition coils **(see illustration 23.6a).** Clean the coil(s) with a dampened cloth and dry them thoroughly.

2   Inspect each coil, for cracks, damage and carbon tracking. If damage exists, replace the coil.

25   Cooling system servicing (draining, flushing and refilling) (every 100,000 miles)

Warning 1: *Wait until the engine is completely cool before beginning this procedure.*

Warning 2: *Do not allow antifreeze to come in contact with your skin or painted surfaces of the vehicle. Rinse off spills immediately with plenty of water. Antifreeze is highly toxic if ingested. Never leave antifreeze lying around in an open container or in puddles on the floor; children and pets are attracted by its sweet smell and may drink it. Check with local authorities about disposing of used antifreeze. Many communities have collection centers which will see that antifreeze is disposed of safely. Never dump used antifreeze on the ground or pour it into drains.*

Caution: *Do not mix coolants of different colors. Doing so might damage the cooling system and/or the engine. The manufacturer specifies a gold colored coolant to be used in these systems. Read the warning label in the engine compartment for additional information.*

Note: *Non-toxic antifreeze is now manufactured and available at local auto parts stores, but even this type must be disposed of properly.*

1   Periodically, the cooling system should be drained, flushed and refilled to replenish the antifreeze mixture and prevent formation of rust and corrosion, which can impair the performance of the cooling system and cause engine damage. When the cooling system is serviced, all hoses and the expansion tank cap should be checked and replaced if necessary.

Draining

Refer to illustrations 25.4

2   Apply the parking brake and block the wheels. If the vehicle has just been driven, wait several hours to allow the engine to cool down before beginning this procedure.

3   Once the engine is completely cool, remove the expansion tank cap.

4   Move a large container under the radiator drain to catch the coolant. Attach a length of hose to the drain fitting to direct the coolant into the container, then open the drain fitting (a pair of pliers may be required to turn it) **(see illustration).**

5   While the coolant is draining, check the condition of the radiator hoses, heater hoses and clamps (refer to Section 10 if necessary). Replace any damaged clamps or hoses.

Flushing

Refer to illustration 25.8

6   Once the system has completely drained, remove the thermostat housing from the engine (see Chapter 3), then reinstall the housing without the thermostat. This will allow the system to be thoroughly flushed.

7   Disconnect the upper hose from the radiator.

8   Place a garden hose in the upper radiator inlet and flush the system until the water

23.10a Apply a thin coat of anti-seize compound to the spark plug threads

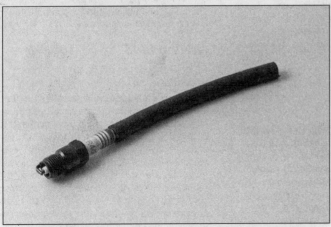

23.10b A length of snug-fitting rubber hose will save time and prevent damaged threads when installing the spark plugs

25.4 The radiator drain fitting is located at the lower corner of the radiator

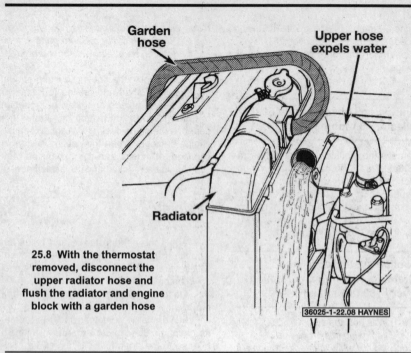

Garden hose

Upper hose expels water

Radiator

25.8 With the thermostat removed, disconnect the upper radiator hose and flush the radiator and engine block with a garden hose

36025-1-22.08 HAYNES

runs clear at the upper radiator hose **(see illustration)**.

9   Severe cases of radiator contamination or clogging will require removing the radiator (see Chapter 3) and reverse flushing it. This involves inserting the hose in the bottom radiator outlet to allow the clean water to run against the normal flow, draining out through the top. A radiator repair shop should be consulted if further cleaning or repair is necessary.

10   When the coolant is regularly drained and the system refilled with the correct coolant mixture there should be no need to employ chemical cleaners or descalers.

Refilling

11   Close and tighten the radiator drain.

12   Place the heater temperature control in the maximum heat position.

13   Slowly add new coolant (a 50/50 mixture of water and antifreeze) to the expansion tank until the level is at the COLD FILL RANGE mark on the expansion tank.

14   Leave the expansion tank cap off and run the engine in a well-ventilated area until the thermostat opens (coolant will begin flowing through the radiator and the upper radiator hose will become hot).

15   Turn the engine off and let it cool. Add more coolant mixture to bring the coolant level between the COLD FILL RANGE mark on the expansion tank.

16   Squeeze the upper radiator hose to expel air, then add more coolant mixture if necessary. Replace the expansion tank cap.

17   Start the engine, allow it to reach normal operating temperature and check for leaks. Also, set the heater and blower controls to the maximum setting and check to see that the heater output from the air ducts is warm. This is a good indication that all air has been purged from the cooling system.

26 Positive Crankcase Ventilation (PCV) valve replacement (every 100,000 miles)

Note: *For additional information on the PCV system refer to Chapter 6.*

1   On 4.0L V6 models, the PCV valve is located on the valve cover. On 4.6L V8 models, there is no actual PCV valve.

2   Disconnect the PCV valve electrical connector, if equipped.

3   Disconnect the PCV valve hose from the PCV valve, then rotate the PCV valve counterclockwise and remove the valve.

5   Installation is the reverse of removal.

27 Automatic transmission fluid change (see the Maintenance schedule for service intervals)

Warning: *This procedure is potentially dangerous and is best left to a professional shop with a safe lifting apparatus. The vehicle must be kept level while being safely raised high enough for access to the check/fill plug on the transmission.*

Note: *When the work is done at a dealership, the factory scan tool is used to read the transmission fluid temperature. To perform the job at home, you will need a special tool to measure the temperature of the oil pan/transmission fluid using infra-red technology or temperature probes with digital readout displays.*

Draining

1   Drive the vehicle to warm the transmission fluid to normal operating temperature, then position the vehicle on a hoist or support it securely on four jackstands.

2   Position a drain pan under the transmission, then unscrew the transmission drain plug (the larger, outer plug on the transmission fluid pan) and allow the the fluid to drain.

3   After the fluid has drained, reinstall the drain plug and tighten it to the torque listed in this Chapter's Specifications.

Adding fluid

Refer to illustrations 27.5a, 27.5b, 27.5c, 27.6a and 27.6b

4   Correct automatic transmission fluid level is extremely important for proper transmission operation. Low fluid level can lead to slipping or loss of drive, while overfilling can cause foaming and loss of fluid. **Note:** *Adding/checking the fluid on these vehicles isn't easy. The transmission is considered by the manufacturer to be a "sealed" unit, to which fluid doesn't need to be added unless a leak is evident. There is no conventional dipstick in the engine compartment, but rather a level-inspection plug accessible only from below the transaxle. Make sure you have a new seal for the plug before you begin. If fluid does have to be added, a special pump will be required.*

27.5a Unscrew the check/fill plug from the center of the drain plug - Note: *You may have to hold the drain plug with a wrench to prevent it from turning*

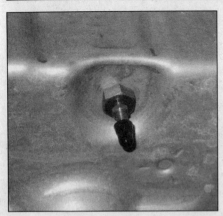

27.5b Install the fitting into the threads in the center of the drain plug

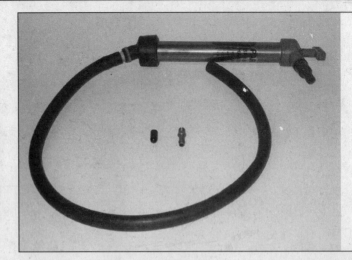

27.5c You will need a special hand pump and fitting to add fluid to the transmission

5   On these transmissions, the fluid is added through the check/fill plug **(see illustration).** Two special tools are available for this purpose; a fitting and a pressure pump **(see illustration).** A hand pump equipped with a rubber hose can be used to pump the fluid up into the pan. The pump and rubber hose is attached to the fitting and fluid is pumped up into the transmission pan **(see illustration).**

6   Slowly add two or three quarts of fluid **(see illustration).** Periodically check the level by removing the pump and rubber hose from the fitting and observing for any excess fluid dripping down **(see illustration).** Until transmission fluid begins to drip from the check/fill plug, the transmission fluid level continues to be low. Once fluid flows, then the transmission fluid level has reached the top of the measured tube positioned inside the oil pan. Wait until the excess fluid has dripped out of the pan. Install the check/fill plug and tighten it securely.

7   Start the engine and allow it to run until the fluid level in the transmission is 80 to 120-degrees F. **Note:** *The manufacturer states that a scan tool must be used to monitor the* temperature of the transmission fluid. If you don't have a scan tool that is capable of reading the transmission fluid temperature, one of the methods described in the Note at the beginning of this Section can be used or, as a last resort, you can feel the fluid pan with your hand. The transmission pan should feel warm to the touch, but not hot enough to cause pain.

8   Remove the check/fill plug again and see if any fluid drains out. If fluid does drain out, wait until the flow stops, then reinstall the check/fill plug and tighten it to the torque listed in this Chapter's Specifications. If no fluid flows out, add fluid by repeating Step 6.

9   The old oil drained from the transmission cannot be reused in its present state and should be discarded. Check with your local refuse disposal company, disposal facility or environmental agency to see if they will accept the oil for recycling. Don't pour used oil into drains or on the ground. After the oil has cooled, it can be drained into a suitable container (capped plastic jugs, topped bottles, milk cartons, etc.) for transport to one of these disposal sites.

28 Spark plug wire check and replacement (V6 engines)(every 60,000 miles or 48 months)

1   The spark plug wires should be checked at the recommended intervals or whenever new spark plugs are installed.

2   Begin this procedure by making a visual check of the spark plug wires while the engine is running. In a darkened garage (make sure there is adequate ventilation) or at night, start the engine and observe each plug wire. Be careful not to come into contact with any moving engine parts. If possible, use an insulated or non-conductive object to wiggle each wire. If there is a break in the wire, you will see arcing or a small blue spark coming from the damaged area. Secondary ignition voltage increases with engine speed and sometimes a damaged wire will not produce an arc at idle speed. Have an assistant press the accelerator pedal to raise the engine speed to approximately 2000 rpm. Check the spark plug wires for arcing as stated previously. If arcing is noticed, replace all spark plug wires.

27.6a Carefully pump the transmission fluid into the oil pan

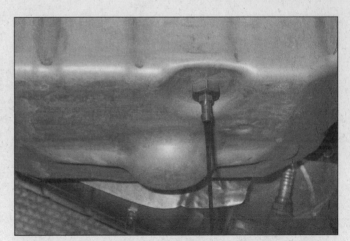

27.6b Remove the pump and observe the fluid drip from the check/fill plug - wait for the excess fluid to drip out then install the plug

Notes

Chapter 2 Part A
V6 engine

Contents

Specifications

General

Displacement	4.0 liters (244 cubic inches)
Bore and stroke	3.953 x 3.320 inches
Cylinder numbers (front-to-rear)	
Left (driver's) side	4-5-6
Right side	1-2-3
Firing order	1-4-2-5-3-6

Cylinder locations
and coil terminal
identification

4.0L
V6 Engine
1-4-2-5-3-6

36025-specs.b HAYNES

Camshafts

Lobe lift (intake and exhaust)	0.259 inch
Allowable lobe lift loss	0.005 inch
Endplay	0.0003 to 0.007 inch
Journal diameter (all)	1.099 to 1.101 inches
Bearing inside diameter (all)	1.102 to 1.104 inches
Journal-to-bearing (oil) clearance	
Standard	0.002 to 0.004 inch
Service limit	0.006 inch

Torque specifications

Ft-lbs (unless otherwise indicated)

Accessory bracket bolts	35
Camshaft sprocket bolt*	63
Camshaft bearing cap bolts	
Step 1	53 in-lbs
Step 2	144 in-lbs
Crankshaft pulley bolt**	
2005 models	
Step 1	33
Step 2	Tighten an additional 85 degrees
2006 models	
Step 1	35
Step 2	Tighten an additional 85 degrees
2007 models	
Step 1	37
Step 2	Tighten an additional 90 degrees
Cylinder head bolts**	
Step 1 (12 mm bolts)	9
Step 2 (12 mm bolts)	18
Step 3 (8 mm bolts)	24
Step 4 (12 mm bolts)	Rotate an additional 90 degrees
Step 5 (12 mm bolts)	Rotate an additional 90 degrees
Drivebelt tensioner bolt	35
Flywheel/driveplate bolts	
Step 1	120 in-lbs
Step 2	52
Exhaust manifold nuts	17
Exhaust pipe-to-manifold nuts	30
Engine front cover bolts	168 in-lbs
Engine mounts	
Engine mount nuts	46
Engine mount bracket-to-engine block bolts	59
Engine mount-to-chassis bolts	52
Intake manifold bolts	89 in-lbs
Jackshaft sprocket bolts	
Front	
Step 1	33
Step 2	Tighten an additional 90 degrees
Rear	
Step 1	15
Step 2	Tighten an additional 90 degrees
Jackshaft chain tensioner bolts	80 in-lbs
Jackshaft chain guide bolts	168 in-lbs
Jackshaft thrust plate bolts	96 in-lbs
Oil pump drive-to-engine block mounting bolt	168 in-lbs
Oil pump screen cover and tube bolt	89 in-lbs
Oil pump-to-block bolts	168 in-lbs
Oil pan (sheet metal)-to-crankcase reinforcement section bolts	80 in-lbs
Crankcase reinforcement section-to-block	
Perimeter bolts/nuts	89 in-lbs
Threaded inserts	27 in-lbs
Rear lower block cradle-to-transmission bolts	35
Rear lower block cradle-to-engine block (rear main oil seal)	71 in-lbs
Lower block cradle bolts	
Step 1	132 in-lbs
Step 2	25

Torque specifications

	Ft-lbs (unless otherwise indicated)
Timing chain tensioner (right side or left side timing chain).....................	32
Timing chain tensioner bolts (main timing chain)	80 in-lbs
Timing chain cassette bolts	
Right side	
Upper bolt	108 in-lbs
Left side	
Upper bolt	108 in-lbs
Lower bolt	168 in-lbs
Valve cover bolts	89 in-lbs

*The right camshaft uses a left-hand threaded bolt
** Bolt(s) must be replaced.

1 General information

This Part of Chapter 2 is devoted to in-vehicle repair procedures for the 4.0L SOHC (Singe Overhead Camshaft) V6 engine as well as procedures such as timing chain(s) and sprocket(s), cylinder head and oil pan (crankcase reinforcement section) removal, which require removal of the engine from the vehicle. Information concerning engine removal and installation, balance shaft removal and installation and engine overhaul can be found in Part C of this Chapter.

This engine uses a jackshaft, in place of a camshaft on a conventional pushrod engine, to drive the camshaft timing chains and the oil pump. The left cylinder bank camshaft is driven by a chain (from the jackshaft) at the front of the engine and the right cylinder bank camshaft is driven by a chain from the rear of the jackshaft. The balance shaft assembly is driven by a chain from the crankshaft sprocket.

The Specifications included in this Part of Chapter 2 apply only to the procedures contained in this Part.

2 Repair operations possible with the engine in the vehicle

Many major repair operations can be accomplished without removing the engine from the vehicle.

Clean the engine compartment and the exterior of the engine with some type of degreaser before any work is done. It will make the job easier and help keep dirt out of the internal areas of the engine.

Depending on the components involved, it may be helpful to remove the hood to improve access to the engine as repairs are performed (refer to Chapter 11 if necessary). Cover the fenders to prevent damage to the paint. Special pads are available, but an old bedspread or blanket will also work.

If vacuum, exhaust, oil or coolant leaks develop, indicating a need for gasket or seal replacement, the repairs can generally be made with the engine in the vehicle.

Exterior engine components, such as the intake and exhaust manifolds, the water pump, the starter motor, the alternator and the fuel system components can be removed for repair with the engine in place.

The cylinder head(s) and the timing chain(s) removal procedures should be performed with the engine removed from the vehicle and bolted to an engine stand. Camshaft removal is also recommended with the engine out. Although professional mechanics perform these procedures in-vehicle, certain components at the rear of the engine are extremely difficult to gain access to. It is recommended for the home mechanic to remove the engine before attempting these repair procedures.

3 Top Dead Center (TDC) for number one piston - locating

Refer to illustration 3.5

Note: *These engines are not equipped with a distributor. Piston position must be determined by checking for compression at the number one spark plug hole, then aligning the ignition timing marks as described in Step 5.*

1 Top Dead Center (TDC) is the highest point in the cylinder that each piston reaches as it travels up-and-down during crankshaft rotation. Each piston reaches TDC on the compression stroke and again on the exhaust stroke, but TDC generally refers to piston position on the compression stroke.

2 Positioning the piston(s) at TDC is an essential part of certain repair procedures discussed in this manual.

3 Before beginning this procedure, be sure to place the transmission in Park or Neutral and apply the parking brake. Remove the ignition coils (V8 engine only) and the spark plugs (see Chapter 1). Disable the ignition system by disconnecting the wiring harness connector from the ignition coil pack (V6 models) (see Chapter 5). Disable the fuel system by removing the fuel pump relay (see Chapter 4, Section 2).

4 In order to bring any piston to TDC, the crankshaft must be turned using one of the methods outlined below. When looking at the front of the engine, normal crankshaft rotation is clockwise.

a) *The preferred method is to turn the crankshaft with a socket and ratchet attached to the bolt threaded into the front of the crankshaft.*

b) *A remote starter switch, which may save some time, can also be used. Follow the instructions included with the switch. Once the piston is close to TDC, use a socket and ratchet as described in the previous paragraph.*

c) *If an assistant is available to turn the ignition switch to the Start position in short bursts, you can get the piston close to TDC without a remote starter switch. Make sure your assistant is out of the vehicle, away from the ignition switch, then use a socket and ratchet as described in Paragraph a) to complete the procedure.*

5 Install a compression gauge in the number one cylinder spark plug hole. Turn

3.5 Timing marks - align the pointer on the crankshaft position sensor (A) with the mark in the crankshaft pulley (B)

4.6 Location of the right-side valve cover mounting bolts

the crankshaft using one of the methods described in the previous Step until compression begins to register on the gauge, then turn it slowly until the TDC notch is aligned with the pointer on the crankshaft position sensor (**see illustration**). **Note:** *There are marks for TDC and for 10-degrees BTDC. Make sure you are on the TDC mark.*

6   After the number one piston has been positioned at TDC on the compression stroke, TDC for any of the remaining pistons can be located by turning the crankshaft and following the firing order. On V6 engines, divide the crankshaft pulley into three equal sections with chalk marks at each point, each indicating 120-degrees of crankshaft rotation. On V8 engines divide the crankshaft pulley into fourths, with the marks 90-degrees apart. Rotating the engine clockwise to the next mark will bring the next piston in the firing order sequence to TDC.

4   Valve covers - removal and installation

Removal

1   Disconnect the cable from the negative battery terminal (see Chapter 5, Section 1).
2   Disconnect the spark plug wires from the valve cover (see Chapter 5).

Right valve cover

Refer to illustration 4.6

3   Disconnect the PCV electrical connector and the PCV tube from the valve cover.
4   Disconnect the heater hose retainer from the side of the valve cover.
5   Remove the wiring harness bracket bolt and the mounting bracket from the rear of the valve cover.
6   Remove the valve cover bolts (**see illustration**). Lift the valve cover off. Tap gently with a soft-face hammer if necessary to break the gasket seal.

Left valve cover

Refer to illustration 4.12

7   Remove the ignition coil assembly (see Chapter 5).
8   Remove the fuel rail supply line bracket assembly (see Chapter 4).
9   Disconnect the PCV hose from the valve cover (see Chapter 6).
10   Disconnect the electrical connector at the camshaft position sensor (see Chapter 6).
11   Disconnect the fuel rail pressure and temperature sensor connector (see Chapter 6).
12   Remove the valve cover bolts (**see illustration**) and lift the valve cover off. Tap it gently with a soft-face hammer if necessary to break the gasket seal.

Installation

13   Check the valve cover gasket for damage or hardness. If it's in good condition it can be re-used.
14   Clean the gasket surfaces on the cylinder head and valve cover. Use a shop rag and lacquer thinner or acetone to wipe off all residue and gasket material from the sealing surfaces.

15   The remainder of installation is the reverse of the removal Steps. Tighten the valve cover bolts evenly, starting with the center bolts and working out, to the torque listed in this Chapter's Specifications.

5   Rocker arms and lash adjusters - removal, inspection and installation

Note: *There are two methods of removing the rocker arms and lash adjusters on this engine. The method recommended by the manufacturer accomplishes the removal of the rocker arms without the removal of the camshaft(s), using a special valve spring compressor made specifically for this engine. The valve spring compressor uses the camshaft as a pivot point and, with a ratchet or breaker bar attached, pushes down on the valve spring to release tension on the rocker arm. The alternative method requires the removal of the camshaft (see Section 12). Either method will achieve the same results, but it is much easier using the manufacturers special tool, if it can be located.*

4.12 Location of the left-side valve cover mounting bolts

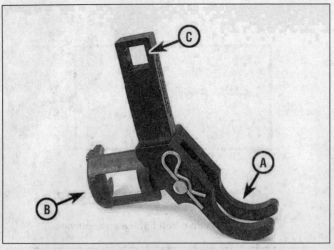

5.2a The special valve spring compressor hooks under the camshaft at (A), pushes on the valve spring retainer at (B), and is operated by a ratchet or breaker bar placed at (C)

5.2b Compress the valve spring just enough to allow the rocker arm to be removed

Removal

Refer to illustrations 5.2a, 5.2b and 5.3

1   Remove the valve cover(s) (see Section 4).

2   Install the special valve spring compressor and compress the spring just enough to remove the rocker arm **(see illustrations)**. **Caution:** *Camshaft rocker arms and hydraulic lash adjusters MUST be reinstalled in the same location they were removed from. Label and store all components to avoid confusion during reassembly.*

3   Remove the hydraulic lash adjuster **(see illustration)**. If there are many miles on the vehicle, the adjusters may have become varnished and difficult to remove. Apply a little penetrating oil around the lash adjuster to help loosen the varnish.

Inspection

Refer to illustrations 5.4 and 5.6

4   Inspect each adjuster carefully for signs

of wear or damage **(see illustration)**. Since the lash adjusters can become clogged as mileage accumulates, we recommend replacing them if you're concerned about their condition or if the engine is exhibiting valvetrain noise.

5   A thin wire or paper clip can be placed in the oil hole to move the plunger and make sure it's not stuck. **Note:** *The lash adjuster must have no more than 1.5 mm of total plunger travel.* It's recommended that if replacement of any of the adjusters is necessary, that the entire set be replaced. This will avoid the need to repeat the repair procedure as the others require replacement in the future.

6   Inspect the rocker arms for signs of wear or damage **(see illustration)**.

Installation

7   Before installing the lash adjusters, bleed them of air. Stand the adjusters upright in a container of oil. Use a thin wire or paper clip to work the plunger up and down. This "primes"

5.3 Remove the lash adjusters and store them and the rocker arms in an organized manner so the components will be returned to their original locations

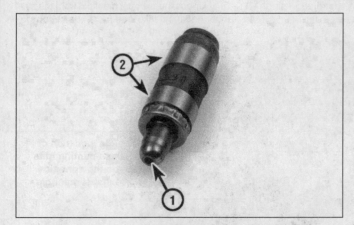

5.4 Inspect the lash adjuster for signs of excessive wear or damage, such as pitting, scoring or signs of overheating (bluing or discoloration) - the areas of wear are the rocker arm pivot point (1) and the side surfaces where the lifter body contacts the cylinder head bore (2)

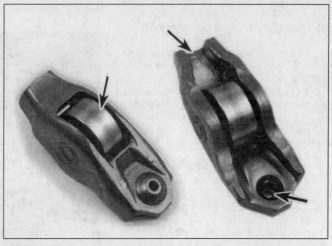

5.6 Check the rocker arm roller, the valve stem contact point and lash adjuster contact point (arrows)

6.13 When replacing the intake manifold, install new O-ring gaskets around each runner

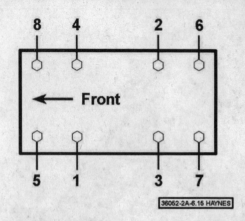

6.15 Intake manifold bolt tightening sequence

the adjuster and removes the air. Leave the adjusters in the oil until ready to install.
8   Lubricate the valve stem tip, rocker arm, and lash adjuster bore with clean engine oil.
9   Install the lash adjusters and, with the valve spring depressed as in Step 2, install each rocker arm.
10   The remainder of installation is the reverse of the removal procedure.
11   When starting the engine after replacing the adjusters, there will normally be some noise until all the air is bled from the lash adjusters. After the engine is warmed-up, raise the speed from idle to 3,000 rpm for one minute. Stop the engine and let it cool down. All of the noise should be gone when it is restarted.

6   Intake manifold - removal and installation

Warning: *Wait until the engine is completely cool before beginning this procedure.*

Removal

1   Disconnect the cable from the negative battery terminal (see Chapter 5, Section 1).
2   Detach the air intake duct from the throttle body.
3   Clamp off the coolant hoses to the throttle body, then disconnect the hoses at the throttle body.
4   Detach the spark plug wires for the right-side cylinder bank from the ignition coil. Remove the ignition coil and the coil bracket and set it aside (see Chapter 5). Also detach the right-side spark plug wire retainer from the intake manifold.
5   Disconnect the EGR module connector (see Chapter 6).
6   Disconnect the IAC valve, the TPS and any other electrical connectors that may interfere with manifold removal.
7   Disconnect the brake booster vacuum hose and the EVAP canister purge solenoid vacuum hose (see Chapter 6).
8   Disconnect the EGR pipe from the EGR

module (see Chapter 6).
9   Disconnect the PCV tube, vapor tube, and any other vacuum hoses from the intake manifold. Mark each hose with tape to insure correct reassembly.
10   Remove the intake manifold mounting bolts.
11   Lift the intake manifold from the engine.

Installation

Refer to illustrations 6.13 and 6.15

12   Clean the manifold mating surfaces. Remove oil and dirt with a cloth and solvent, such as lacquer thinner.
13   Install new O-ring gaskets around each of the six intake runners **(see illustration)**. **Note:** *No sealant is required.*
14   Install the manifold to the cylinder heads, making sure the bolt holes are aligned. Install the bolts and tighten them finger-tight. **Note:** *Do not move the manifold back and forth, or the O-ring seals could move out of position, causing vacuum leaks.*
15   Tighten the bolts to the torque listed in this Chapter's Specifications, following the correct torque sequence **(see illustration)**.
16   The remainder of installation is the reverse of the removal steps.

17   Check the coolant level, adding as necessary (see Chapter 1).
18   Run the engine and check for oil, coolant and vacuum leaks.

7   Exhaust manifolds - removal and installation

Note: *If the exhaust system fasteners are rusted or "frozen," apply penetrating oil to them and let it soak in for awhile, then attempt to unscrew them.*

Removal

Refer to illustrations 7.5a and 7.5b

1   Disconnect the cable from the negative battery terminal (see Chapter 5, Section 1).
2   Disconnect the EGR pipe from the left manifold and the EGR module (see Chapter 6).
3   Raise the vehicle and support it on jackstands.
4   Detach the exhaust pipe from the manifold.
5   Unbolt the exhaust manifold from the cylinder head and take it off **(see illustrations)**.

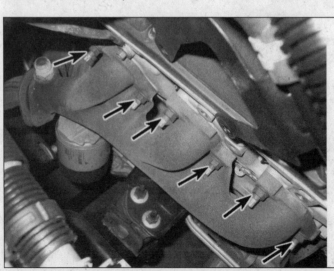

7.5a Location of the mounting nuts on the right-side exhaust manifold

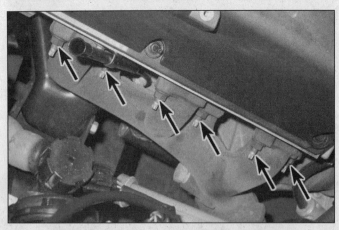

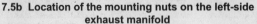

7.5b Location of the mounting nuts on the left-side exhaust manifold

7.7 Place the new gasket over the studs on the cylinder head

Installation

Refer to illustration 7.7

6   Using a scraper, thoroughly clean the mating surfaces on the cylinder head, manifold and exhaust pipe. Remove residue with a solvent such as acetone or lacquer thinner.

7   Check that the mating surfaces are perfectly flat and not damaged in any way. Warped or damaged manifolds may require machining. Install the new gasket to the cylinder head studs and place the manifold on the cylinder head **(see illustration)**. Tighten the bolts evenly to the torque listed in this Chapter's Specifications.

8   Connect the exhaust pipe to the manifold and tighten the nuts evenly to the torque listed in this Chapter's Specifications.

9   The remainder of installation is the reverse of the removal steps.

10   Run the engine and check for exhaust leaks.

8   Cylinder heads - removal and installation

Caution: *The engine must be completely cool*

when the heads are removed. Failure to allow the engine to cool off could result in head warpage.

Note 1: *This is a difficult procedure and it is recommended that the engine be removed from the vehicle and placed on an engine stand. It is possible to remove the cylinder heads in-vehicle, but this method will require all the necessary special tools and an expert skill level in automotive repairs.*

Note 2: *The rocker arms and valve lash adjusters must be removed before performing any repair procedures on the cylinder heads, timing chains and jackshaft chains. Refer to Section 5 and remove all the rocker arms and valve lash adjusters.*

Removal

Refer to illustration 8.7

1   Disconnect the cable from the negative battery terminal (see Chapter 5, Section 1).

2   Drain the cooling system (see Chapter 1).

3   Remove the engine from the vehicle and mount it on an engine stand (see Chapter 2C).

4   Remove the intake manifold (see Section 6).

5   Label and disconnect the electrical connectors from the fuel injectors (see Chapter 4).

6   Remove the exhaust manifolds (see Section 7).

7   Remove the drivebelt (see Chapter 1) and the drivebelt tensioner **(see illustration)**.

8   Remove the rocker arms and valve lash adjusters (see Section 5).

Right cylinder head

Refer to illustration 8.9

9   Remove the alternator (see Chapter 5) and the lower alternator mounting bracket **(see illustration)**.

10   Remove the upper radiator hose, the coolant tube bracket and the coolant bypass hose (see Chapter 3).

11   Disconnect the engine coolant temperature (ECT) sensor (see Chapter 6). Remove the bolts that retain the engine wiring harness to the engine brackets.

12   Remove the thermostat housing (see Chapter 3).

13   Perform Steps 17 through 19. If you are working in-vehicle, the sprocket is at the rear of the right cylinder head. Remove the large plug covering the rear of the jackshaft **(see**

8.7 The drivebelt tensioner is retained by one bolt

8.9 Remove the bolts from the alternator bracket and set it aside

8.16 Location of the accessory bracket mounting bolts

8.17 Remove the hydraulic chain tensioner - left side shown, right side is located at the rear of the cylinder head near the timing chain

illustration 10.10a). **Note:** *Obtain a new plug for reassembly. The plug is not reusable.* Remove the Torx bolt and spacer through the opening. A special offset tool is used to remove the sprocket bolt, but it is possible (though difficult) to remove it with a conventional wrench in-vehicle. Secure the chain for the rear sprocket with a rubber band as in Step 19. Remove the lower cassette bolt and the cassette and chain. **Caution:** *The sprocket bolt for the right cylinder head is a left-hand thread.*

Left cylinder head

Refer to illustrations 8.16, 8.17, 8.18 and 8.19

14   Disconnect the radio capacitor from the side of the ignition coil pack.
15   Remove the ignition coil assembly (see Chapter 5).
16   Remove the accessory bracket mounting bolts and position it off to the side **(see illustration)**. **Note:** *The accessory bracket*

must be removed and positioned off to the side with the power steering pump lines and air conditioning compressor lines attached. It will be necessary to remove the coil bracket bolts and carefully move the entire accessory bracket assembly toward the front of the vehicle.

17   Remove the hydraulic chain tensioner **(see illustration)**.
18   Remove the bolt from the camshaft sprocket, then remove the Torx bolt below it in the head, one of two that secure the camshaft chain "cassette" to the head **(see illustration)**.
19   Remove the lower bolt holding the chain cassette to the head. Keep the slack out of the chain (to prevent dropping it below) and secure it to the cassette with a large rubber band **(see illustration)**. Then remove the cassette, chain and sprockets.

Both cylinder heads

20   If you are working in-vehicle, the

sprocket is at the rear of the right cylinder head. Remove the large plug covering the rear of the jackshaft **(see illustration 10.10a)**. **Note:** *Obtain a new plug for reassembly. The plug is not reusable.* Remove the Torx bolt and spacer through the opening. A special, offset tool is used to remove the sprocket bolt, but it is possible (though difficult) to remove it with a conventional wrench in-vehicle. Secure the chain for the rear sprocket with a rubber band as in Step 19. Remove the lower cassette bolt and the cassette and chain. **Caution:** *The sprocket bolt for the right cylinder head is a left-hand thread.*

21   Remove the cylinder head bolts, following the reverse of the recommended tightening sequence **(see illustration 8.30)**. Loosen the bolts in sequence 1/4-turn at a time. **Note:** *There are two 8 mm external Torx bolts (one on each side of the chain opening in the head) and eight 12 mm internal Torx bolts. Loosen the 8 mm bolts first, then the 12 mm bolts. These are torque-to-yield bolts and should be*

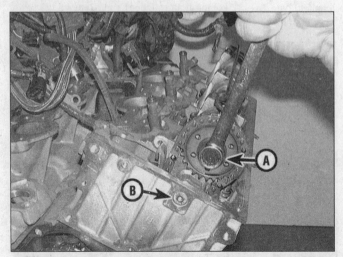

8.18 The chain will hold the camshaft and sprocket in position while you remove the sprocket bolt (A) - then remove the upper cassette bolt (B)

8.19 Slip a heavy-duty rubber band (A) over the chain, just below the sprocket, then remove the cassette bolt (B)

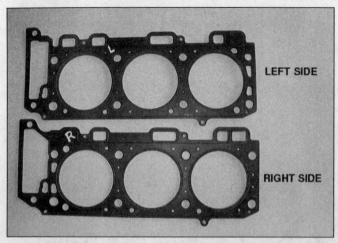

8.29 Make sure the cylinder head gaskets are installed in the correct location - they are not interchangeable

8.30 Cylinder head bolt tightening sequence - bolts (A) are 8 mm bolts, the rest are 12 mm bolts

discarded; do not reuse them. If the head is to be completely overhauled, refer to Section 11 for removal of the camshafts.

22 Use a pry bar at the corners of the head-to-block mating surface to break the gasket seal. Do not pry between the cylinder head and engine block in the gasket sealing area.

23 Lift the cylinder head(s) off the engine. If resistance is felt, place a wood block against the end and strike the wood block with a hammer.

24 Store the cylinder heads on wood blocks to prevent damage to the gasket sealing surfaces.

25 Remove the old cylinder head gasket(s). Before removing, note which gasket goes on which side, they are different and cannot be interchanged.

Installation

Refer to illustrations 8.29, 8.30, 8.34, 8.35 and 8.37

26 The mating surfaces of the cylinder heads and block must be perfectly clean when the heads are installed. Use a gasket scraper to remove all traces of carbon and old gasket material, then clean the mating surfaces with lacquer thinner or acetone. If there's oil on

the mating surfaces when the cylinder heads are installed, the gaskets may not seal correctly and leaks may develop. When working on the engine block, cover the open areas of the engine with shop rags to keep debris out during repair and reassembly. Use a vacuum cleaner to remove any debris that falls into the cylinders.

27 Check the engine block and cylinder head mating surfaces for nicks, deep scratches and other damage.

28 Use a tap of the correct size to chase the threads in the cylinder head bolt holes. Dirt, corrosion, sealant and damaged threads will affect torque readings.

29 Make sure the new gaskets are on the correct cylinder banks, and located on the dowels in the block. They are not interchangeable **(see illustration)**.

30 Carefully position the cylinder heads on the engine block without disturbing the gaskets. Install new cylinder head bolts and following the recommended sequence **(see illustration)**, tighten the bolts. First, install the eight 12 mm bolts and tighten, in the order shown, to the torque listed in Step 1 and 2 of this Chapter's Specifications. Then install and tighten the two 8 mm bolts to the torque

listed in Step 3 of this Chapter's Specifications. Finally, tighten the 12 mm bolts to the torque listed in Steps 4 and 5 of this Chapter's Specifications. Mark a stripe on each of the 12 mm head bolts to help keep track of the bolts that have been tightened the additional 90-degrees. **Note:** *The method used for the head bolt tightening procedure is referred to as "torque-angle" or "torque-to-yield." A special torque angle gauge (available at most auto parts stores) is available to attach to a breaker bar and socket for better accuracy during the tightening procedure.*

31 Pull up the slack chain, align the camshaft sprocket with the camshaft and install the sprocket bolt. **Caution:** *Do not tighten the sprocket bolt at this time; the sprocket must rotate freely on the camshaft!* Do this on both cylinder heads and install the bolts that hold the chain cassettes to the heads.

32 Install the special camshaft chain tensioning tool into the chain tensioner location on the left cylinder head.

33 Make sure the engine is located at the correct position for cylinder number 1 at TDC (see Section 3). Rotate the engine clockwise to obtain number 1 TDC if necessary.

34 Clamp the special crankshaft holding tool over the crankshaft damper (flush with the rear edge of the damper) and with the straight end against the bottom of the block **(see illustration)**. Tighten the bolt on the holding tool.

35 Install the camshaft holding tool on the rear of the left camshaft **(see illustration 8.37)**. Rotate the camshaft with a wrench to align the off-center slot on the rear of the camshaft with the tool. The slot is positioned down when aligned with the tool. Tighten the tool bolts securely. Install the camshaft sprocket holding tool on the left camshaft sprocket **(see illustration)**. Tighten the holding tool bolts to hold the camshaft sprocket stationary, then tighten the camshaft sprocket bolt to the torque listed in this Chapter's Specifications.

36 Remove the tensioning tool from the left cylinder head and install the hydraulic tensioner. Install the tensioning tool into the right cylinder head tensioner location.

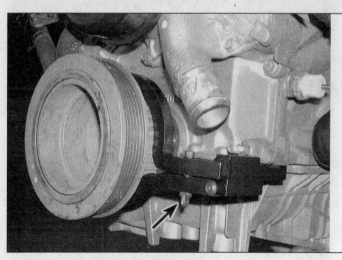

8.34 Clamp the special crankshaft holding tool around the damper as shown, then tighten the bolt - this locks the crankshaft at TDC

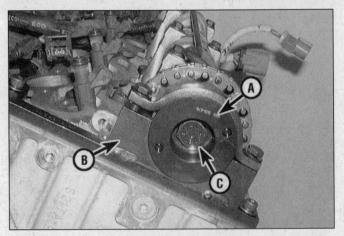

8.35 Install the camshaft sprocket holding tool to the cylinder head, inserting the pins on the aligning tool (A) into the holes in the sprocket - tighten the holder bolts (B), then tighten the sprocket bolt (C)

8.37 Camshaft positioning tool at the front of the right camshaft - align the tool's projection with the slot in the end of the camshaft

Remove the camshaft and camshaft sprocket holding tools and the crankshaft holding tool.

37   Install the camshaft holding tool onto the front of the right camshaft **(see illustration)**. Rotate the camshaft to align the slots in the end of the camshaft with the tool. Tighten the holding tool bolts securely. Install the camshaft sprocket holding tool onto the rear of the camshaft and tighten the sprocket bolt to the torque listed in this Chapter's Specifications. **Caution:** *The right camshaft sprocket bolt is left-hand thread.*

38   Remove the chain tensioning tool and install the hydraulic tensioner. Remove the camshaft and crankshaft holding tools.

39   Install the rocker arms and hydraulic valve lash adjusters (see Section 5).

40   The remaining installation steps are the reverse of removal.

41   Change the engine oil and filter and refill the cooling system (Chapter 1), then start the engine and check carefully for oil and coolant leaks.

9   Crankshaft pulley and front oil seal - removal and installation

Removal

Refer to illustrations 9.7, 9.8 and 9.9

1   Disconnect the cable from the negative battery terminal (see Chapter 5, Section 1).

2   Remove the drivebelt (see Chapter 1).

3   Remove the engine cooling fan/shroud assembly (see Chapter 3).

4   Raise the vehicle and support it securely on jackstands.

5   Remove the power steering fluid cooler mounting bolts and position the assembly to the side without disconnecting the fluid lines.

6   Remove the crankshaft pulley bolt.

7   Use a breaker bar and socket to remove the crankshaft pulley center bolt **(see illustration)**. Discard the bolt and obtain a new one for installation. **Note:** *It will be necessary to lock the pulley in position using a strap or chain*

wrench. Be sure to wrap a shop rag around the pulley before installing the special tool.

8   Using a bolt-type puller, pull the pulley from the crankshaft **(see illustration)**. **Note:** *Because the pulley is recessed, an adapter may be needed between the puller bolt and the crankshaft.*

9   Use a seal puller to remove the crankshaft front oil seal **(see illustration)**.

10   Clean the seal bore and check it for nicks or gouges. Also examine the area of the hub that rides in the seal for signs of abnormal wear or scoring. For many popular engines, a repair sleeve is available to restore a smooth finish to the sealing surface. Check with your auto parts store for availability of this sleeve.

Installation

Refer to illustration 9.11

11   Coat the lip of the new seal with clean engine oil and drive it into the bore with a socket or section of pipe slightly smaller in

9.7 To remove the crankshaft pulley, remove the center bolt

9.8 Remove the pulley with a puller that bolts to the pulley hub; an adapter may be needed between the puller bolt and the crankshaft snout to prevent damaging the threads in the end of the crankshaft

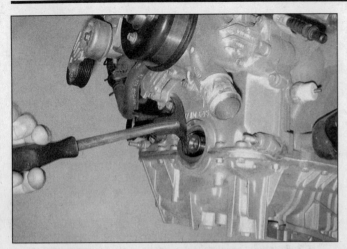

9.9 Use a seal puller to remove the old crankshaft seal, taking care not to damage the crankshaft or the seal bore in the cover

9.11 Drive the new seal in with a large socket or short section of appropriate-size pipe

diameter than the seal **(see illustration)**. The open side of the seal faces into the engine.

12 Using clean engine oil, lubricate the sealing surface of the hub. Install the crankshaft pulley/damper with a special installation tool, available at most auto parts stores. Do not use a hammer to install the pulley/damper. Install a new center bolt and tighten it to the torque listed in this Chapter's Specifications. **Note:** *You must use a **new** pulley bolt.*

13 The remainder of the installation is the reverse of the removal procedure.

10.4 Remove the engine front cover bolts - the water pump may remain attached to the cover, if desired

10 Timing chain and sprockets - removal, inspection and installation

Note 1: *This is a difficult procedure, involving special tools and the removal of the engine from the vehicle. This procedure covers the removal of the timing chain, the cassettes and the sprockets. It will be necessary to refer to Section 8 for the camshaft timing procedure and the photos of the special tools and their placement on the cylinder heads. Read through both Sections and obtain the necessary tools before beginning the procedure.*

Note 2: *The rocker arms and valve lash adjusters must be removed before performing any repair procedures on the cylinder heads, timing chains and jackshaft chains. Refer to Section 5 and remove all the rocker arms and valve lash adjusters.*

Removal

Refer to illustrations 10.4, 10.5, 10.6, 10.7, 10.8, 10.10a, 10.10b and 10.10c

1   Refer to Chapter 2, Part C and remove the engine from the vehicle. The balance of this procedure is written assuming you have the engine out and mounted on an engine stand.

2   Remove the rocker arms and the valve lash adjusters (see Section 5).

3   Refer to Section 12 and remove the

lower oil pan cover, oil pump pickup tube and the crankcase reinforcement section.

4   Remove the bolts/nuts and the front cover from the engine block **(see illustration)**.

5   This engine uses a jackshaft (in place of a camshaft on a conventional pushrod engine) to drive the camshaft timing chains and the

oil pump **(see illustration)**. The left cylinder bank camshaft is driven by a chain at the front of the engine and the right cylinder bank camshaft is driven by a chain from the rear of the jackshaft. A balance shaft assembly is driven by a chain from the crankshaft sprocket.

6   Remove the bolts retaining the jackshaft

10.5 Timing chain components

1   *Jackshaft sprocket*
2   *Jackshaft chain*
3   *Chain/cassette for left camshaft*
4   *Crankshaft sprocket*

10.6 Remove the bolts retaining the jackshaft chain tensioner (A) and the chain guide (B)

10.7 Remove the jackshaft sprocket bolt, then remove the sprocket and chain

chain tensioner and the jackshaft chain guide **(see illustration)**.

7   Remove the bolt in the center of the jackshaft sprocket, then remove the sprocket with the jackshaft chain **(see illustration)**.

8   If not previously removed, remove the left-side upper cassette bolt **(see illustration 8.18)**, the camshaft sprocket **(see Section 8)**, and the lower cassette bolt. Then remove the cassette and chain with the jackshaft and camshaft sprockets **(see illustration)**.

9   To remove the right camshaft chain, remove the camshaft sprocket bolt and upper cassette bolt (see Section 8).

10   Remove the large plug covering the rear of the jackshaft **(see illustration)**. **Note:** *Obtain a new plug for reassembly. The plug is not reusable.* Remove the Torx bolt and spacer through the opening, then remove the lower cassette bolt and the right-side cassette and chain **(see illustrations)**.

Installation

Refer to illustration 10.16

11   If the crankshaft has been rotated during this procedure, make sure the number

one piston is at the top of its stroke (TDC) (see Section 3). The crankshaft keyway should point straight up. **Note:** *Do not rotate the crankshaft counterclockwise, or incorrect engine timing will result.*

12   Install the right camshaft timing chain cassette to the block. Install the rear jackshaft sprocket bolt and tighten it to the torque listed in this Chapter's Specifications. Install the rear jackshaft plug.

13   Install the left camshaft cassette and tighten the bolts to the torque listed in this Chapter's Specifications.

14   Drape the jackshaft chain over the jackshaft sprocket and engage the chain onto the crankshaft sprocket. Install the jackshaft sprocket onto the jackshaft and tighten the jackshaft sprocket bolt to the torque listed in this Chapter's Specifications.

15   Install the jackshaft chain guide and tensioner **(see illustration 10.6)**. Tighten the bolts to the torque listed in this Chapter's Specifications.

16   Clean the front surface of the engine block and front cover with lacquer thinner and install a new gasket **(see illustration)**. Install

10.8 Remove the bolt and remove the left camshaft chain cassette and chain

the front cover, tightening the bolts to the torque listed in this Chapter's Specifications.

17   Set the camshaft timing for the left camshaft sprocket and the right camshaft sprocket using the special tools (see Section 8).

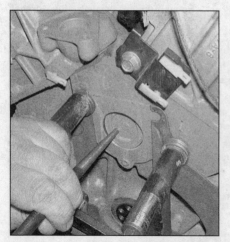

10.10a Remove the rear jackshaft plug by tapping it sideways, then use pliers to extract it

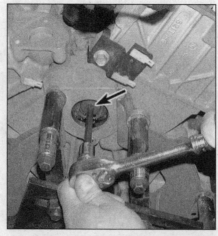

10.10b Remove this bolt and spacer . . .

10.10c . . . then remove this bolt and the right-side chain and cassette

10.16 Install the new front cover gasket

11.7a Areas to look for excessive wear or damage on the camshafts are the bearing surfaces and camshaft lobes

18   The remainder of installation is the reverse of the removal Steps.
19   Fill the crankcase with oil and change the filter, then add engine coolant (see Chapter 1).
20   Run the engine and check for oil or coolant leaks.

11   Camshafts - removal, inspection and installation

Note: *This is a difficult procedure, involving special tools and the removal of the engine from the vehicle. This procedure covers the removal of the camshafts and the camshaft bearing caps. It will be necessary to refer to Section 8 for the camshaft timing procedure and the photos of the special tools and their placement on the cylinder heads. Read through both Sections and obtain the necessary tools before beginning the procedure.*

Removal

1   Remove the valve covers (see Section 4).
2   Follow the procedure in Section 5 for removing the rocker arms and valve lash adjusters and see Section 8 for the procedure to disconnect the camshaft

sprockets and timing chains from the camshafts.
3   Mount a dial indicator to the front of the cylinder head and measure the camshaft endplay of each camshaft. If the clearance is greater than the value listed in this Chapter's Specifications, replace the camshaft and/or the cylinder head.
4   Loosen the bearing cap bolts in 1/4-turn increments, following the reverse sequence of the tightening procedure **(see illustration 11.13)**, until they can be removed by hand.
5   Remove the bearing caps and lift the camshaft off the cylinder head. Don't mix up the camshafts or any of the components. They must all go back to their original locations, and on the same cylinder head they were removed from.
6   Repeat this procedure for removal of the other camshaft.

Inspection

Refer to illustrations 11.7a, 11.7b, 11.9a and 11.9b

7   Visually examine the cam lobes and bearing journals for score marks, pitting, galling and evidence of overheating (blue, discolored areas). Look for flaking of the hardened surface of each lobe **(see illustrations)**.

8   Using a micrometer, measure the diameter of each camshaft journal and the lift of each camshaft lobe (see Chapter 2B, **illustrations 12.10a, 12.10b and 12.10c**). Compare your measurements with the Specifications listed at the front of this Chapter, and if the measurement of any one of these is less than specified, replace the camshaft.
9   Check the oil clearance for each camshaft bearing as follows:

a)   *Clean the bearing surfaces and the camshaft journals with lacquer thinner or acetone.*
b)   *Carefully lay the camshaft(s) in place in the cylinder head. Don't install the rocker arms or lash adjusters and don't use any lubrication.*
c)   *Lay a strip of Plastigage on each journal* **(see illustration)**.
d)   *Install the camshaft caps.*
e)   *Tighten the caps, a little at a time, to the torque listed in this Chapter's Specifications.* **Note:** *Don't turn the camshaft while the Plastigage is in place.*
f)   *Remove the bolts and detach the caps.*
g)   *Compare the width of the crushed Plastigage (at its widest point) to the scale on the Plastigage envelope* **(see illustration)**.

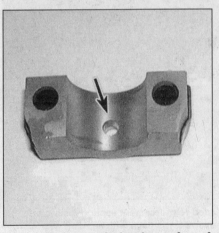

11.7b Also inspect the bearing surface of each camshaft cap

11.9a Lay a strip of Plastigage on each camshaft journal

11.9b Compare the width of the crushed Plastigage to the scale on the envelope to determine the oil clearance

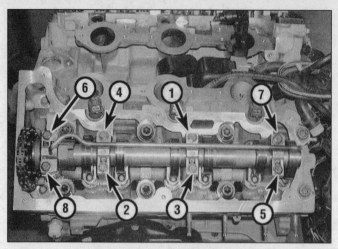

11.13 Camshaft bearing cap tightening sequence

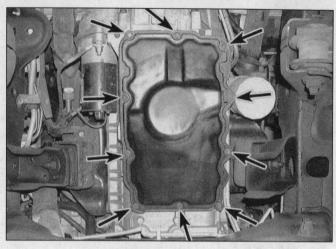

12.2 Lower oil pan bolts

h) *If the clearance is greater than specified, and the diameter of any journal is less than specified, replace the camshaft. If the journal diameters are within specifications but the oil clearance is too great, the cylinder head is worn and must be replaced.*

10   Scrape off the Plastigage with your fingernail or the edge of a credit card - don't scratch or nick the journals or bearing surfaces.

Installation

Refer to illustration 11.13

11   If the lash adjusters and/or rocker arms have been removed, install them in their original locations (see Section 5).
12   Apply moly-based engine assembly lubricant to the camshaft lobes and bearing journals, then install the camshaft(s).
13   Install the camshaft caps in the correct locations, and following the correct bolt tightening sequence **(see illustration)**, tighten the bolts to the torque listed in this Chapter's Specifications.
14   Refer to Section 10 for the timing pro-

cedure for each camshaft, using the special tools.
15   The remainder of installation is the reverse of the removal procedure.

---

12   Oil pan - removal and installation

Note: *The complete oil pan assembly is comprised of two sections. The lower section is a sheet metal oil pan, while the upper section is a large aluminum casting that serves as a structural reinforcement for the lower part of the crankcase. The lower oil pan and the oil pump pick-up/screen may be removed with the engine in-vehicle, but the engine must be removed from the vehicle to remove the reinforcement section for access to the oil pump and crankshaft.*

Removal

Refer to illustrations 12.2 and 12.5

1   Remove the engine from the vehicle (see Chapter 2, Part C).
2   Remove the lower oil pan fasteners **(see**

illustration) and remove the sheet metal oil pan.
3   Remove the two bolts and the oil pump screen cover and tube.
4   To remove the reinforcement section, begin by removing the eight bolts in the center of the section **(see illustration 12.16)**.
5   Remove the two rear reinforcement section-to-block bolts **(see illustration)**, then the outside bolts on each side.
6   Carefully remove the reinforcement section from the block. **Caution:** *Do not pry between the reinforcement section and the block on any gasket sealing surface.*

Installation

Refer to illustrations 12.8, 12.9a, 12.9b and 12.16

Caution: *Since two of the oil pan bolts are attached to the transmission, spacers are used between the oil pan and transmission. Failure to check the clearance and select the shims correctly can cause oil pan damage or oil leaks when the pan is installed. The engine must be removed from the engine stand for this check.*
7   Using a gasket scraper, thoroughly clean

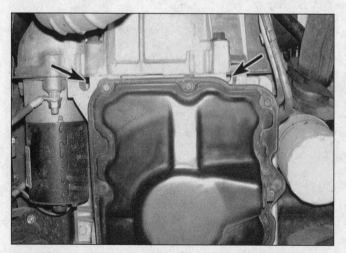

12.5 Location of the two rearmost reinforcement bolts

12.8 Turn the threaded reinforcement section inserts out (counterclockwise) several turns each with an Allen wrench to ensure they do not contact the main cap bolt head when the section is initially installed

12.9a Apply RTV sealant to the four places indicated on the front cover . . .

12.9b . . . and the two places indicated on the rear main cap

all old gasket material from the engine block, reinforcement section and oil pan. Remove residue and oil film with a solvent such as acetone or lacquer thinner.

8　The eight center bolts of the reinforcement section are threaded into the main cap bolt heads. Threaded inserts are installed in the reinforcement section to create support for the bolts without stressing the reinforcement section. Use an Allen wrench to back the inserts out until they are in the "index" position before installing the reinforcement section to the engine block **(see illustration)**.

9　Apply RTV sealant to the rear main cap and the bottom of the front cover **(see illustrations)**. On the front cover, there are four spots to seal, where the cover meets the block, and on either side of the front crankshaft seal, where the rubber end gasket will meet the side gaskets.

10　Position the new gaskets on the engine block. Install the two side gaskets, then the rubber end seals, making sure that the ends of the side gaskets fit into the notches on the rubber end seals. Add a small amount of RTV sealant over the joints between the side gaskets and end seals. Install the reinforcement section. **Note:** *Install the reinforcement section within five minutes of applying the sealant.*

11　Install, but do not tighten, the reinforcement section side rail bolts, including the two rear bolts.

12　Place a straightedge across the transmission mounting surface on the cylinder block and one of the mounting pads for the reinforcement section-to-transmission bolts. Measure the clearance between the mounting pad and straightedge with a feeler gauge.

13　Move the straightedge to the other mounting pad and measure the clearance. The clearance should not exceed 0.010-inch (0.25 mm). If the reinforcement section protrudes over the rear edge of the engine block, tap it toward the front of the engine with a soft-faced hammer until it's flush (it may pro-

trude a maximum of 0.002-inch [0.05 mm]). If the clearance is excessive, tap it to the rear of the engine.

14　Tighten the perimeter bolts/nuts, including the two rear lower block cradle-to-transmission bolts and the two rear lower block cradle-to-engine block bolts, to the torque listed in this Chapter's Specifications.

15　Tighten the reinforcement section threaded inserts to the torque listed in this Chapter's Specifications.

16　Install the interior reinforcement section bolts (lower cradle block bolts), using new seals on the two front bolts (the front two bolts are distinguished by their silver color). Tighten them in sequence to the torque listed in this Chapter's Specifications **(see illustration)**.

17　Repeat the reinforcement section-to-engine block clearance check (see Step 13).

18　Install the oil pump pickup tube, place a new oil pan gasket in position and install the oil pan and bolts. Tighten the bolts to the torque listed in this Chapter's Specifications.

13　Oil pump - removal and installation

Removal

Refer to illustration 13.2

1　Remove the lower and upper oil pan sections (see Section 12). **Note:** *This involves removing the engine from the vehicle (see Part C of this Chapter).*

2　Remove the oil pump bolts and take the pump off the engine **(see illustration)**.

3　Pull the oil pump driveshaft out of the engine.

Installation

Refer to illustration 13.5

4　Fill one of the pump ports with clean engine oil and rotate the pump by hand to prime it.

5　Insert the driveshaft into the engine,

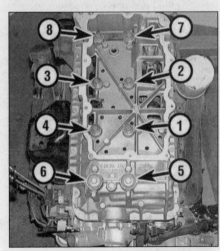

12.16 Crankcase reinforcement section bolt tightening sequence

13.2 Remove the two oil pump bolts

13.5 Insert the oil pump driveshaft into the pump, then guide the driveshaft into the engine, making sure the travel limit clip is in place

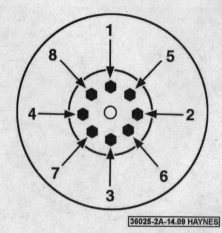

36025-2A-14.09 HAYNES

14.9 Flywheel/driveplate bolt tightening sequence

pointed end first, until it engages the oil pump drive **(see illustration)**. Make sure the travel limit clip is in place, and closer to the camshaft position sensor end. It fits tightly on a non-machined section of the driveshaft.

6   Install the oil pump on the block, using a new gasket. Make sure the pump engages the driveshaft. Install the oil pump bolts and tighten to the torque listed in this Chapter's Specifications.

7   The remainder of installation is the reverse of removal.

8   Run the engine and make sure oil pressure comes up to normal quickly. If it doesn't, stop the engine and find out the cause. Severe engine damage can result from running an engine with insufficient oil pressure!

14   Flywheel/driveplate - removal and installation

Removal

1   Raise the vehicle and support it securely on jackstands, then refer to Chapter 7 and remove the transmission. If it's leaking, now would be a very good time to replace the front pump seal/O-ring (automatic transmission only).

2   Remove the pressure plate and clutch disc (Chapter 8) (manual transmission equipped vehicles). Now is a good time to check/replace the clutch components and pilot bearing.

3   Use a center punch or paint to make alignment marks on the flywheel/driveplate and crankshaft to ensure correct alignment during reinstallation.

4   Remove the bolts that secure the flywheel/driveplate to the crankshaft. If the crankshaft turns, wedge a screwdriver in the ring gear teeth to jam the flywheel.

5   Remove the flywheel/driveplate from the crankshaft. Since the flywheel is fairly heavy, be sure to support it while removing the last bolt. **Caution:** *The ring gear teeth may be*

sharp. Wear gloves to protect your hands. Automatic transmission equipped vehicles have a spacer between the crankshaft and the driveplate.

Installation

Refer to illustration 14.9

6   Clean the flywheel to remove grease and oil. Inspect the surface for cracks, rivet grooves, burned areas and score marks. Light scoring can be removed with emery cloth. Check for cracked and broken ring gear teeth. Lay the flywheel on a flat surface and use a straightedge to check for warpage.

7   Clean and inspect the mating surfaces of the flywheel/driveplate and the crankshaft. If the crankshaft rear seal is leaking, replace it before reinstalling the flywheel/driveplate.

8   Position the flywheel/driveplate against the crankshaft. Be sure to align the marks made during removal. Note that some engines have an alignment dowel or staggered bolt holes to ensure correct installation. Before installing the bolts, apply thread locking compound to the threads.

9   Wedge a screwdriver in the ring gear teeth to keep the flywheel/driveplate from turning and tighten the bolts to the torque listed in this Chapter's Specifications. Follow the correct torque sequence **(see illustration)** and work up to the final torque in three or four steps.

10   The remainder of installation is the reverse of the removal procedure.

15   Rear main oil seal - replacement

1   The one-piece rear main oil seal is pressed into the engine block and the crankcase reinforcement section. Remove the transmission (see Chapter 7), the clutch components, if equipped (see Chapter 8) and the flywheel (see Section 14).

2   Pry out the old seal with a special seal removal tool or a flat blade screwdriver. **Cau-**

tion: *To prevent an oil leak after the new seal is installed, be very careful not to scratch or otherwise damage the crankshaft sealing surface or the bore in the engine block.*

3   Clean the crankshaft and seal bore in the block thoroughly and de-grease these areas by wiping them with a rag soaked in lacquer thinner or acetone. Lubricate the lip of the new seal and the outer diameter of the crankshaft with engine oil. Make sure the edges of the new oil seal are not rolled over.

4   Position the new seal onto the crankshaft. **Note:** *When installing the new seal, if so marked, the words THIS SIDE OUT on the seal must face out, toward the rear of the engine.* Use a special rear main oil seal installation tool or a socket with the exact diameter of the seal to drive the seal in place. Make sure the seal is not off-set; it must be flush along the entire circumference of the seal carrier.

5   The remainder of installation is the reverse of removal.

16   Engine mounts - check and replacement

Check

1   Engine mounts seldom require attention, but broken or deteriorated mounts should be replaced immediately or the added strain placed on the driveline components may cause damage or wear.

2   During the check, the engine must be raised slightly to remove the weight from the mounts.

3   Raise the vehicle and support it securely on jackstands, then position a jack under the engine oil pan. Place a wood plank between the jack head and the oil pan, then carefully raise the engine just enough to take the weight off the mounts. **Warning:** *DO NOT place any part of your body under the engine when it's supported only by a jack!* **Note:** *If your jack and piece of wood will fit under the vehicle, it*

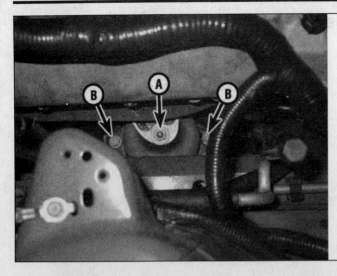

16.13a Right-side engine mount details

A   Mount upper nut
B   Mount-to-chassis bolts

16.13b Left-side engine mount details

A   Mount upper nut
B   Mount-to-chassis bolts

won't be necessary to raise the vehicle.

4   Check the mounts to see if the rubber is cracked or hardened. Sometimes the rubber will split right down the center.

5   Have an assistant raise the jack and see if the upper portion of the mount raises off the lower portion. If it does, replace the mount.

6   Check for relative movement between the mount plates and the engine or frame (use a large screwdriver or pry bar to attempt to move the mounts).

7   If movement is noted, lower the engine and tighten the mount fasteners.

Replacement

Refer to illustrations 16.13a and 16.13b

8   Disconnect the cable from the negative battery terminal (see Chapter 5, Section 1).

9   Detach the air intake duct from the throttle body. If you're replacing the left-side mount, remove the air filter housing to provide more working room (see Chapter 4).

10   Remove the throttle body (see Chapter 4).

11   Remove the accessory bracket bolt from the front of the left cylinder head.

12   Install an engine support fixture or engine hoist, with the chain secured with the bolt removed in the previous Step. Make sure the chain is anchored properly and the bolt is tightened securely. As an alternative, a floor jack can be used, along with a piece of wood, positioned under the oil pan.

13   Remove the fasteners holding the mount to the mount bracket and the chassis **(see illustrations)**.

14   Unbolt the mount bracket from the engine, then remove the mount.

15   Installation is the reverse of removal. Use thread-locking compound on the mount fasteners and tighten them to the torque listed in this Chapter's Specifications.

Notes

Chapter 2 Part B
V8 engine

Contents

Specifications

General

Displacement	281 cubic inches (4.6 liters)
Cylinder numbers (front to rear)	
Right side	1-2-3-4
Left (driver's) side	5-6-7-8
Firing order	1-3-7-2-6-5-4-8

Camshaft

Lobe lift	0.217 inch
Allowable lobe lift loss	0.005 inch
Endplay	0.0002 to 0.0090 inch
Journal diameter (all)	1.126 to 1.127 inches
Bearing inside diameter (all)	1.128 to 1.129 inches
Journal-to-bearing (oil) clearance	
Standard	0.001 to 0.003 inch
Service limit	0.002 inch maximum

Torque specifications

Ft-lbs (unless otherwise indicated)

Accessory drivebelt pulley bolt(s)	18
Camshaft sprocket bolt	
Step 1	30
Step 2	Tighten an additional 90-degrees
Camshaft cap bolts	89 in-lbs
Coolant crossover assembly bolts	89 in-lbs
Cylinder head bolts*	
Step 1	30
Step 2	Tighten an additional 90-degrees
Step 3	Tighten an additional 90-degrees
Crankshaft pulley-to-crankshaft bolt	
Step 1	66
Step 2	Loosen one full turn
Step 3	37
Step 4	Tighten an additional 90-degrees
Subframe mounting bolts	85
Drivebelt tensioner bolts	18

4.6L V8 ENGINE
1-3-7-2-6-5-4-8

36061-1-specs.C HAYNES

Cylinder locations

Torque specifications (continued)

	Ft-lbs (unless otherwise indicated)
Engine mounts	
Mount-to-bracket nuts	46
Mount bracket-to-engine block bolts	41
Mount-to-chassis bolts	41
Exhaust manifold-to-cylinder head nuts	18
Exhaust pipe-to-exhaust manifold nuts	30
Flywheel/driveplate bolts	59
Intake manifold-to-cylinder head bolts	89 in-lbs
Oil pan bolts	
Step 1	18 in-lbs
Step 2	15
Step 3	Tighten an additional 60-degrees
Oil filter adapter bolts	18
Oil pump-to-engine block mounting bolts	89 in-lbs
Oil pick-up screen-to-engine block bolt	18
Oil pick-up tube-to-oil pump bolts	89 in-lbs
Rear main seal retainer-to-engine block bolts	89 in-lbs
Timing chain cover bolts **(see illustration 8.21)**	
Step 1	15
Step 2	Tighten an additional 60-degrees
Timing chain stationary guide bolts	89 in-lbs
Timing chain tensioner bolts	18
Valve cover bolts	89 in-lbs

Use new cylinder head bolts

1   General information

This Part of Chapter 2 is devoted to in-vehicle repair procedures for the 4.6L Three-valve, Single Overhead Camshaft (SOHC) engine, as well as cylinder head removal, which requires removal of the engine from the vehicle. Information concerning engine removal and installation and overhaul can be found in Part C of this Chapter.

This engine has aluminum cylinder heads, an aluminum block and three valves per cylinder.

This engine is an "interference" design. In the event the timing chain breaks, the pistons will contact the valves and cause damage.

2   Repair operations possible with the engine in the vehicle

Many major repair operations can be accomplished without removing the engine from the vehicle.

If possible, clean the engine compartment and the exterior of the engine with some type of pressure washer before any work is started. It will make the job easier and help keep dirt out of the internal areas of the engine.

It may help to remove the hood to improve access to the engine as repairs are performed (refer to Chapter 11).

If vacuum, exhaust, oil or coolant leaks develop, indicating a need for gasket or seal replacement, the repairs can generally be made with the engine in the vehicle. The intake and exhaust manifold gaskets, timing chain cover gasket, oil pan gasket and crankshaft oil seals are all accessible with the engine in place.

Exterior engine components, such as the intake and exhaust manifolds, the oil pan, the water pump, the starter motor, the alternator and the fuel system components can be removed for repair with the engine in place. Replacement of the timing chain and sprockets and oil pump is also possible with the engine in the vehicle.

In extreme cases caused by a lack of necessary equipment, repair or replacement of piston rings, pistons, connecting rods and rod bearings is also possible with the engine in the vehicle. However, this practice is not recommended because of the cleaning and preparation work that must be done to the components involved.

3   Top Dead Center (TDC) for number one piston - locating

Refer to illustration 3.1

Refer to Chapter 2, Part A for the TDC locating procedure, but use the illustration provided with this Section for the appropriate reference marks and the following exceptions:

a) *Remove the ignition coils (see Chapter 5).*

b) *Remove the spark plugs and install a compression gauge in the number one cylinder. Turn the crankshaft clockwise with a socket and breaker bar.*

c) *When the piston approaches TDC, compression will be noted on the compression gauge. Continue turning the crankshaft until the notch in the crankshaft pulley is aligned with the TDC mark on*

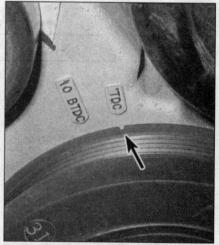

3.1 When placing the engine at Top Dead Center (TDC), align the notch in the crankshaft pulley with the TDC indicator on the timing chain cover

the front cover **(see illustration)**. *At this point, number one cylinder is at TDC on the compression stroke.*

d) *After the number one piston has been positioned at TDC on the compression stroke, TDC for any of the remaining cylinders can be located by turning the crankshaft in 90-degree increments and following the firing order (refer to the Specifications). Divide the crankshaft pulley into four equal sections with chalk marks at four points, each indicating 90-degrees of crankshaft rotation. Rotating the engine 90-degrees past TDC number 1 will put the engine at TDC for cylinder number 3.*

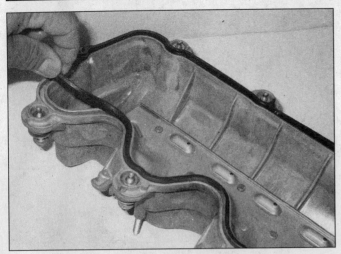

4.16 Make sure the gasket is pushed all the way into the groove in the valve cover

4.17 Apply a dab of RTV sealant to the mating joints between the timing chain cover and the cylinder head before installing the valve cover

4 Valve covers - removal and installation

Removal

1 Disconnect the cable from the negative battery terminal (see Chapter 5, Section 1).
2 Disconnect the PCV hose quick-connect fittings from the valve cover(s).
3 Remove the ignition coils (see Chapter 5).
4 Disconnect the fuel injector harness from the injectors and position the harness off to the side (see Chapter 4).

Right (passenger's side) valve cover

5 Unplug the electrical connector from the Variable Camshaft Timing (VCT) control solenoid (see Chapter 6).
6 Remove the engine harness connectors from the valve cover.
7 Remove the PCV hose from the valve cover and the intake manifold (see Chapter 6).
8 Remove the battery and the battery tray (see Chapter 5).

9 Loosen the valve cover bolts and remove the valve cover. Follow the reverse of the tightening sequence **(see illustration 4.19a)**. If it's stuck, tap it with a hammer and block of wood.

Left (driver's side) valve cover

10 Remove the air filter housing and intake air duct (see Chapter 4).
11 Remove the oil dipstick tube mounting bolt and position the tube off to the side.
12 Unplug the electrical connector from the Variable Camshaft Timing (VCT) control solenoid (see Chapter 6).
13 Disconnect the EVAP tube from the intake manifold (see Chapter 6).
14 Loosen the valve cover bolts and remove the valve cover. Follow the reverse of the tightening sequence **(see illustration 4.19b)**. If it's stuck, tap it with a hammer and block of wood.

Installation

Refer to illustrations 4.16, 4.17, 4.19a and 4.19b

15 The mating surfaces of each cylinder

head and valve cover must be perfectly clean when the valve covers are installed. Remove all traces of sealant, and clean the mating surfaces with lacquer thinner or acetone. If there's old sealant or oil on the mating surfaces when the valve cover is installed, oil leaks may develop.
16 Install a new valve cover gasket in the cover's groove. Make sure the gasket is pushed all the way into the groove in the valve cover **(see illustration)**. The use of sealant in the groove will help hold the gasket in place.
17 At the mating joint (two spots per cylinder head) between the timing chain cover and cylinder head, apply a dab of RTV sealant before installing the valve cover **(see illustration)**.
18 Carefully position the valve cover on the cylinder head and install the bolts. **Note:** *Install the covers within five minutes of applying the RTV sealant.*
19 Tighten the fasteners in two steps to the torque listed in this Chapter's Specifications. Follow the correct torque sequence **(see illustrations)**.
20 The remaining installation steps are the reverse of removal.

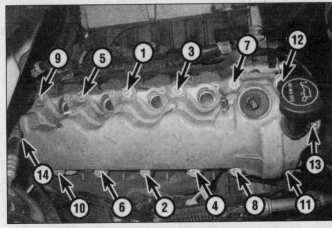

4.19a Right valve cover bolt tightening sequence

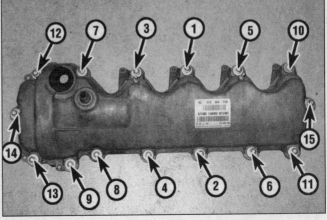

4.19b Left valve cover bolt tightening sequence

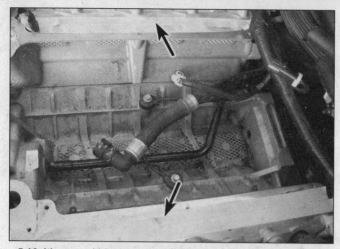

5.16 It's a good idea to cover the intake ports with duct tape to prevent debris from falling in

5.18 Location of the intake manifold O-ring seals

21   Start the engine and check for oil leaks as the engine warms up.

5   Intake manifold - removal and installation

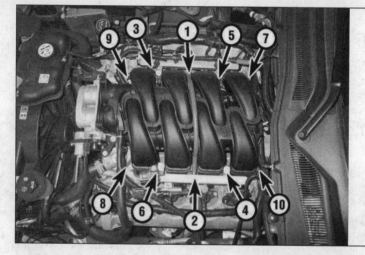

5.20 Intake manifold bolt tightening sequence

Removal

Refer to illustration 5.16

1   Relieve the fuel system pressure (see Chapter 4).
2   Disconnect the cable from the negative battery terminal (see Chapter 5, Section 1).
3   Remove the drivebelt (see Chapter 1).
4   Detach the air intake air duct from the throttle body (see Chapter 4).
5   Disconnect the brake booster vacuum line from the intake manifold.
6   Disconnect the electrical connectors from the electronic throttle body (see Chapter 4).
7   Detach the PCV hoses from the intake manifold.
8   Disconnect the quick-connect coupling on the EVAP line at the intake manifold (see Chapter 6).
9   Disconnect the Fuel Rail Pressure and Temperature (FRPT) sensor connector (see Chapter 6).
10   Remove the fuel rail and injectors (see Chapter 4).
11   Disconnect the Charge Motion Control Valve (CMCV) electrical connector (see Chapter 6) located at the rear of the intake manifold. **Note:** *It may be easier to do this after the manifold has been unbolted and moved forward.*
12   Loosen the intake manifold bolts in 1/4-turn increments, following the reverse order of the tightening sequence **(see illustration 5.20)**, until they can be removed by hand.
13   Move the intake manifold assembly forward and disconnect the cylinder head temperature (CHT) sensor jumper harness connector.
14   Lift the intake manifold from the cylinder

heads. If it's stuck, it shouldn't take more than a little bit of force to jar it loose.
15   Remove the intake manifold gaskets (O-rings) and clean all traces of dirt and gasket or sealant material from the sealing surfaces of the cylinder heads and intake manifold. **Caution:** *The mating surfaces of the cylinder heads, engine block and intake manifold must be perfectly clean. Since the cylinder heads are aluminum and the intake manifold is plastic, aggressive scraping can cause damage! Use only a plastic-tipped scraper, not a metal one.*
16   While the manifold is removed, cover the intake ports with shop rags or duct tape to keep debris out of the engine **(see illustration)**. Use a vacuum cleaner to remove any debris that falls into the intake ports in the cylinder heads.

Installation

Refer to illustrations 5.18 and 5.20

17   If you are replacing the manifold, transfer all components to the new unit.
18   Attach the new O-rings to the intake manifold **(see illustration)**.
19   Carefully set the intake manifold in place. Don't move the manifold fore-and-aft after it

contacts the cylinder heads, as the O-rings might be disturbed.
20   Install the intake manifold bolts and, following the recommended tightening sequence **(see illustration)**, tighten them to the torque listed in this Chapter's Specifications.
21   The remaining installation steps are the reverse of removal. **Caution:** *Make sure that the electrical harness does not interfere with the movement of the Charge Motion Control Valve actuator rods through the entire range of operation.*

6   Exhaust manifolds - removal and installation

Removal

1   Disconnect the cable from the negative battery terminal (see Chapter 5, Section 1).
2   Remove the alternator (see Chapter 5).
3   Raise the vehicle and support it securely on jackstands.
4   Remove the inner splash shield from the fenderwell (see Chapter 11).
5   Working under the vehicle, apply penetrating oil to the catalytic converter-to-mani-

6.12 Left-side exhaust manifold mounting nuts (engine removed for clarity)

7.5 Be sure to wrap a shop rag or a piece of drivebelt material around the pulley before installing the strap wrench

fold studs and nuts and to the exhaust manifold mounting nuts (they're usually corroded or rusty).

6   Remove the catalytic converter nuts from the exhaust manifolds and move the catalytic converter assemblies to the rear of the vehicle (see Chapter 6).

7   Connect an engine support fixture or engine hoist as shown in Chapter 2C, Section 7, and remove the engine mount upper nuts (see Section 18).

Left (driver's side) exhaust manifold

Refer to illustration 6.12

8   Remove the intake duct and the air filter housing (see Chapter 4).

9   Disconnect the rear heated oxygen sensor connector (see Chapter 6).

10   Disconnect the intermediate shaft from the power steering gear. **Warning:** *Don't allow the steering column to rotate after the intermediate shaft has been disconnected, as the airbag clockspring could be damaged. Either remove the ignition key to lock the steering column, or pass the seat belt through the steering wheel and clip it into place.*

11   Raise the engine 1-1/2 inches to provide clearance for exhaust manifold removal. If you're using an engine hoist, place a block of wood between the mount and the mount bracket, just in case the hoist hydraulic ram fails.

12   Remove the eight mounting nuts from the exhaust manifold **(see illustration)**. Remove the exhaust manifold.

Right (passenger's side) exhaust manifold

13   Remove the starter (see Chapter 5).

14   Disconnect the front and rear heated oxygen sensor connectors (see Chapter 6).

15   Detach the transmission cooler tube bracket and position the tube off to the side.

16   Raise the engine 1-1/2 inches to provide clearance for exhaust manifold removal. If you're using an engine hoist, place a block of wood between the mount and the mount

bracket, just in case the hoist hydraulic ram fails.

17   Remove the eight mounting nuts from the exhaust manifold. Remove the exhaust manifold.

Installation

18   Check the exhaust manifolds for cracks. Make sure the stud threads are clean and undamaged. The exhaust manifold and cylinder head mating surfaces must be clean before the exhaust manifolds are reinstalled - use a gasket scraper to remove all carbon deposits.

19   Position a new gasket in place and slip the exhaust manifold over the studs on the cylinder head. Install the mounting nuts.

20   When tightening the mounting nuts, work from the rear to the front, alternating between top and bottom rows. Tighten the bolts in three equal steps to the torque listed in this Chapter's Specifications.

21   The remaining installation steps are the reverse of removal.

22   Start the engine and check for exhaust leaks.

7   Crankshaft pulley and front oil seal - removal and installation

Removal

Refer to illustrations 7.5, 7.6 and 7.7

1   Disconnect the cable from the negative battery terminal (see Chapter 5, Section 1).

2   Remove the drivebelt (see Chapter 1).

3   Remove the engine cooling fan and shroud assembly (see Chapter 3).

4   Raise the vehicle and support it securely on jackstands.

5   Use a breaker bar and socket to remove the crankshaft pulley center bolt **(see illustration)**. **Note:** *It will be necessary to lock the pulley in position using a strap or chain wrench. Be sure to wrap a shop rag or a piece of drivebelt material around the pulley before installing the special tool.*

6   Using a bolt-type puller, pull the pulley

7.6 Remove the crankshaft pulley with a puller that bolts to the crankshaft pulley hub

7.7 Prying out the oil seal with a special seal removal tool - be careful to not damage the crankshaft or the bore while using the special tool

from the crankshaft **(see illustration)**. **Caution:** *Because the pulley is recessed, an adapter may be needed between the puller bolt and the crankshaft (to prevent damage to the bore and threads in the end of the crankshaft.*

7   Use a seal puller to remove the crankshaft front oil seal **(see illustration)**.

8   Clean the seal bore and check it for nicks or gouges. Also, examine the area of the hub that rides in the seal for signs of abnormal wear or scoring. For many popular engines, a repair sleeve is available to restore a smooth finish to the sealing surface. Check with your auto parts store for availability of this sleeve.

Installation

Refer to illustrations 7.9 and 7.10

9   Coat the lip of the new seal with clean engine oil and drive it into the bore with a seal driver or a large socket slightly smaller in diameter than the seal **(see illustration)**. The open side of the seal faces into the engine.
10   Lubricate the oil seal contact surface of the crankshaft pulley hub **(see illustration)** with multi-purpose grease or clean engine oil. Apply a dab of RTV sealant to the keyway slot in the crankshaft pulley before installation.
11   Install the crankshaft pulley with a special installation tool, available at most auto parts stores. Do not use a hammer to install the pulley. Install the center bolt and tighten it to the torque listed in this Chapter's Specifications.
12   The remainder of the installation is the reverse of the removal procedure.

**8   Timing chain cover - removal and
        installation**

Warning: *Wait until the engine is completely cool before beginning this procedure.*

Removal

Refer to illustration 8.16

1   Disconnect the cable from the negative battery terminal (see Chapter 5, Section 1).
2   Drain the engine oil and engine coolant (see Chapter 1). Also remove the oil filter.
3   Remove the expansion tank and the engine cooling fan (see Chapter 3).
4   Detach the coolant hoses from the coolant crossover housing.

7.9 There is a special tool for installing the front oil seal into the timing chain cover - if the tool is not available, a large socket (the same diameter of the seal) can be used to drive the seal into place

5   Remove the drivebelt (see Chapter 1) and the water pump pulley (see Chapter 3).
6   Remove the crankshaft pulley (see Section 7).
7   Remove both valve covers (see Section 4).
8   Remove the nut from the transmission fluid cooler line bracket and remove the bracket from the lower timing chain cover.
9   Remove the mounting bolts and position the power steering pump off to the side (see Chapter 10).
10   Remove the drivebelt tensioner (see Chapter 1) and the drivebelt idler pulleys from the timing chain cover.
11   Disconnect the electrical connector to the crankshaft position sensor (see Chapter 6).
12   Remove the camshaft position sensors (see Chapter 6).
13   Remove the radio ignition interference capacitors from the left and right sides of the upper timing chain cover.
14   Remove the four front oil pan bolts (see

7.10 Inspect the crankshaft pulley for signs of damage or excessive wear

1   Oil seal surface
2   Woodruff keyway

Section 14).
15   Remove the timing chain cover-to-engine block bolts. Note the locations of studs and different length bolts so they can be reinstalled in their original locations.
16   Separate the timing chain cover from the engine block **(see illustration)**. If it's stuck, tap it gently with a soft-face hammer to break the gasket bond. **Caution:** *DO NOT use excessive force or you may crack the cover. If the cover is difficult to remove, make sure all of the bolts have been removed.*

Installation

Refer to illustrations 8.18, 8.19 and 8.21

17   Clean the mating surfaces of the timing chain cover, engine block and cylinder heads to remove all traces of old gasket material, oil and dirt. Final cleaning should be with lacquer thinner or acetone. **Warning:** *Be careful when cleaning any of the aluminum components. Use of a metal scraper could cause scratches or gouges that could lead to an oil leak later.*

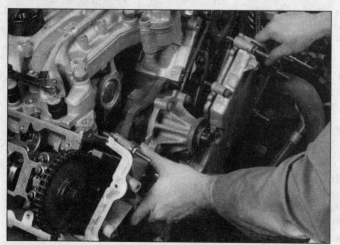

8.16 Separate the timing chain cover from the engine, using a soft-faced hammer if necessary to break the gasket seal

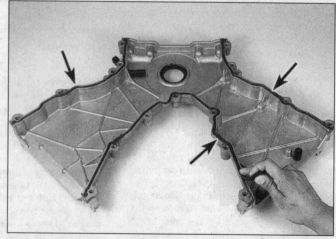

8.18 Install three new gaskets into the grooves in the back of the timing chain cover

8.19 Apply a bead of RTV sealant to the mating junctions of the oil pan-to-engine block and cylinder head-to-engine block

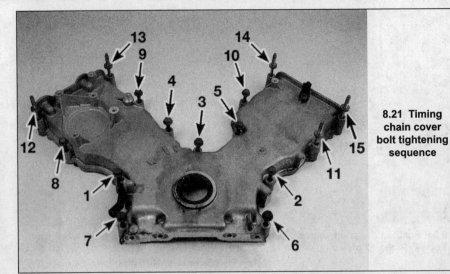

8.21 Timing chain cover bolt tightening sequence

18   Install the three new gaskets into the grooves of the timing chain cover (see illustration).

19   Apply a 1/8-inch bead of RTV sealant to the junctions of the oil pan-to-engine block and the cylinder head-to-engine block (see illustration). Apply a small dab of RTV where the timing chain cover and engine block meet, and where the timing chain cover and cylinder head meet at the valve cover surface.

20   Lubricate the timing chains and the lip of the crankshaft front oil seal with clean engine oil.

21   Install the timing chain cover on the engine, within five minutes of applying the RTV sealant. Position the bottom/front edge of the timing chain cover flush with the front edge of the oil pan and "tilt" the top of the cover into place against the engine. Do not press the cover straight in against the engine or the sealant may be scraped off the front of the oil pan and cause a leak. Tighten the timing chain cover-to-engine block bolts in

the recommended sequence (see illustration), to the torque and sequence listed in this Chapter's Specifications. Tighten the oil pan-to-timing chain cover bolts to the torque listed in this Chapter's Specifications.

22   Install the remaining parts in the reverse order of removal.

23   Install a new oil filter and add the proper type and quantity of engine oil (see Chapter 1). Also refill the cooling system (see Chapter 1). Run the engine and check for leaks.

---

9   Timing chains, tensioners and sprockets - removal, inspection and installation

Caution 1: *These engines are difficult to work on and require special tools for many procedures. On any procedure involving timing chains, camshafts or cylinder head removal, the steps must be read carefully and disassembly must proceed using the special tools, otherwise damage to the engine could result.*

Caution 2: *Because this is an "interference" engine design, if a chain has broken, there will be damage to the valves (and possibly the pistons) and will require removal of the cylinder heads.*

Removal

Refer to illustrations 9.3, 9.4, 9.5a, 9.5b, 9.8 and 9.9

1   Disconnect the cable from the negative battery terminal (see Chapter 5, Section 1).

2   Remove the spark plugs (see Chapter 1) and position the number one cylinder on TDC (see Section 3).

3   Remove the timing chain cover (see Section 8). Two long timing chains connect camshafts to the crankshaft (see illustration).

4   Remove the crankshaft position sensor toothed wheel (see illustration) by sliding it off the end of the crankshaft nose. Note the direction of the teeth on the wheel to insure correct reassembly (the wheel is marked "front").

5   Rotate the crankshaft and position the

9.3 These V8 engines have two long timing chains with two main tensioners

9.4 Remove the crankshaft position sensor toothed wheel

9.5a Position the crankshaft with the keyway in the 12 o'clock position . . .

9.5b . . . and the camshaft lobes for cylinder number one positioned like this (up position)

A   Intake lobes
B   Exhaust lobe

9.8 Location of the left side timing chain tensioner mounting bolts

crankshaft keyway in the 12 o'clock position **(see illustrations)**.
6   Remove the rocker arms from the "Stage 1" designated cylinders (see Step 3 in Section 10).
7   Rotate the crankshaft CLOCKWISE and position the crankshaft keyway in the 6 o'clock position. Remove the remaining rocker arms from the cylinder heads.
8   Remove the left side timing chain

tensioner **(see illustration)** and the timing chain tensioner guide.
9   Remove the right side timing chain tensioner **(see illustration)** and the timing chain tensioner guide.
10   Remove the right timing chain from the

crankshaft and camshaft sprockets.
11   Remove the left timing chain from the crankshaft and camshaft sprockets, by slipping the chain off the camshaft sprocket and under the crankshaft sprocket.
12   Remove the timing chain stationary guides.
13   The camshafts can now be removed, if desired (see Section 12).

Inspection

Refer to illustrations 9.15a and 9.15b
14   Inspect the individual sprocket teeth and keyways for wear and damage. Check the chain for cracked plates and pitted or worn rollers. Check the wear surface of the chain guides for wear and damage. Replace any worn or defective parts with new ones. **Caution:** *If excessive plastic material is missing from the chain guides, the oil pan should be removed and cleaned of all debris (see Section 13). Check the oil pick-up tube and screen. Replace the assembly if it is clogged.*
15   Check the timing chain tensioners:

a) *Check the condition of the tensioner seal* **(see illustration)**. *Make sure the seal is intact and not broken, chipped or damaged.*

9.9 Location of the right side timing chain tensioner mounting bolts

9.15a Check the condition of the tensioner seal

9.15b Install a special tool (A) onto the plunger and tensioner side rail (B) to lock the plunger in the retracted position

9.19 This tool slips over the end of the crankshaft and engages with the dowel pin on the right (passenger's) side of the engine

9.21a Colored link on the timing chain aligned with the mark on the crankshaft sprocket (right timing chain shown, left timing chain similar)

b) Check the condition of the plunger. Depress the plunger to make sure it moves freely.

c) For installation, compress the tensioner in a vise and lock it in the retracted position using a special retainer clip **(see illustration)**.

Installation

Refer to illustrations 9.19, 9.21a, 9.21b and 9.23

16   Install the timing chain stationary guides, for both sides, and tighten the bolts to the torque listed in this Chapter's Specifications.

17   Install the crankshaft sprocket on the crankshaft with the timing marks facing forward.

18   Install the camshafts (see Section 12) if they were removed. **Caution:** *Do NOT install the rocker arms at this time. The rocker arms will be installed as a final step when the timing chains and tensioners have been installed.*

19   Using the special tool **(see illustration)**, position the crankshaft keyway in the 10 o'clock position. This is the TDC number 1 position.

20   The timing chains should have three bright (copper plated) or colored links on each chain. If no colored links are present, purchase timing chains equipped with colored links.

21   Loop the left side timing chain under the crankshaft sprocket and align the bright link with the alignment mark on the crankshaft sprocket **(see illustration)**. Install the timing chain over the camshaft sprocket, aligning the two bright links with the L on the sprocket **(see illustration)**. **Note:** *The slack side of the chain should be below the tensioner arm dowel on the block.*

22   Install the left timing chain tensioner guide. Before assembling the tensioner with the chain guide, compress the tensioner and lock it in this position with a retainer clip **(see illustration 9.15b)**. Install the tensioner, tighten the bolts to the torque listed in this Chapter's Specifications and remove the retainer clip to apply force against the tensioner chain guide so the tensioner fully extends against the chain guide and all slack is removed from the chain.

23   Loop the right side timing chain under the crankshaft sprocket and align the bright link with the alignment mark on the crankshaft sprocket **(see illustration 9.21a)**. Install the timing chain over the camshaft sprocket, aligning the two bright links with the I on the camshaft sprocket **(see illustration)**. **Note 1:** *The slack side of the chain should be above the tensioner arm dowel on the block.* **Note 2:** *The camshaft sprocket alignment*

mark can be an I, 0, 0R or an arrow.

24   Install the right timing chain tensioner guide. Before assembling the tensioner with the chain guide, compress the tensioner and lock it in this position with a retainer clip **(see illustration 9.15b)**. Install the tensioner, tighten the bolts to the torque listed in this Chapter's Specifications and remove the retainer clip to apply force against the tensioner chain guide so the tensioner fully extends against the chain guide and all slack is removed from the chain.

25   Recheck all the timing marks to make sure they are still in alignment.

26   Install the rocker arms (see Section 10).

27   Slowly rotate the crankshaft in the normal direction of rotation (clockwise) at least two revolutions and again bring the engine to TDC. If you feel any resistance, stop and find out why.

28   The remainder of installation is the reverse of removal.

9.21b Align the L on the left camshaft sprocket (A) with the two colored links (B) on the timing chain

9.23 Align the I on the right camshaft sprocket (A) with the two colored links (B) on the timing chain

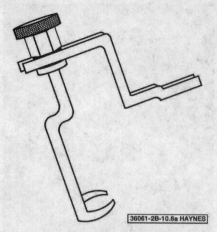

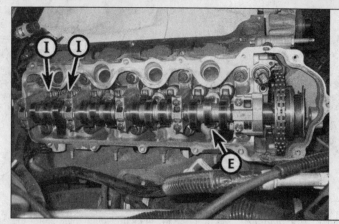

10.3b With the crankshaft keyway positioned at 12 o'clock (just past TDC for cylinder number one), the exhaust rocker arm can be removed from cylinder number 1, the intake rocker arms from cylinder number 4 . . .

10.3a Use a special tool to compress the valve springs and remove the rocker arms

10 Rocker arms and valve lash adjusters - removal, inspection and installation

Caution: *These engines are difficult to work on and require special tools for many procedures. On any procedure involving timing chains, camshafts or cylinder head removal, the steps must be read carefully and disassembly must proceed using the special tools otherwise damage to the engine could result.*

Removal

Refer to illustrations 10.3a, 10.3b and 10.3c

1   Remove the valve cover(s) (see Section 4).

2   Because of the interference design of these engines, the rocker arms must be removed before the timing chains and camshafts are removed. This will be accomplished in two stages. The first stage, certain valve springs on designated cylinders must be compressed and the rocker arms removed to prevent any valve-to-piston contact during the repair procedure. Position the number one cylinder slightly advanced from TDC number 1 and follow the procedure carefully using the special tools (Step 3). The second stage, the engine will be rotated 180-degrees CLOCK-

WISE (crankshaft keyway in the 6 o'clock position) and the remaining rocker arms removed (Step 4).

3   **Stage 1:** Position the engine at TDC number 1 (see Section 3), then continue to rotate the engine until the crankshaft keyway is in the 12 o'clock position and remove the rocker arms from the designated cylinders:

a)   *Working on the right cylinder head, install a special valve spring compressing tool onto the indicated valve springs* **(see illustration)**, *compress the springs and remove the rocker arms* **(see illustration). Caution:** *Do not allow the valve spring keepers to fall off the valve stems or the valves will release and drop down into the cylinders.*

b)   *Working on the left cylinder head, install a special valve spring compressing tool onto the indicated valve springs* **(see illustration 10.3a)**, *compress the springs and remove the rocker arms* **(see illustration). Caution:** *Do not allow the valve spring keepers to fall off the valve stems or the valves will release and drop down into the cylinders.*

4   **Stage 2:** Rotate the engine 180-degrees CLOCKWISE until the crankshaft keyway is in the 6 o'clock position. Install the valve spring compressing tool onto the remaining valve springs, compress the springs and remove the remaining rocker arms. Camshaft rocker arms MUST be reinstalled with the same camshaft

lobe that they were removed from. Label and store all components to avoid confusion during reassembly.

5   Remove the timing chains (see Section 9) and the camshafts from the cylinder heads (see Section 12) to access the hydraulic lash adjusters.

6   Remove the hydraulic lash adjusters. The hydraulic lash adjusters MUST be reinstalled in their original locations. Label and store all components to avoid confusion during reassembly.

Inspection

Refer to illustrations 10.7 and 10.9

7   Inspect each adjuster carefully for signs of wear or damage. The areas of possible wear are the ball tip that contacts the rocker arm and the sides of the adjuster that contact the bore in the cylinder head **(see illustration)**. Since the lash adjusters frequently become clogged as mileage increases, we recommend replacing them if you're concerned about their condition or if the engine is exhibiting valve "tapping" noises.

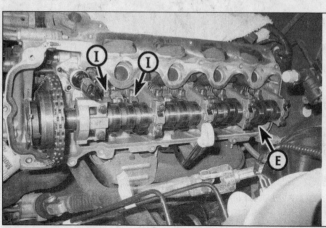

10.3c . . . the intake rocker arms from cylinder number 5 and the exhaust rocker arm from cylinder number 8

10.7 Inspect the lash adjuster for signs of excessive wear or damage, such as pitting, scoring or signs of overheating (bluing or discoloration), where the tip contacts the rocker arm (1) and the side surfaces that contact the lifter bore in the cylinder head (2)

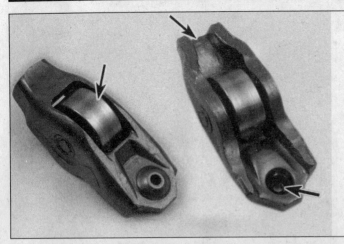

10.9 Check the roller surface of the rocker arms and the areas where the valve stem and lash adjuster contact the rocker arm

8   A thin wire or paper clip can be placed in the oil hole to move the plunger and make sure it's not stuck. **Note:** *The lash adjuster must have no more than 1/16-inch of total plunger travel.* It's recommended that if replacement of any of the adjusters is necessary, that the entire set be replaced. This will avoid the need to repeat the repair procedure as the others require replacement in the future.

9   Inspect the rocker arms for signs of wear or damage. The areas of wear are the ball socket that contacts the lash adjuster and the roller where the follower contacts the camshaft **(see illustration).**

Installation

10   Before installing the lash adjusters, bleed as much air as possible out of them. Stand the adjusters upright in a container of oil. Use a thin wire or paper clip to work the plunger up and down. This "primes" the adjuster and removes most of the air. Leave the adjusters in the oil until ready to install (just be sure not to mix them up).

11   Lubricate the valve stem tip, rocker arm, and lash adjuster bore with clean engine oil.

12   Install the lash adjusters and, with the valve spring depressed as in Step 3, install

each rocker arm.

13   The remainder of installation is the reverse of the removal procedure.

14   When re-starting the engine after replacing the adjusters, the adjusters will normally make some "tapping" noises, until all the air is bled from them. After the engine is warmed-up, raise the speed from idle to 3,000 rpm for one minute. Stop the engine and let it cool down. All of the noise should be gone when it is restarted.

11   Variable Camshaft Timing (VCT) system - general information

General information

1   The Variable Camshaft Timing (VCT) system consists of a VCT solenoid mounted on the front of each cylinder head over the camshaft, a camshaft position (CMP) sensor with a trigger wheel and a variable camshaft timing sprocket mounted on each camshaft. When oil flow is sent to the timing sprockets by way of the VCT solenoids, the phasing of the camshafts is altered, thereby advancing or retarding the actuation of the valves. The Variable Camshaft Timing (VCT) system is

described in detail in Chapter 6.

2   The Variable Camshaft Timing (VCT) system should be diagnosed by a dealer service department or other qualified automotive repair facility. The VCT system is controlled and monitored by the Powertrain Control Module (PCM), thereby requiring a specialized scan tool for diagnostics.

12   Camshaft(s) - removal, inspection and installation

Caution: *These engines are difficult to work on and require special tools for many procedures. On any procedure involving timing chains, camshafts or cylinder head removal, the steps must be read carefully and disassembly must proceed using the special tools otherwise damage to the engine could result.*

Removal

Refer to illustrations 12.3 and 12.4

1   Remove the valve covers (see Section 4), and the timing chain cover (see Section 8).

2   Remove the timing chains (see Section 9) and the rocker arms (see Section 10).

3   Measure the thrust clearance (endplay) of the camshaft(s) with a dial indicator **(see illustration)**. If the clearance is greater than the value listed in this Chapter's Specifications, replace the camshaft and/or the cylinder head.

4   Remove the camshaft sprockets. It will be necessary to lock the camshaft in position using a special tool before loosening the camshaft sprocket bolt **(see illustration)**. **Caution:** *Don't mix up the sprockets. Make an identification mark on each sprocket to insure correct reassembly.*

5   Working on the right side (passenger's side) cylinder head, remove the camshaft cap bolts. It's IMPORTANT to loosen the bearing cap bolts only 1/4-turn at a time, following the reverse of the tightening sequence **(see illustrations 12.16)**, until they can be removed by

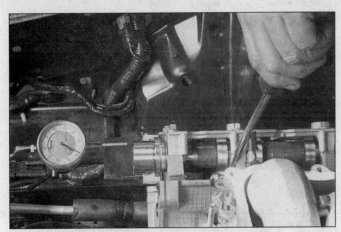

12.3 Camshaft endplay can be checked by setting up a dial indicator off the front of the camshaft and prying the camshaft gently forward and back

12.4 A special tool that is bolted to the cylinder head locks the camshaft sprocket in position

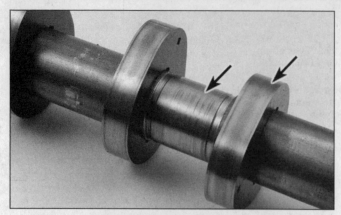

12.9 Areas to look for excessive wear or damage on the camshafts are the bearing surfaces and the camshaft lobes (also inspect the bearing surfaces of the camshaft bearing caps for signs of excessive wear, damage or overheating)

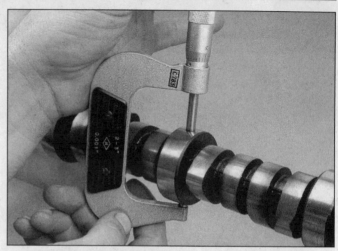

12.10a Measuring the camshaft bearing journal diameter

hand. **Note:** *Be sure to lift the front thrust camshaft cap straight up and even to prevent side-loading.*

6   Working on the left side (driver's side) cylinder head, remove the camshaft cap bolts. It's IMPORTANT to loosen the bearing

cap bolts only 1/4-turn at a time, following the reverse of the tightening sequence **(see illustrations 12.16)**, until they can be removed by hand. **Note:** *Be sure to lift the front thrust camshaft cap straight up and even to prevent side-loading.*

7   Remove the caps and lift the camshaft off the cylinder head. You may have to lightly tap the camshaft caps to jar them loose. Don't mix up the camshafts or any of the components. They must all go back on the same positions, and on the same cylinder head they were removed from.

8   Repeat this procedure for removal of the remaining camshaft.

Inspection

Refer to illustrations 12.9, 12.10a, 12.10b, 12.10c, 12.11a, 12.11b and 12.13

9   Visually examine the cam lobes and bearing journals for score marks, pitting, galling and evidence of overheating (blue, discolored areas). Look for flaking of the hardened surface of each lobe **(see illustration)**.

10   Using a micrometer, measure the diameter of each camshaft journal and the lift of each camshaft lobe **(see illustrations)**. Compare your measurements with the Specifications listed at the front of this Chapter, and if the diameter of any one of these is less than specified, replace the camshaft.

11   Check the oil clearance for each camshaft journal as follows:

12.10b Measure the camshaft lobe at its greatest dimension

12.10c Subtract the camshaft lobe diameter at its smallest dimension to obtain the lobe lift specification

12.11a Lay a strip of Plastigage on each of the camshaft journals

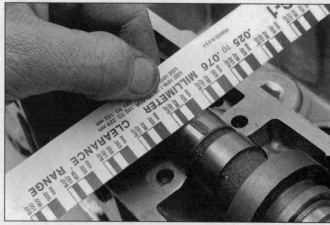

12.11b Compare the width of the crushed Plastigage to the scale on the envelope to determine the oil clearance

12.13 Oil is delivered to the timing chain tensioner by a feed tube and reservoir in the cylinder head

1   Tensioner oil feed tube
2   Reservoir

a) Clean the bearing surfaces and the camshaft journals with lacquer thinner or acetone.
b) Carefully lay the camshaft(s) in place in the cylinder head. Don't install the rocker arms and don't use any lubrication.
c) Lay a strip of Plastigage on each journal **(see illustration)**.
d) Install the camshaft bearing caps.
e) Tighten the cap bolts, a little at a time, to the torque listed in this Chapter's Specifications. **Note:** *Don't turn the camshaft while the Plastigage is in place.*
f) Remove the bolts and detach the caps.
g) Compare the width of the crushed Plastigage (at its widest point) to the scale on the Plastigage envelope **(see illustration)**.
h) If the clearance is greater than specified, and the diameter of any journal is less than specified, replace the camshaft. If the journal diameters are within specifications but the oil clearance is too great, the cylinder head is worn and must be replaced.

12   Scrape off the Plastigage with your fingernail or the edge of a credit card - don't scratch or nick the journals or bearing surfaces.
13   Finally, be sure to check the timing chain tensioner oil feed tube and reservoir before installing the camshaft caps **(see illustration)**. It must be absolutely clean and free of all obstructions or it will affect the operation of the timing chain tensioner.

Installation

Refer to illustration 12.16

14   If the lash adjusters and/or camshaft followers have been removed, install them in their original locations (see Section 10).
15   Apply moly-base grease or camshaft installation lube to the camshaft lobes and

bearing journals, then install the camshaft(s).
16   Install the camshaft caps in the correct locations, and loosely install all the bolts. Refer to Section 9 to align the camshaft sprockets before tightening the cap bolts. Following the correct bolt-tightening sequence **(see illustration)**, tighten the bolts in 1/4-turn increments to the torque listed in this Chapter's Specifications.
17   Set the camshafts at TDC before reinstalling the timing chain(s) (see Section 9).
18   Install the timing chain(s), tensioners and timing chain cover (see Section 9).
19   The remainder of installation is the reverse of the removal procedure.

13   Cylinder heads - removal and installation

Caution 1: *These engines are difficult to work on and require special tools for many procedures. On any procedure involving timing chains, camshafts or cylinder head removal, the steps must be read carefully and disassembly must proceed using the special tools otherwise damage to the engine could result.*
Caution 2: *The engine must be completely cool when the cylinder heads are removed. Failure to allow the engine to cool off could result in cylinder head warpage.*
Caution 3: *The manufacturer states that the cylinder head cannot be machined in the*

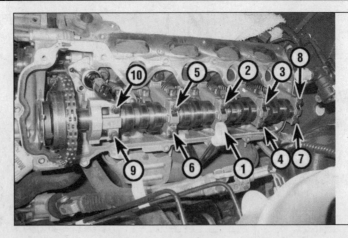

12.16 Camshaft cap bolt tightening sequence

event of warping. If the cylinder head warpage exceeds 0.001 inch (0.0254 mm) it must be replaced.
Note: *Due to limited working space, the following procedure is written using the manufacturer's recommended method, with the engine removed from the vehicle.*

Removal

Refer to illustration 13.5

1   Relieve the fuel system pressure (see Chapter 4)
2   Disconnect the cable from the negative battery terminal (see Chapter 5, Section 1).
3   Drain the cooling system (see Chapter 1).
4   Remove the valve covers (see Section 4).
5   Remove the intake manifold (see Section 5) and the coolant crossover housing **(see illustration)**.
6   Remove the engine (see Chapter 2C).
7   Remove the exhaust manifolds (see Section 6).
8   Remove the timing chain, tensioners and sprockets (see Section 9).
9   Remove the camshafts (see Section 12).
10   Following the reverse of the tightening sequence **(see illustrations 13.19a)**, use a breaker bar to remove the cylinder head bolts. Loosen the bolts in sequence, 1/4-turn at a time.

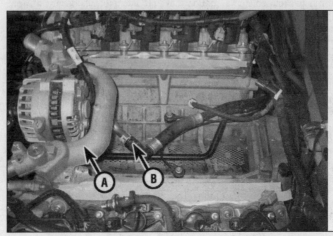

13.5 Location of the coolant crossover housing (A) and the heater hose connection (B)

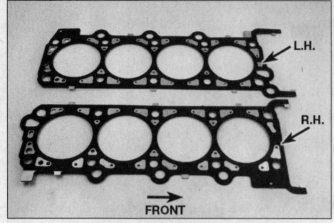

13.17 Identify the left and right cylinder head gaskets, the shapes are different and cannot be interchanged

13.18 Position the gaskets on the correct cylinder banks, then push them down over the alignment dowels

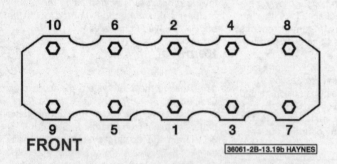

13.19a Cylinder head bolt tightening sequence on the 4.6L engine

13.19b Mark each cylinder head bolt with a paint stripe and, using a breaker bar and socket, tighten the bolts in sequence to the correct torque angle

11   Lift the cylinder head(s) off the engine. If resistance is felt, place a wood block against the end and strike the wood block with a hammer. **Caution:** *The cylinder heads are aluminum; store them on wood blocks to prevent damage to the gasket sealing surfaces.*

12   If the head sticks, use a prybar at the corners of the cylinder head-to-engine block mating surface to break the cylinder head gasket seal. Do not pry between the cylinder head and engine block in the gasket sealing area.

13   Remove the cylinder head gasket(s). Before removing, note which gasket goes on which side (they are different and cannot be interchanged).

Installation

Refer to illustrations 13.17, 13.18, 13.19a, and 13.19b

Caution: *New cylinder head bolts must be used for reassembly. Failure to use new bolts may result in cylinder head gasket leakage and engine damage.*

14   The mating surfaces of the cylinder heads and engine block must be perfectly clean when the cylinder heads are installed.

Use a gasket scraper to remove all traces of carbon and old gasket material, then clean the mating surfaces with lacquer thinner or acetone. If there's oil on the mating surfaces when the cylinder heads are installed, the gaskets may not seal correctly and leaks may develop. When working on the engine block, cover the open areas of the engine with shop rags to keep debris out during repair and reassembly. Use a vacuum cleaner to remove any debris that falls into the cylinders. **Caution:** *Do not use abrasive wheels or sharp metal scrapers on the heads or block surface; use a plastic scraper and chemical gasket remover, or the head gasket surfaces could develop leaks.*

15   Check the engine block and cylinder head mating surfaces for nicks, deep scratches and other damage.

16   Use a tap of the correct size to chase the threads in the cylinder head bolt holes. Dirt, corrosion, sealant and damaged threads will affect torque readings.

17   Make sure the new gaskets are installed on the correct cylinder banks **(see illustration)**. They are not interchangeable.

18   Position the new gasket(s) over the alignment dowels **(see illustration)** in the

engine block.

19   Carefully position the cylinder heads on the engine block without disturbing the gaskets. Install the NEW cylinder head bolts (the cylinder head bolts are torque-to-yield design and they cannot be reused). Following the recommended sequence **(see illustrations)**, tighten the cylinder head bolts, in five steps, to the torque and angle of rotation listed in this Chapter's Specifications. **Note:** *The method used for the cylinder head bolt tightening procedure is referred to as "torque-angle" or "torque-to-yield" method. Follow the procedure exactly.* Tighten the bolts in the first step using a torque wrench, then use a breaker bar and a special torque-angle gauge (available at most auto parts stores) to tighten the bolts the required angle. If the adapter is not available, mark each bolt with a paint stripe to aid in the torque angle process **(see illustration)**.

20   The remaining installation steps are the reverse of removal.

21   Install the engine (see Chapter 2C).

22   Change the engine oil and filter and refill the cooling system (see Chapter 1), then start the engine and check carefully for oil and coolant leaks.

14.6b Location of the upper front subframe mounting nuts (left side shown, right side similar)

14.6a Location of the rearmost subframe mounting bolts (right side shown, left side similar)

14 Oil pan - removal and installation

Removal

Refer to illustrations 14.6a, 14.6b and 14.8

1 Disconnect the cable from the negative battery terminal (see Chapter 5, Section 1).

2 Loosen the front wheel lug nuts, raise the front of the vehicle and support it securely on jackstands. Remove the front wheels.

3 Drain the engine oil and remove the oil filter (see Chapter 1).

4 Remove the intake air duct and the air filter housing, and the throttle body (see Chapter 4). Also remove the alternator (see Chapter 5).

5 Install an engine support fixture and remove the engine mounts (see Section 18).

6 Install a floor jack under the front of the subframe. Make reference marks along the subframe and the unibody structure, remove the subframe mounting bolts **(see illustrations)** and lower the assembly approximately

two inches.

7 Remove the engine oil dipstick.

8 Remove the oil pan mounting bolts **(see illustration)**.

9 Carefully separate the oil pan from the engine block. Don't pry between the engine block and oil pan or damage to the sealing surfaces may result and oil leaks could develop. Instead, dislodge the oil pan with a large rubber mallet or a wood block and a hammer.

Installation

Refer to illustration 14.12

10 Use a gasket scraper or putty knife to remove all traces of old gasket material and sealant from the pan and engine block. **Caution:** *Be careful not to gouge the oil pan or block, or oil leaks could develop later.*

11 Clean the mating surfaces with lacquer thinner or acetone. Make sure the bolt holes in the engine block are clean.

12 Apply a bead of RTV sealant to the four corner seams where the rear seal retainer meets the engine block, and where the front cover meets the engine block **(see illustration)**.

13 Carefully position the oil pan against the engine block and install the bolts finger tight. Make sure the gaskets haven't shifted, then tighten the bolts to the torque listed in this Chapter's Specifications. Start at the center of the oil pan and work out toward the ends in a spiral pattern.

14 Install the subframe bolts and tighten them to the torque listed in this Chapter's Specifications.

15 The remaining steps are the reverse of removal. **Caution:** *Don't forget to refill the engine with oil before starting it (see Chapter 1).*

16 Start the engine and check carefully for oil leaks at the oil pan. Drive the vehicle and check again.

15 Oil pump - removal and installation

Refer to illustrations 15.2, 15.4 and 15.5
Note: *The oil pump is available as a complete replacement unit only. No service parts or repair specifications are available from the manufacturer.*

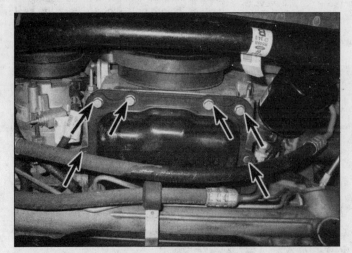

14.8 Remove the bolts from around the perimeter of the oil pan - front oil pan bolts shown, others hidden from view

14.12 Apply a bead of RTV sealant at the junctions of the front cover-to-engine block and the rear-seal retainer-to-engine block before installing the oil pan

15.2 Remove the two bolts retaining the pickup tube to the oil pump

15.4 Remove the oil pump mounting bolts and detach the oil pump from the engine block

Removal

1   Remove the oil pan (see Section 14).
2   Remove the two bolts that attach the oil pump pick-up tube to the oil pump **(see illustration)**.
3   Remove the timing chain cover, timing chains, chain guides and crankshaft sprocket (see Sections 8 and 9).
4   Remove the oil pump mounting bolts

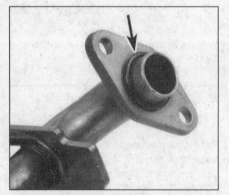

15.5 Before bolting the pickup tube back into the oil pump, inspect the O-ring and replace it if necessary

16.1 Flywheel and driveplate bolt tightening sequence

(see illustration) and separate the pump from the engine block.

Installation

5   Inspect the O-ring gasket on the pick-up tube **(see illustration)**. If it's damaged, replace it.
6   Install the oil pump to the engine and tighten the bolts to the torque listed in this Chapter's Specifications. **Note:** *Prime the oil pump prior to installation. Pour clean oil into the pick-up port and turn the pump by hand.*
7   The remainder of installation is the reverse of removal procedure.
8   Fill the engine with the correct type and quantity of oil. Start the engine and check for leaks.

---

16 Flywheel/driveplate - removal and installation

Refer to illustration 16.1

    This procedure is essentially the same as for the V6 engine. Refer to Part A and follow the procedure outlined there. However, use the correct tightening sequence **(see illustration)** and bolt torque value listed in this Chapter's Specifications.

---

17 Rear main oil seal - replacement

Refer to illustration 17.7

Note: *Rear main oil seal replacement is a time-consuming job requiring several special tools; read through the procedure and obtain the necessary tools before beginning.*
1   Disconnect the cable from the negative battery terminal (see Chapter 5, Section 1).
2   Raise the vehicle and support it securely on jackstands.
3   Remove the transmission (see Chapter 7).
4   Remove the flywheel/driveplate (Section 16).
5   Use a special tool and remove the crankshaft rear oil slinger.
6   Use a seal puller and remove the rear main oil seal from the retainer. **Caution:** *To prevent an oil leak after the new seal is installed, be very careful not to scratch or otherwise damage the crankshaft sealing surface or the bore in the engine block.*
7   Clean the crankshaft and seal bore in the retainer thoroughly and de-grease these areas by wiping them with a rag soaked in lacquer thinner or acetone. Check the seal contact surface on the crankshaft very carefully for scratches or nicks that could damage the new seal lip and cause oil leaks **(see**

17.7 Inspect the seal contact surface on the crankshaft for signs of excessive wear or grooves (seal retainer removed for clarity)

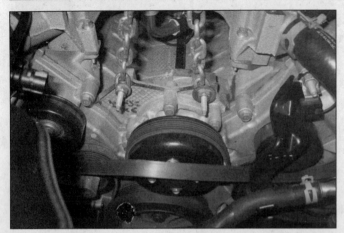

18.12 With the throttle body and alternator removed from the engine, attach the engine support fixture or hoist chains to the alternator mounting studs. Be sure to use washers and tighten the nuts securely

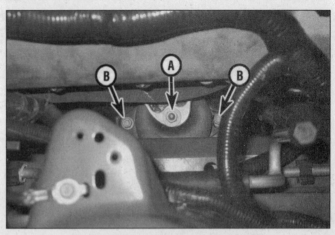

18.13a Right-side engine mount details

A   Mount upper nut   B   Mount-to-chassis bolts

illustration). Lubricate the lip of the new seal and the outer diameter of the crankshaft with engine oil. Make sure the edges of the new oil seal are not rolled over.

8   Position the new seal on the crankshaft. **Note:** *When installing the new seal, if so marked, the words THIS SIDE OUT on the seal must face out, toward the rear of the engine.* Use a special rear main oil seal installation tool to drive the seal in place. Make sure the seal is not offset; it must be flush along the entire circumference of the seal retainer.

9   Use a special tool and install the crankshaft rear oil slinger.

10   The remainder of installation is the reverse of removal.

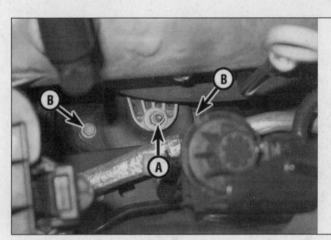

18.13b Left-side engine mount details

A   Mount upper nut
B   Mount-to-chassis bolts

18   Engine mounts - check and replacement

Check

1   Engine mounts seldom require attention, but broken or deteriorated mounts should be replaced immediately or the added strain placed on the driveline components may cause damage or wear.

2   During the check, the engine must be raised slightly to remove the weight from the mounts.

3   Raise the vehicle and support it securely on jackstands, then position a jack under the engine oil pan. Place a wood plank between the jack head and the oil pan, then carefully raise the engine just enough to take the weight off the mounts. **Warning:** *DO NOT place any part of your body under the engine when it's supported only by a jack!* **Note:** *If your jack and piece of wood will fit under the vehicle, it*

won't be necessary to raise the vehicle.

4   Check the mounts to see if the rubber is cracked or hardened. Sometimes the rubber will split right down the center.

5   Have an assistant raise the jack and see if the upper portion of the mount raises off the lower portion. If it does, replace the mount.

6   Check for relative movement between the mount plates and the engine or frame (use a large screwdriver or pry bar to attempt to move the mounts).

7   If movement is noted, lower the engine and tighten the mount fasteners.

Replacement

Refer to illustrations 18.12, 18.13a and 18.13b

8   Disconnect the cable from the negative battery terminal (see Chapter 5, Section 1).

9   Detach the air intake duct from the throttle body. If you're replacing the left-side mount, remove the air filter housing to provide

more working room (see Chapter 4).

10   Remove the throttle body (see Chapter 4).

11   Remove the alternator (see Chapter 5).

12   Install an engine support fixture or engine hoist, with the chains connected to the alternator mounting studs. Make sure the chain is anchored properly and the nuts are tightened securely **(see illustration).** As an alternative, a floor jack can be used, along with a piece of wood, positioned under the oil pan.

13   Remove the fasteners holding the mount to the mount bracket and the chassis **(see illustrations).** If you're removing the left-side mount, also unbolt the mount bracket from the engine.

14   Raise the engine approximately 1-1/2 inches and remove the mount.

15   Installation is the reverse of removal. Use thread-locking compound on the mount fasteners and tighten them to the torque listed in this Chapter's Specifications.

Notes

Chapter 2 Part C
General engine overhaul procedures

Contents

Specifications

General

Displacement
V6 engine	4.0 liters (244 cubic inches)
V8 engine	4.6 liters (281 cubic inches)

Bore and stroke
V6 engine	3.953 X 3.320 inches
V8 engine	3.550 X 3.540 inches
Cylinder compression	Lowest cylinder must be within 75 percent of highest cylinder

Oil pressure (minimum oil pressure with engine at operating temperature)
V6 engine	15 psi @ 2,000 rpm
V8 models	75 psi @ 2,000 rpm

Torque specifications
Ft-lbs (unless otherwise indicated)

Connecting rod bearing cap nuts/bolts*
V6 engine	
Step 1	15
Step 2	Tighten an additional 90-degrees
V8 engine	
Step 1	32
Step 2	Tighten an additional 105-degrees
Jackshaft thrust plate bolts	96 in-lbs
Balance shaft assembly bolts	20
Balance shaft chain guide bolts	89 in-lbs
Balance shaft tensioner bolts	21
Main bearing cap bolts	
V6 engine	
Step 1	26
Step 2	Tighten an additional 57-degrees
V8 engine (tighten the bolts in the order listed here)	
Main bolts **(see illustration 10.19a)**	
Step 1 (Bolts 1 through 20)	89 in-lbs
Step 2 (Bolts 1 through 10)	18
Step 3 (Bolts 11 through 20)	30
Step 4 (Bolts 1 through 20)	Tighten an additional 90-degrees
Side bolts (horizontal bolts) **(see illustration 10.19b)**	
Step 5	30
Step 6	Tighten an additional 90-degrees

Use new connecting rod bolts/nuts on V6 models and connecting rod bolts on V8 models

1   General information - engine overhaul

Refer to illustrations 1.1, 1.2, 1.3, 1.4, 1.5 and 1.6

Included in this portion of Chapter 2 are general information and diagnostic testing procedures for determining the overall mechanical condition of your engine.

The information ranges from advice concerning preparation for an overhaul and the purchase of replacement parts and/or components to detailed, step-by-step procedures covering removal and installation.

The following Sections have been written to help you determine whether your engine needs to be overhauled and how to remove and install it once you've determined it needs to be rebuilt. For information concerning in-vehicle engine repair, see Chapter 2A or 2B.

The Specifications included in this Part are general in nature and include only those necessary for testing the oil pressure and checking the engine compression. Refer to Chapter 2A or 2B for additional engine Specifications.

It's not always easy to determine when, or if, an engine should be completely overhauled, because a number of factors must be considered.

High mileage is not necessarily an indication that an overhaul is needed, while low mileage doesn't preclude the need for an overhaul. Frequency of servicing is probably the most important consideration. An engine that's had regular and frequent oil and filter changes, as well as other required maintenance, will most likely give many thousands of miles of reliable service. Conversely, a neglected engine may require an overhaul very early in its service life.

Excessive oil consumption is an indication that piston rings, valve seals and/or valve guides are in need of attention. Make sure that oil leaks aren't responsible before deciding that the rings and/or guides are bad. Perform a cylinder compression check to determine the extent of the work required (see Section 3). Also check the vacuum readings under various conditions (see Section 4).

1.1 An engine block being bored. An engine rebuilder will use special machinery to recondition the cylinder bores

Check the oil pressure with a gauge installed in place of the oil pressure sending unit and compare it to this Chapter's Specifications (see Section 2). If it's extremely low, the bearings and/or oil pump are probably worn out.

Loss of power, rough running, knocking or metallic engine noises, excessive valve train noise and high fuel consumption rates may also point to the need for an overhaul, especially if they're all present at the same time. If a complete tune-up doesn't remedy the situation, major mechanical work is the only solution.

An engine overhaul involves restoring the internal parts to the specifications of a new engine. During an overhaul, the piston rings are replaced and the cylinder walls are reconditioned (rebored and/or honed) **(see illustrations 1.1 and 1.2)**. If a rebore is done

1.2 If the cylinders are bored, the machine shop will normally hone the engine on a machine like this

by an automotive machine shop, new oversize pistons will also be installed. The main bearings, connecting rod bearings and camshaft bearings are generally replaced with new ones and, if necessary, the crankshaft may be reground to restore the journals **(see illustration 1.3)**. Generally, the valves are serviced as well, since they're usually in less-than-perfect condition at this point. While the engine is being overhauled, other components, such as the distributor, starter and alternator, can be rebuilt as well. The end result should be similar to a new engine that will give many trouble free miles. **Note:** *Critical cooling system components such as the hoses, drivebelts, thermostat and water pump should be replaced with new parts when an engine is overhauled.*

1.3 A crankshaft having a main bearing journal ground

1.4 A machinist checks for a bent connecting rod, using specialized equipment

1.5 A bore gauge being used to check the main bearing bore

1.6 Uneven piston wear like this indicates a bent connecting rod

The radiator should be checked carefully to ensure that it isn't clogged or leaking (see Chapter 3). *If you purchase a rebuilt engine or short block, some rebuilders will not warranty their engines unless the radiator has been professionally flushed. Also, we don't recommend overhauling the oil pump - always install a new one when an engine is rebuilt.*

Overhauling the internal components on today's engines is a difficult and time-consuming task which requires a significant amount of specialty tools and is best left to a professional engine rebuilder **(see illustrations 1.4, 1.5 and 1.6)**. A competent engine rebuilder will handle the inspection of your old parts and offer advice concerning the reconditioning or replacement of the original engine, never purchase parts or have machine work done on other components until the block has been thoroughly inspected by a professional machine shop. As a general rule, time is the primary cost of an overhaul, especially since the vehicle may be tied up for a minimum of two weeks or more. Be aware that some engine builders only have the capability to rebuild the engine you bring them while other rebuilders have a large inventory of

rebuilt exchange engines in stock. Also be aware that many machine shops could take as much as two weeks time to completely rebuild your engine depending on shop workload. Sometimes it makes more sense to simply exchange your engine for another engine that's already rebuilt to save time.

2 Oil pressure check

Refer to illustrations 2.2a and 2.2b

1   Low engine oil pressure can be a sign of an engine in need of rebuilding. A "low oil pressure" indicator (often called an "idiot light") is not a test of the oiling system. Such indicators only come on when the oil pressure is dangerously low. Even an oil pressure gauge in the instrument panel is only a relative indication, although much better for driver information than a warning light. A better test is with a mechanical (not electrical) oil pressure gauge.
2   Locate the oil pressure sending unit on the engine block:
a) *On V6 engines, the oil pressure sending unit is located on the front left side of the engine block near the front cover* **(see illustration)**.

b) *On V8 engines, the oil pressure sending unit is located near the lower left side of the engine block on the oil filter adapter* **(see illustration)**.
3   Unscrew and remove the oil pressure sending unit and screw in the hose for your oil pressure gauge. If necessary, install an adapter fitting. Use Teflon tape or thread sealant on the threads of the adapter and/or the fitting on the end of your gauge's hose.
4   Connect an accurate tachometer to the engine, according to the tachometer manufacturer's instructions.
5   Check the oil pressure with the engine running (normal operating temperature) at the specified engine speed, and compare it to this Chapter's Specifications. If it's extremely low, the bearings and/or oil pump are probably worn out.

3 Cylinder compression check

Refer to illustration 3.6

1   A compression check will tell you what mechanical condition the upper end of your engine (pistons, rings, valves, head gaskets) is in. Specifically, it can tell you if the compression is down due to leakage caused by worn piston rings, defective valves and seats or a blown head gasket. **Note:** *The engine must be at normal operating temperature and the battery must be fully charged for this check.*
2   Begin by cleaning the area around the spark plugs before you remove them (compressed air should be used, if available). The idea is to prevent dirt from getting into the cylinders as the compression check is being done.
3   Disable the ignition system (see Chapter 5). On V6 models, disconnect the primary (low voltage) wires from the coil pack (assembly). On V8 models, disconnect the primary (low voltage) wires from each ignition coil assembly.
4   Remove all of the spark plugs from the engine (see Chapter 1).
5   Remove the fuel pump relay from the fuse/relay center in the engine compartment (see Chapter 4, Section 2).

2.2a On V6 engines, the oil pressure sending unit is located on the left side of the engine block near the front cover

2.2b On V8 engines, the oil pressure sending unit is located on the oil filter adapter

3.6 Use a compression gauge with a threaded fitting for the spark plug hole, not the type that requires hand pressure to maintain the seal

6   Install a compression gauge in the spark plug hole **(see illustration)**.
7   While depressing the accelerator pedal to the floor (wide open throttle), crank the engine over at least seven compression strokes and watch the gauge. The compression should build up quickly in a healthy engine. Low compression on the first stroke, followed by gradually increasing pressure on successive strokes, indicates worn piston rings. A low compression reading on the first stroke, which doesn't build up during successive strokes, indicates leaking valves or a blown head gasket (a cracked head could also be the cause). Deposits on the undersides of the valve heads can also cause low compression. Record the highest gauge reading obtained.
8   Repeat the procedure for the remaining cylinders and compare the results to this Chapter's Specifications.
9   Add some engine oil (about three squirts from a plunger-type oil can) to each cylinder, through the spark plug hole, and repeat the test.
10   If the compression increases after the oil is added, the piston rings are definitely worn. If the compression doesn't increase significantly, the leakage is occurring at the valves or head gasket. Leakage past the valves may be caused by burned valve seats and/or faces or warped, cracked or bent valves.
11   If two adjacent cylinders have equally low compression, there's a strong possibility that the head gasket between them is blown. The appearance of coolant in the combustion chambers or the crankcase would verify this condition.
12   If one cylinder is slightly lower than the others, and the engine has a slightly rough idle, a worn lobe on the camshaft could be the cause.
13   If the compression is unusually high, the combustion chambers are probably coated with carbon deposits. If that's the case, the cylinder head(s) should be removed and decarbonized.
14   If compression is way down or varies

greatly between cylinders, it would be a good idea to have a leak-down test performed by an automotive repair shop. This test will pinpoint exactly where the leakage is occurring and how severe it is.

4   Vacuum gauge diagnostic checks

Refer to illustrations 4.4 and 4.6

A vacuum gauge provides inexpensive but valuable information about what is going on in the engine. You can check for worn rings or cylinder walls, leaking head or intake manifold gaskets, incorrect carburetor adjustments, restricted exhaust, stuck or burned valves, weak valve springs, improper ignition or valve timing and ignition problems.
Unfortunately, vacuum gauge readings are easy to misinterpret, so they should be used in conjunction with other tests to confirm the diagnosis.
Both the absolute readings and the rate of needle movement are important for accurate interpretation. Most gauges measure vacuum in inches of mercury (in-Hg). The following references to vacuum assume the diagnosis is being performed at sea level. As elevation increases (or atmospheric pressure

4.4 A simple vacuum gauge can be handy in diagnosing engine condition and performance

decreases), the reading will decrease. For every 1,000 foot increase in elevation above approximately 2,000 feet, the gauge readings will decrease about one inch of mercury.
Connect the vacuum gauge directly to the intake manifold vacuum, not to ported (throttle body) vacuum **(see illustration)**. Be sure no hoses are left disconnected during the test or false readings will result.

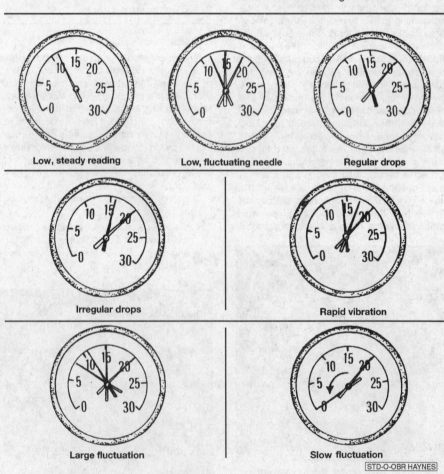

Low, steady reading   Low, fluctuating needle   Regular drops

Irregular drops   Rapid vibration

Large fluctuation   Slow fluctuation

STD-O-OBR HAYNES

4.6 Typical vacuum gauge readings

Before you begin the test, allow the engine to warm up completely. Block the wheels and set the parking brake. With the transmission in Park, start the engine and allow it to run at normal idle speed. **Warning:** *Keep your hands and the vacuum gauge clear of the fans.*

Read the vacuum gauge; an average, healthy engine should normally produce about 17 to 22 in-Hg with a fairly steady needle **(see illustration)**. Refer to the following vacuum gauge readings and what they indicate about the engine's condition:

1   A low steady reading usually indicates a leaking gasket between the intake manifold and cylinder head(s) or throttle body, a leaky vacuum hose, late ignition timing or incorrect camshaft timing. Check ignition timing with a timing light and eliminate all other possible causes, utilizing the tests provided in this Chapter before you remove the timing chain cover to check the timing marks.

2   If the reading is three to eight inches below normal and it fluctuates at that low reading, suspect an intake manifold gasket leak at an intake port or a faulty fuel injector.

3   If the needle has regular drops of about two-to-four inches at a steady rate, the valves are probably leaking. Perform a compression check or leak-down test to confirm this.

4   An irregular drop or down-flick of the needle can be caused by a sticking valve or an ignition misfire. Perform a compression check or leak-down test and read the spark plugs.

5   A rapid vibration of about four in-Hg vibration at idle combined with exhaust smoke indicates worn valve guides. Perform a leak-down test to confirm this. If the rapid vibration occurs with an increase in engine speed, check for a leaking intake manifold gasket or head gasket, weak valve springs, burned valves or ignition misfire.

6   A slight fluctuation, say one inch up and down, may mean ignition problems. Check all the usual tune-up items and, if necessary, run the engine on an ignition analyzer.

7   If there is a large fluctuation, perform a compression or leak-down test to look for a weak or dead cylinder or a blown head gasket.

8   If the needle moves slowly through a wide range, check for a clogged PCV system, incorrect idle fuel mixture, throttle body or intake manifold gasket leaks.

9   Check for a slow return after revving the engine by quickly snapping the throttle open until the engine reaches about 2,500 rpm and let it shut. Normally the reading should drop to near zero, rise above normal idle reading (about 5 in-Hg over) and return to the previous idle reading. If the vacuum returns slowly and doesn't peak when the throttle is snapped shut, the rings may be worn. If there is a long delay, look for a restricted exhaust system (often the muffler or catalytic converter). An easy way to check this is to temporarily disconnect the exhaust ahead of the suspected part and redo the test.

5   Engine rebuilding alternatives

The do-it-yourselfer is faced with a number of options when purchasing a rebuilt engine. The major considerations are cost, warranty, parts availability and the time required for the rebuilder to complete the project. The decision to replace the engine block, piston/connecting rod assemblies and crankshaft depends on the final inspection results of your engine. Only then can you make a cost effective decision whether to have your engine overhauled or simply purchase an exchange engine for your vehicle.

Some of the rebuilding alternatives include:

Individual parts - If the inspection procedures reveal that the engine block and most engine components are in reusable condition, purchasing individual parts and having a rebuilder rebuild your engine may be the most economical alternative. The block, crankshaft and piston/connecting rod assemblies should all be inspected carefully by a machine shop first.

Short block - A short block consists of an engine block with a crankshaft and piston/connecting rod assemblies already installed. All new bearings are incorporated and all clearances will be correct. The existing camshafts, valve train components, cylinder head and external parts can be bolted to the short block with little or no machine shop work necessary.

Long block - A long block consists of a short block plus an oil pump, oil pan, cylinder head, valve cover, camshaft and valve train components, timing sprockets and chain or gears and timing cover. All components are installed with new bearings, seals and gaskets incorporated throughout. The installation of manifolds and external parts is all that's necessary.

Low mileage used engines - Some companies now offer low mileage used engines which is a very cost effective way to get your vehicle up and running again. These engines often come from vehicles which have been in totaled in accidents or come from other countries which have a higher vehicle turn over rate. A low mileage used engine also usually has a similar warranty like the newly remanufactured engines.

Give careful thought to which alternative is best for you and discuss the situation with local automotive machine shops, auto parts dealers and experienced rebuilders before ordering or purchasing replacement parts.

6   Engine removal - methods and precautions

Refer to illustrations 6.1, 6.2, 6.3 and 6.4

If you've decided that an engine must be removed for overhaul or major repair work, several preliminary steps should be taken. Read all removal and installation procedures carefully prior to committing to this job.

Locating a suitable place to work is extremely important. Adequate work space, along with storage space for the vehicle, will

6.1 After tightly wrapping water-vulnerable components, use a spray cleaner on everything, with particular concentration on the greasiest areas, usually around the valve cover and lower edges of the block. If one section dries out, apply more cleaner

6.2 Depending on how dirty the engine is, let the cleaner soak in according to the directions and hose off the grime and cleaner. Get the rinse water down into every area you can get at; then dry important components with a hair dryer or paper towels

6.3 Get an engine stand sturdy enough to firmly support the engine while you're working on it. Stay away from three-wheeled models; they have a tendency to tip over more easily, so get a four-wheeled unit

6.4 A clutch alignment tool is necessary if you plan to install a rebuilt engine equipped with a manual transmission

be needed. If a shop or garage isn't available, at the very least a flat, level, clean work surface made of concrete or asphalt is required.

Cleaning the engine compartment and engine before beginning the removal procedure will help keep tools clean and organized **(see illustrations 6.1 and 6.2)**.

An engine hoist will also be necessary. Make sure the hoist is rated in excess of the combined weight of the engine and transmission. Safety is of primary importance, considering the potential hazards involved in removing the engine from the vehicle.

If you're a novice at engine removal, get at least one helper. One person cannot easily do all the things you need to do to remove a big heavy engine and transmission assembly from the engine compartment. Also helpful is to seek advice and assistance from someone who's experienced in engine removal.

Plan the operation ahead of time. Arrange for or obtain all of the tools and equipment you'll need prior to beginning the job **(see illustrations 6.3 and 6.4)**. Some of the equipment necessary to perform engine removal and installation safely and with rela-

tive ease are (in addition to a vehicle hoist and an engine hoist) a heavy duty floor jack (preferably fitted with a transmission jack head adapter), complete sets of wrenches and sockets as described in the front of this manual, wooden blocks, plenty of rags and cleaning solvent for mopping up spilled oil, coolant and gasoline.

Plan for the vehicle to be out of use for quite a while. A machine shop can do the work that is beyond the scope of the home mechanic. Machine shops often have a busy schedule, so before removing the engine, consult the shop for an estimate of how long it will take to rebuild or repair the components that may need work.

7   Engine - removal and installation

Refer to illustrations 7.8, 7.11, 7.28a and 7.28b

Warning 1: *Gasoline is extremely flammable, so take extra precautions when you work on any part of the fuel system. Don't smoke or allow open flames or bare light bulbs near the work area, and don't work in a garage where a gas-type appliance (such as a water heater or clothes dryer) is present. Since gasoline is carcinogenic, wear fuel-resistant gloves when there's a possibility of being exposed to fuel and, if you spill any fuel on your skin, rinse it off immediately with soap and water. Mop up any spills immediately and do not store*

fuel-soaked rags where they could ignite. The fuel system is under constant pressure, so, if any fuel lines are to be disconnected, the fuel pressure in the system must be relieved first (see Chapter 4 for more information). When you perform any kind of work on the fuel system, wear safety glasses and have a Class B type fire extinguisher on hand.

Warning 2: *The air conditioning system is under high pressure. DO NOT loosen any fittings or remove any components until after the system has been discharged. Air conditioning refrigerant should be properly discharged into an EPA-approved container at a dealer service department or an automotive air conditioning repair facility. Always wear eye protection when disconnecting air conditioning system fittings.*

Warning 3: *The engine must be completely cool before beginning this procedure.*

Removal

1   Have the air conditioning system discharged by an automotive air conditioning technician.

2   Relieve the fuel system pressure (see Chapter 4).

3   Disconnect the cable from the negative battery terminal (see Chapter 5, Section 1).

4   Remove the hood (see Chapter 11). Cover the fenders and cowl using special pads. An old bedspread or blanket will also work.

5   On V8 models, remove the cowl vent screen (see Chapter 11), then remove the ground strap bolt.

6   Remove the air filter housing and intake duct (see Chapter 4).

7   Disconnect the upper and lower electrical connectors at the Powertrain Control Module (PCM) (see Chapter 6).

8   Separate the upper part of the underhood fuse/relay box from the lower part, then remove the bolt and detach the large electrical connector for the harness that leads to the engine **(see illustration)**. On V6 models, also remove the bolt and disconnect the battery cable from the underhood fuse/relay box (see Chapter 5).

9   Remove the power steering fluid reservoir and position it off to the side.

10   Remove the battery and the battery tray (see Chapter 5).

11   Clearly label and disconnect all vacuum

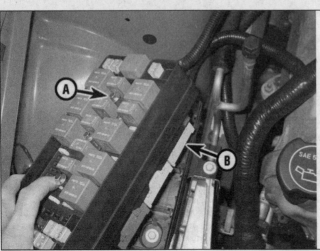

7.8 After separating the upper part of the underhood fuse/relay box from the lower part, remove this bolt (A) and unplug the large electrical connector (B) from underneath

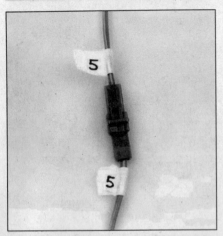

7.11 Label both ends of each wire and hose before disconnecting it

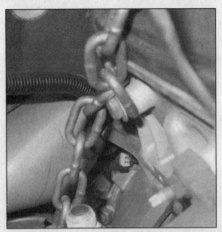

7.28a On V6 engines, attach the hoist chain to the threaded boss on each cylinder head

lines, emissions hoses, wiring harness connectors, ground straps and fuel lines between the engine and the chassis. Masking tape and/or a touch up paint applicator work well for marking items **(see illustration)**. Take instant photos or sketch the locations of components and brackets.

12　Remove the drivebelt and the drivebelt tensioner (see Chapter 1).

13　Remove the power steering pump (see Chapter 10) without disconnecting the power steering fluid lines and position the assembly off to the side.

14　If you're working on a V8 model, remove the intake manifold (see Chapter 2B).

15　Drain the cooling system (see Chapter 1).

16　Detach the radiator hoses from the engine. If you're working on a V8 model, remove the thermostat housing along with the lower hose.

17　If you're working on a V8 model, remove the coolant crossover housing.

18　Disconnect the heater hoses at the firewall (see Chapter 3).

19　Raise the vehicle and support it securely on jackstands.

20　Install an engine support fixture to support the rear of the engine while the transmission is being removed. As an alternative, a floor jack and piece of wood placed underneath the oil pan can be used.

21　Drain the engine oil and remove the oil filter (see Chapter 1).

22　Remove the transmission (see Chapter 7A or 7B).

23　On manual transmission models, remove the clutch pressure plate and disc (see Chapter 8).

24　Remove the air conditioning compressor (see Chapter 3) without disconnecting the refrigerant lines and position the assembly off to the side.

25　Unplug the upstream oxygen sensor electrical connector(s).

26　Detach the exhaust pipes from the exhaust manifolds (see Chapter 2A or 2B).

27　Remove the starter motor (see Chapter 5).

28　Roll the engine hoist into position and attach the chain or sling to the engine:

a) *If you're working on a V6 engine, connect the hoist chain to the threaded bosses on each cylinder head* **(see illustration)**. *If no bosses are present, remove an exhaust manifold bolt from each cylinder head and, using a longer bolt and washer(s), connect the chain to the heads.*

b) *If you're working on a V8 engine, connect the hoist chain to the threaded bosses on the engine block, at the front and rear of the valley between the cylinder banks* **(see illustration)**.

c) **Warning:** *Regardless of which engine you're working on, be sure the bolts securing the hoist chain are tightened securely, using spacers and/or washers on either side of the chain link, if necessary, to ensure this.*

Take up the slack in the sling or chain, but don't lift the engine. **Warning:** *DO NOT place any part of your body under the engine when it's supported only by a hoist or other lifting device.*

29　Remove the engine mount upper nuts (see Chapter 2B).

30　Recheck to be sure nothing is still connecting the engine to the vehicle. Disconnect anything still remaining.

31　Raise the engine slightly and inspect it thoroughly once more to make sure that *nothing* is still attached, then slowly raise the engine out of the engine compartment. Check carefully to make sure nothing is hanging up.

32　It may be necessary to tilt or turn the engine as it is being raised. **Warning:** *Don't place any part of your body under the engine or between the engine and the vehicle.*

33　Remove the flywheel/driveplate (see Chapter 2A or 2B) and mount the engine on an engine stand.

34　Inspect the engine mounts (see Chapter 2B) and transmission mount (see Chapter 7A). If they're worn or damaged, replace them.

Installation

35　Install the flywheel/driveplate (see Chapter 2A or 2B).

36　If you're working on a vehicle with manual transmission, install the clutch disc and pressure plate (see Chapter 8). Now is a good time to install a new clutch.

37　Carefully lower the engine into the engine compartment and engage it with the engine mounts.

38　Install the transmission (see Chapter 7A or 7B).

39　Install the transmission-to-engine bolts and tighten them securely. **Caution:** *DO NOT use the bolts to force the transmission and engine together!*

40　Reinstall the remaining components in the reverse order of removal.

41　Add coolant, oil and transmission fluid as needed (see Chapter 1).

42　Reconnect the battery, run the engine and check for leaks and proper operation of all accessories, then install the hood and test drive the vehicle. The Powertrain Control Module (PCM) must relearn its idle and fuel trim strategy for optimum driveability and performance, which may take a few trips.

43　Have the air conditioning system recharged by the shop that discharged it.

7.28b On V8 engines, attach the hoist chain to the threaded holes at each end of the engine block, in the valley between the cylinder banks

8   Engine overhaul - disassembly sequence

1   It's much easier to remove the external components if it's mounted on a portable engine stand. A stand can often be rented quite cheaply from an equipment rental yard. Before the engine is mounted on a stand, the flywheel/driveplate should be removed from the engine.

2   If a stand isn't available, it's possible to remove the external engine components with it blocked up on the floor. Be extra careful not to tip or drop the engine when working without a stand.

3   If you're going to obtain a rebuilt engine, all external components must come off first, to be transferred to the replacement engine. These components include:

> Flywheel (models with manual transmission)
> Driveplate (models with automatic transmission)
> Ignition system components
> Emissions-related components
> Engine mounts and mount brackets
> Engine rear cover (spacer plate between flywheel/driveplate and engine block)
> Intake/exhaust manifolds
> Fuel injection components
> Oil filter
> Spark plugs
> Thermostat and housing assembly
> Water pump

Note: *When removing the external components from the engine, pay close attention to details that may be helpful or important during installation. Note the installed position of gaskets, seals, spacers, pins, brackets, washers, bolts and other small items.*

4   If you're going to obtain a short block (assembled engine block, crankshaft, pistons and connecting rods), then remove the timing chain, cylinder head, oil pan, oil pump pick-up tube, oil pump and water pump from your engine so that you can turn in your old short block to the rebuilder as a core. See *Engine rebuilding alternatives* for additional information regarding the different possibilities to be considered.

9   Jackshaft and balance shaft (V6 models) - removal, inspection and installation

Removal

Refer to illustrations 9.3 and 9.5

1   Disconnect the cable from the negative battery terminal (see Chapter 5, Section 1).

2   The camshafts, timing chains and sprockets and the jackshaft chain and sprocket must be removed before extracting the jackshaft (see Chapter 2, Part A).

3   From above, at the top-rear of the block, remove the bolt holding the oil pump drive and pull out the drive assembly **(see illustration)**.

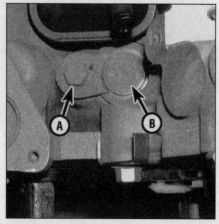

9.3 Remove the oil pump drive assembly retaining bolt and clamp (A) and pull the drive assembly (B) straight up and out of the engine block

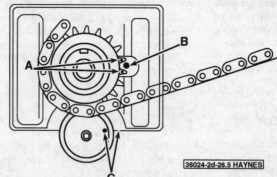

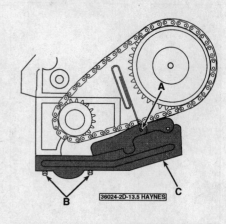

9.5 Press the balance shaft tensioner in and insert a pin (A) - remove the balance shaft tensioner bolts (B) and remove the tensioner assembly (C)

9.14 To align the balance shaft for chain installation, turn the sprocket until the marks on the sprocket (A) align with the hole in the assembly (B) and the lower shaft mark (C) aligns with the mating line of the assembly and engine block

4   Remove the two jackshaft thrust plate bolts, and any spacer that may be in place. Installing a long bolt in the front of the jackshaft to use as a handle, guide the jackshaft gently out of the block, without nicking the journals on the jackshaft bearings in the block.

5   Compress the balance shaft tensioner by hand and insert a drill bit or pin to hold it in place, then remove the tensioner bolts and the tensioner **(see illustration)**.

6   Slide the balance shaft chain and crankshaft sprocket from the crankshaft and balance shaft sprocket. **Note:** *Do not remove the bolt from the balance shaft sprocket.*

7   Rotate the engine on the engine stand so that the bottom of the block is up. Remove the two rear bolts and the balance shaft assembly. **Note:** *Two of the balance shaft assembly mounting bolts are the ones removed to take off the tensioner.*

Inspection

8   The jackshaft rides in a bearing at the front of the block and a bushing at the rear. Check the jackshaft bearings in the block for wear and damage. Look for galling, pitting and discolored areas.

9   Jackshaft bearing replacement requires special tools and expertise that place it outside the scope of the home mechanic. Take the block to an automotive machine shop to ensure the job is done correctly.

Installation

Jackshaft

10   Lubricate the jackshaft bearing journals with engine assembly lubricant.

11   Slide the jackshaft into the engine, using a long bolt screwed into the front of the jackshaft as a handle. Support the shaft near the block and be careful not to scrape or nick the bearings. Install the jackshaft thrust plate and tighten the bolts to the torque listed in this Chapter's Specifications.

Balance shaft

Refer to illustration 9.14

12   Lubricate the front bearing and rear journal of the balance shaft with engine oil and insert the balance shaft assembly carefully onto the block. The block should be turned bottom-end up on the engine stand, and the crankshaft should be positioned with cylinder number 1 at TDC.

13   Install and tighten the balance shaft bolts to the torque listed in this Chapter's Specifications.

14   Turn the balance shaft sprocket until the two dots on the sprocket align over the hole in

10.1 Before you try to remove the pistons, use a ridge reamer to remove the raised material (ridge) from the top of the cylinders

10.3 Checking the connecting rod endplay (side clearance)

10.4 If the connecting rods and caps are not marked, mark the caps to the rods by cylinder number (for example, this would be No. 4 connecting rod)

the front of the assembly, and the timing mark on the front of the lower shaft aligns with the balance shaft-to-block mating line **(see illustration)**. Install a 4 mm pin or drill bit into the hole to hold it in this position. **Note**: *It may take quite a few turns of the balance shaft sprocket to get the shaft and sprocket timing marks sets to align, because of the reduction gearing in the assembly.*

15   Install the balance shaft chain and the crankshaft sprocket. Install the balance shaft chain guide and tensioner (with the tensioner pinned) **(see illustration 9.5)**. Tighten the chain guide and tensioner bolts to the torque listed in this Chapter's Specifications. Remove the pin from the tensioner and remove the alignment pin from the balance shaft sprocket.

16   Refer to Chapter 2A and install the camshaft timing chain cassettes, jackshaft chain and jackshaft sprocket.

---

10   Pistons and connecting rods - removal and installation

Removal

Refer to illustrations 10.1, 10.3 and 10.4

Note: *Prior to removing the piston/connecting rod assemblies, remove the cylinder heads and oil pan (see Chapter 2A or 2B).*

1   Use your fingernail to feel if a ridge has formed at the upper limit of ring travel (about 1/4-inch down from the top of each cylinder). If carbon deposits or cylinder wear have produced ridges, they must be completely removed with a special tool **(see illustration)**. Follow the manufacturer's instructions provided with the tool. Failure to remove the ridges before attempting to remove the piston/connecting rod assemblies may result in piston breakage.

2   After the cylinder ridges have been removed, turn the engine so the crankshaft is facing up.

3   Before the main bearing cap assembly and connecting rods are removed, check the connecting rod endplay with feeler gauges. Slide them between the first connecting rod and the crankshaft throw until the play is removed **(see illustration)**. Repeat this procedure for each connecting rod. The endplay is equal to the thickness of the feeler gauge(s). Check with an automotive machine shop for the endplay service limit (a typical endplay limit should measure between 0.005 to 0.015 inch [0.127 to 0.369 mm]). If the play exceeds the service limit, new connecting rods will be required. If new rods (or a new crankshaft) are installed, the endplay may fall under the minimum allowable. If it does, the rods will have to be machined to restore it. If necessary, consult an automotive machine shop for advice.

4   Check the connecting rods and caps for identification marks. If they aren't plainly

marked, use paint or marker to clearly identify each rod and cap (1, 2, 3, etc., depending on the cylinder they're associated with) **(see illustration)**.

5   Remove the connecting rod cap bolts (V8 models) or nuts (V6 models) evenly. **Note 1:** *On V6 models, obtain new nuts and bolts for final installation but use the old nuts and bolts for checking the clearances. On V8 models, save the old bolts for checking clearances, but install new connecting rod bolts on the final torque.* **Note 2:** *On V6 models, on cylinders 1, 2 and 3, remove the connecting rod nut at the oil split hole side first. On cylinders 4, 5 and 6, remove the opposite nut first. Loosen the first nut until it protrudes approximately 1/16-inch above the bolt. Tap on the nut to release the bolt from the connecting rod.*

6   Remove the number one connecting rod cap and bearing insert. Don't drop the bearing insert out of the cap.

7   Remove the bearing insert and push the connecting rod/piston assembly out through the top of the engine. Use a wooden or plastic hammer handle to push on the upper bearing surface in the connecting rod. If resistance is felt, double-check to make sure that all of the ridge was removed from the cylinder.

8   Repeat the procedure for the remaining cylinders.

9   After removal, reassemble the connecting rod caps and bearing inserts in their respective connecting rods and install the cap bolts/nuts finger tight. Leaving the old bearing inserts in place until reassembly will help prevent the connecting rod bearing surfaces from being accidentally nicked or gouged.

10   The pistons and connecting rods are now ready for inspection and overhaul at an automotive machine shop.

Piston ring installation

Refer to illustrations 10.13, 10.14, 10.15, 10.19a, 10.19b and 10.22

11   Before installing the new piston rings, the ring end gaps must be checked. It's assumed that the piston ring side clearance has been checked and verified correct.

ENGINE BEARING ANALYSIS

Debris

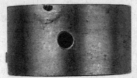

Babbitt bearing embedded with debris from machinings

Microscopic detail of debris

Microscopic detail of gouges

Overplated copper alloy bearing gouged by cast iron debris

Aluminum bearing embedded with glass beads

Microscopic detail of glass beads

Damaged lining caused by dirt left on the bearing back

Misassembly

Result of a lower half assembled as an upper - blocking the oil flow

Excessive oil clearance is indicated by a short contact arc

Polished and oil-stained backs are a result of a poor fit in the housing bore

Result of a wrong, reversed, or shifted cap

Overloading

Damage from excessive idling which resulted in an oil film unable to support the load imposed

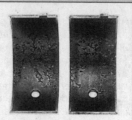

Damaged upper connecting rod bearings caused by engine lugging; the lower main bearings (not shown) were similarly affected

The damage shown in these upper and lower connecting rod bearings was caused by engine operation at a higher-than-rated speed under load

Misalignment

A warped crankshaft caused this pattern of severe wear in the center, diminishing toward the ends

A poorly finished crankshaft caused the equally spaced scoring shown

A tapered housing bore caused the damage along one edge of this pair

A bent connecting rod led to the damage in the "V" pattern

Lubrication

Result of dry start: The bearings on the left, farthest from the oil pump, show more damage

Result of a low oil supply or oil starvation

Severe wear as a result of inadequate oil clearance

Corrosion

Microscopic detail of corrosion

Corrosion is an acid attack on the bearing lining generally caused by inadequate maintenance, extremely hot or cold operation, or inferior oils or fuels

Microscopic detail of cavitation

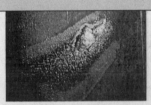

Example of cavitation - a surface erosion caused by pressure changes in the oil film

Damage from excessive thrust or insufficient axial clearance

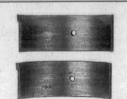

Bearing affected by oil dilution caused by excessive blow-by or a rich mixture

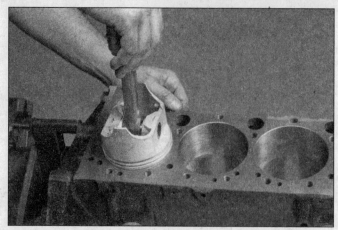

10.13 Install the piston ring into the cylinder then push it down into position using a piston so the ring will be square in the cylinder

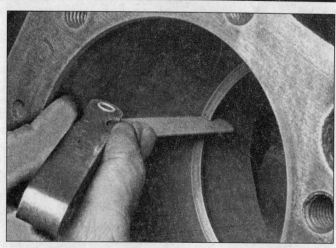

10.14 With the ring square in the cylinder, measure the ring end gap with a feeler gauge

10.15 If the ring end gap is too small, clamp a file in a vise as shown and file the piston ring ends - be sure to remove all raised material

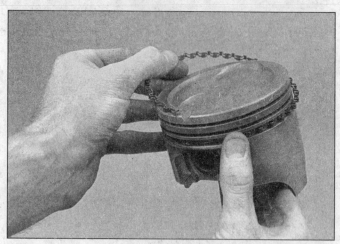

10.19a Installing the oil ring spacer in the piston oil ring groove

12   Lay out the piston/connecting rod assemblies and the new ring sets so the ring sets will be matched with the same piston and cylinder during the end gap measurement and engine assembly.

13   Insert the top (number one) ring into the first cylinder and square it up with the cylinder walls by pushing it in with the top of the piston **(see illustration)**. The ring should be near the bottom of the cylinder, at the lower limit of ring travel.

14   To measure the end gap, slip feeler gauges between the ends of the ring until a gauge equal to the gap width is found **(see illustration)**. The feeler gauge should slide between the ring ends with a slight amount of drag. A typical ring gap should fall between 0.010 and 0.020 inch [0.25 to 0.50 mm] for compression rings and up to 0.030 inch [0.76 mm] for the oil ring steel rails. If the gap is larger or smaller than specified, double-check to make sure you have the correct rings before proceeding.

15   If the gap is too small, it must be enlarged or the ring ends may come in contact with each other during engine operation, which can cause serious damage to the engine. If

necessary, increase the end gaps by filing the ring ends very carefully with a fine file. Mount the file in a vise equipped with soft jaws, slip the ring over the file with the ends contacting the file face and slowly move the ring to remove material from the ends. When performing this operation, file only by pushing the ring from the outside end of the file towards the vise **(see illustration)**.

16   Excess end gap isn't critical unless it's greater than 0.040 inch (1.01 mm). Again, double-check to make sure you have the correct ring type.

17   Repeat the procedure for each ring that will be installed in the first cylinder and for each ring in the remaining cylinders. Remember to keep rings, pistons and cylinders matched up.

18   Once the ring end gaps have been checked/corrected, the rings can be installed on the pistons.

19   The oil control ring (lowest one on the piston) is usually installed first. It's composed of three separate components. Slip the spacer/expander into the groove **(see illustration)**. If an anti-rotation tang is used, make sure it's inserted into the drilled hole in the

ring groove. Next, install the upper side rail in the same manner **(see illustration)**. Don't use a piston ring installation tool on the oil ring side rails, as they may be damaged. Instead, place one end of the side rail into the groove between the spacer/expander and the ring

10.19b DO NOT use a piston ring installation tool when installing the oil control side rails (oil rings)

10.22 Use a piston ring installation tool to install the number 2 and the number 1 (top) compression rings - be sure the directional mark on the piston ring(s) is facing toward the top of the piston

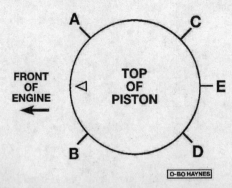

10.30 Position the piston ring end gaps as shown here before installing the piston/connecting rod assemblies into the engine

A *Top compression ring gap*
B *Second compression ring gap*
C *Upper oil ring side rail gap*
D *Lower oil ring side rail gap*
E *Oil ring spacer gap*

land, hold it firmly in place and slide a finger around the piston while pushing the rail into the groove. Finally, install the lower side rail.

20   After the three oil ring components have been installed, check to make sure that both the upper and lower side rails can be rotated smoothly inside the ring grooves.

21   The number two (middle) ring is installed next. It's usually stamped with a mark which must face up, toward the top of the piston. Do not mix up the top and middle rings, as they have different cross-sections. **Note:** *Always follow the instructions printed on the ring package or box - different manufacturers may require different approaches.*

22   Use a piston ring installation tool and make sure the identification mark is facing the top of the piston, then slip the ring into the middle groove on the piston **(see illustration)**. Don't expand the ring any more than necessary to slide it over the piston.

23   Install the number one (top) ring in the same manner. Make sure the mark is facing up. Be careful not to confuse the number one and number two rings.

24   Repeat the procedure for the remaining pistons and rings.

Installation

25   Before installing the piston/connecting rod assemblies, the cylinder walls must be perfectly clean, the top edge of each cylinder bore must be chamfered, and the crankshaft must be in place.

26   Remove the cap from the end of the number one connecting rod (refer to the marks made during removal). Remove the original bearing inserts and wipe the bearing surfaces of the connecting rod and cap with a clean, lint-free cloth. They must be kept spotlessly clean.

Connecting rod bearing oil clearance check

Refer to illustrations 10.30, 10.35, 10.37 and 10.41

27   Clean the back side of the new upper bearing insert, then lay it in place in the connecting rod.

28   Make sure the tab on the bearing fits into the recess in the rod. Don't hammer the bearing insert into place and be very careful not to nick or gouge the bearing face. Don't lubricate the bearing at this time.

29   Clean the back side of the other bearing insert and install it in the rod cap. Again, make sure the tab on the bearing fits into the recess in the cap, and don't apply any lubricant. It's critically important that the mating surfaces of the bearing and connecting rod are perfectly clean and oil free when they're assembled.

30   Position the piston ring gaps at intervals around the piston as shown **(see illustration)**.

31   Lubricate the piston and rings with clean engine oil and attach a piston ring compressor to the piston. Leave the skirt protruding about 1/4-inch to guide the piston into the cylinder. The rings must be compressed until they're flush with the piston.

32   Rotate the crankshaft until the number one connecting rod journal is at BDC (bottom dead center) and apply a liberal coat of engine oil to the cylinder walls.

33   With the weight designation mark on top of the piston facing the front (timing chain end) of the engine, gently insert the piston/connecting rod assembly into the number one cylinder bore and rest the bottom edge of the ring compressor on the engine block. Install the pistons with the cavity mark(s) facing toward the timing chain.

34   Tap the top edge of the ring compressor to make sure it's contacting the block around its entire circumference.

35   Gently tap on the top of the piston with the end of a wooden or plastic hammer handle **(see illustration)** while guiding the end of the connecting rod into place on the crankshaft journal. The piston rings may try to pop out of the ring compressor just before entering the cylinder bore, so keep some downward pressure on the ring compressor. Work slowly, and if any resistance is felt as the piston enters the cylinder, stop immediately. Find out what's hanging up and fix it before proceeding. Do not, for any reason, force the piston into the cylinder - you might break a ring and/or the piston.

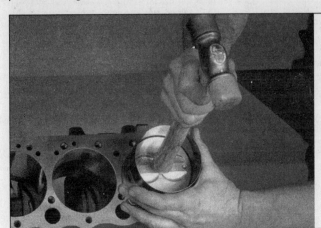

10.35 Use a plastic or wooden hammer handle to push the piston into the cylinder

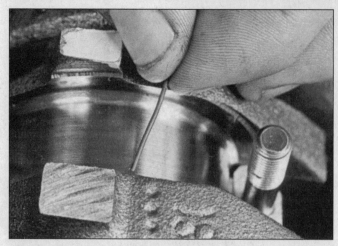

10.37 Place Plastigage on each connecting rod bearing journal parallel to the crankshaft centerline

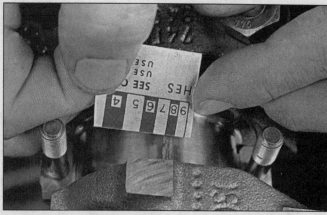

10.41 Use the scale on the Plastigage package to determine the bearing oil clearance - be sure to measure the widest part of the Plastigage and use the correct scale; it comes with both standard and metric scales

36   Once the piston/connecting rod assembly is installed, the connecting rod bearing oil clearance must be checked before the rod cap is permanently installed.

37   Cut a piece of the appropriate size Plastigage slightly shorter than the width of the connecting rod bearing and lay it in place on the number one connecting rod journal, parallel with the journal axis **(see illustration)**.

38   Clean the connecting rod cap bearing face and install the rod cap. Make sure the mating mark on the cap is on the same side as the mark on the connecting rod **(see illustration 10.4)**.

39   Install the old rod bolts (V8 models) or nuts (V6 models), at this time, and tighten them to the torque listed in this Chapter's Specifications. **Note:** *Use a thin-wall socket to avoid erroneous torque readings that can result if the socket is wedged between the rod cap and the bolt. If the socket tends to wedge itself between the fastener and the cap, lift up on it slightly until it no longer contacts the cap. DO NOT rotate the crankshaft at any time during this operation.*

40   Remove the fasteners and detach the rod cap, being very careful not to disturb the Plastigage. Discard the cap bolts at this time as they cannot be reused. **Note:** *You MUST use new connecting rod bolts on V8 engines and connecting rod nuts and bolts on V6 engines.*

41   Compare the width of the crushed Plastigage to the scale printed on the Plastigage envelope to obtain the oil clearance **(see illustration)**. The connecting rod oil clearance is usually about 0.001 to 0.002 inch. Consult an automotive machine shop for the clearance specified for the rod bearings on your engine.

42   If the clearance is not as specified, the bearing inserts may be the wrong size (which means different ones will be required). Before deciding that different inserts are needed, make sure that no dirt or oil was between the bearing inserts and the connecting rod or cap when the clearance was measured. Also, recheck the journal diameter. If the Plasti-

gage was wider at one end than the other, the journal may be tapered. If the clearance still exceeds the limit specified, the bearing will have to be replaced with an undersize bearing. **Caution:** *When installing a new crankshaft always use a standard size bearing.*

Final installation

43   Carefully scrape all traces of the Plastigage material off the rod journal and/or bearing face. Be very careful not to scratch the bearing - use your fingernail or the edge of a plastic card.

44   Make sure the bearing faces are perfectly clean, then apply a uniform layer of clean moly-base grease or engine assembly lube to both of them. You'll have to push the piston into the cylinder to expose the face of the bearing insert in the connecting rod.

45   Slide the connecting rod back into place on the journal, install the rod cap, install the new bolts (V8 models) or bolts and nuts (V6 models) and tighten them to the torque listed in this Chapter's Specifications. **Caution:** *Install new connecting rod cap bolts (and nuts, on V6 models). Do NOT reuse old bolts - they have stretched and cannot be reused. Also on V6 models, make sure the bolt heads sit parallel to the edges of the connecting rods.*

46   Repeat the entire procedure for the remaining pistons/connecting rods.

47   The important points to remember are:

a) *Keep the back sides of the bearing inserts and the insides of the connecting rods and caps perfectly clean when assembling them.*

b) *Make sure you have the correct piston/rod assembly for each cylinder.*

c) *The mark on the piston must face the front (timing chain end) of the engine.*

d) *Lubricate the cylinder walls liberally with clean oil.*

e) *Lubricate the bearing faces when installing the rod caps after the oil clearance has been checked.*

48   After all the piston/connecting rod assemblies have been correctly installed,

rotate the crankshaft a number of times by hand to check for any obvious binding.

49   As a final step, check the connecting rod endplay, as described in Step 3. If it was correct before disassembly and the original crankshaft and rods were reinstalled, it should still be correct. If new rods or a new crankshaft were installed, the endplay may be inadequate. If so, the rods will have to be removed and taken to an automotive machine shop for resizing.

---

11   Crankshaft - removal and installation

Removal

Refer to illustrations 11.1 and 11.3

Note: *The crankshaft can be removed only after the engine has been removed from the vehicle. It's assumed that the flywheel or driveplate, crankshaft pulley, timing chain, oil pan, oil pump body, oil filter and piston/connecting rod assemblies have already been removed. The rear main oil seal retainer must be unbolted and separated from the block before proceeding with crankshaft removal.*

1   Before the crankshaft is removed, measure the endplay. Mount a dial indicator with the indicator in line with the crankshaft and just touching the end of the crankshaft as shown **(see illustration)**.

2   Pry the crankshaft all the way to the rear and zero the dial indicator. Next, pry the crankshaft to the front as far as possible and check the reading on the dial indicator. The distance traveled is the endplay. A typical crankshaft endplay will fall between 0.003 to 0.010 inch (0.076 to 0.254 mm). If it is greater than that, check the crankshaft thrust surfaces for wear after it's removed. If no wear is evident, new main bearings should correct the endplay.

a) *On V6 engines, the crankshaft thrust bearings are located in the upper number 3 crankshaft saddle and the lower number 3 crankshaft main bearing cap.*

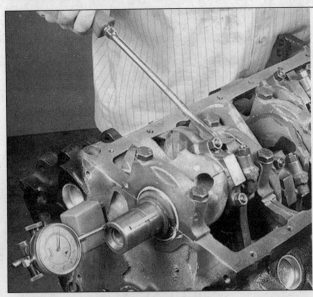

11.1 Checking crankshaft endplay with a dial indicator

b) *On 4.6L V8 engines, the crankshaft thrust washer is located on the back of the number 5 main bearing journal saddle and the thrust bearing is located in the number 5 main bearing cap.*

Note: *When installing thrust washers, make sure the grooves in the washer face the crankshaft.*

3   If a dial indicator isn't available, feeler gauges can be used. Gently pry the crankshaft all the way to the front of the engine. Slip feeler gauges between the crankshaft and the front face of the thrust bearing or washer to determine the clearance **(see illustration)**.

4   Loosen the main bearing cap bolts 1/4-turn at a time each, until they can be removed by hand. On V8 models, follow the reverse of the tightening sequence when loosening the side bolts and the vertical bolts **(see illustrations 11.19a and 11.19b)**.

5   Remove the main bearing caps. Gently tap the main bearing cap with a soft-face hammer. Pull the main bearing caps straight up and off the cylinder block. Try not to drop the bearing inserts if they come out with the assembly.

6   Carefully lift the crankshaft out of the engine. It may be a good idea to have an assistant available, since the crankshaft is quite heavy and awkward to handle. With the bearing inserts in place inside the engine block and main bearing caps, reinstall the main bearing caps onto the engine block and tighten the bolts finger tight. Make sure the caps are in the proper order, and with the arrow pointing toward the front (timing chain end) of the engine.

Installation

7   Crankshaft installation is the first step in engine reassembly. It's assumed at this point that the engine block and crankshaft have been cleaned, inspected and repaired or reconditioned.

8   Position the engine block with the bottom facing up.

9   Remove the bolts and lift off the main bearing caps.

10   If they're still in place, remove the original bearing inserts from the block and from the main bearing cap assembly. Wipe the bearing surfaces of the block and main bearing cap assembly with a clean, lint-free cloth. They must be kept spotlessly clean. This is critical for determining the correct bearing oil clearance.

Main bearing oil clearance check

Refer to illustrations 11.17, 11.19a, 11.19b and 11.21

11   Without mixing them up, clean the back sides of the new upper main bearing inserts (with grooves and oil holes) and lay one in each main bearing saddle in the engine block. Each upper bearing (engine block) has an oil groove and oil hole in it. **Caution:** *The oil holes in the block must line up with the oil holes in the engine block inserts.* The thrust washer or thrust bearing insert must be installed in the correct location. **Note:** *The oil grooves in the thrust washers must face toward the front of the engine.* Clean the back sides of the lower main bearing inserts and lay them in the corresponding location in the main bearing cap assembly. Make sure the tab on the bearing insert fits into the recess in the block or main bearing cap assembly. **Caution:** *Do not hammer the bearing insert into place and don't nick or gouge the bearing faces. DO NOT apply any lubrication at this time.*

12   Clean the faces of the bearing inserts in the block and the crankshaft main bearing journals with a clean, lint-free cloth.

13   Check or clean the oil holes in the crankshaft, as any dirt here can go only one way - straight through the new bearings.

14   Once you're certain the crankshaft is clean, carefully lay it in position in the cylinder block.

15   Before the crankshaft can be permanently installed, the main bearing oil clearance must be checked.

16   Cut several strips of the appropriate size of Plastigage. They must be slightly shorter than the width of the main bearing journal.

17   Place one piece on each crankshaft main bearing journal, parallel with the journal axis as shown **(see illustration)**.

18   Clean the faces of the bearing inserts

11.3 Checking the crankshaft endplay with feeler gauges at the thrust bearing journal

11.17 Place the Plastigage onto the crankshaft bearing journal as shown

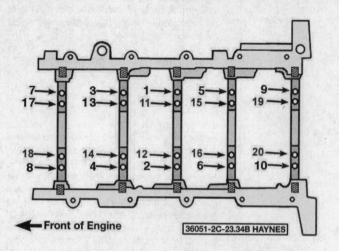

11.19a Main bearing cap bolt tightening sequence on V8 engines

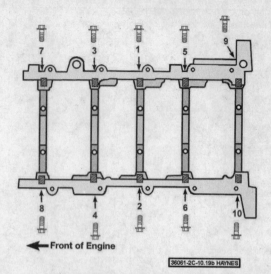

11.19b Main bearing cap side (horizontal) bolt tightening sequence (V8 engines)

in the main bearing cap assembly. Hold the bearing inserts in place and install the assembly onto the crankshaft and cylinder block. DO NOT disturb the Plastigage.

19 Apply clean engine oil to all bolt threads prior to installation, then install all bolts finger-tight. On V6 engines, tighten the main bearing caps starting with the center cap and working out. On V8 engines, tighten the bearing cap assembly bolts in the sequence shown **(see illustrations)** progressing in steps, to the torque listed in this Chapter's Specifications. DO NOT rotate the crankshaft at any time during this operation. **Note:** *The V8 engines are equipped with side bolts, but it isn't necessary to install them for the oil clearance check.*

20 Remove the bolts in the *reverse* order of the tightening sequence and carefully lift the main bearing caps straight up and off the block. Do not disturb the Plastigage or rotate the crankshaft. If the main bearing cap assembly is difficult to remove, tap it gently from side-to-side with a soft-face hammer to loosen it.

21 Compare the width of the crushed Plastigage on each journal to the scale printed on the Plastigage envelope to determine the main bearing oil clearance **(see illustration)**. Check with an automotive machine shop for the crankshaft endplay service limits.

22 If the clearance is not as specified, the bearing inserts may be the wrong size (which means different ones will be required). Before deciding if different inserts are needed, make sure that no dirt or oil was between the bearing inserts and the cap assembly or block when the clearance was measured. If the Plastigage was wider at one end than the other, the crankshaft journal may be tapered. If the clearance still exceeds the limit specified, the bearing insert(s) will have to be replaced with an undersize bearing insert(s). **Caution:** *When installing a new crankshaft always install a standard bearing insert set.*

23 Carefully scrape all traces of the Plastigage material off the main bearing journals and/or the bearing insert faces. Be sure to remove all residue from the oil holes. Use your fingernail or the edge of a plastic card - don't nick or scratch the bearing faces.

Final installation

24 Carefully lift the crankshaft out of the cylinder block.

25 Clean the bearing insert faces in the cylinder block, then apply a thin, uniform layer of moly-base grease or engine assembly lube to each of the bearing surfaces. Be sure to coat the thrust faces as well as the journal face of the thrust bearing.

26 Make sure the crankshaft journals are clean, then lay the crankshaft back in place in the cylinder block.

27 Clean the bearing insert faces and apply the same lubricant to them. Clean the engine block and the main bearing caps thoroughly. The surfaces must be free of oil residue.

28 On V6 engines, apply a bead of RTV

sealant to the engine block on the rear main bearing cap parting line. Be sure the main bearing cap is installed within four minutes after the RTV sealant is applied.

29 Prior to installation, apply clean engine oil to all bolt threads wiping off any excess, then install all bolts finger-tight.

30 Tighten the main bearing cap bolts to the torque listed in this Chapter's Specifications. On V6 engines, start with the center bearing cap bolts and work out. On V8 models, follow the correct torque sequence **(see illustrations 11.19a and 11.19b)**. **Note:** *On V8 engines, tighten the main bearing cap (vertical) bolts first, then tighten the side bolts.*

31 Recheck the crankshaft endplay with a feeler gauge or a dial indicator. The endplay should be correct if the crankshaft thrust faces aren't worn or damaged and if new bearings have been installed.

32 Rotate the crankshaft a number of times by hand to check for any obvious binding. It should rotate with a running torque of 50 in-lbs or less. If the running torque is too high,

11.21 Use the scale on the Plastigage package to determine the bearing oil clearance - be sure to measure the widest part of the Plastigage and use the correct scale; it comes with both standard and metric scales

correct the problem at this time.

33 Install the new rear main oil seal (see Chapter 2A or 2B).

12 Engine overhaul - reassembly sequence

1 Before beginning engine reassembly, make sure you have all the necessary new parts, gaskets and seals as well as the following items on hand:

Common hand tools
A 1/2-inch drive torque wrench
New engine oil
Gasket sealant
Thread locking compound

2 If you obtained a short block it will be necessary to install the cylinder head, the oil pump and pick-up tube, the oil pan, the water pump, the timing chain and cover, and the valve covers (see Chapter 2A or 2B). In order to save time and avoid problems, the external components must be installed in the following general order:

Water pump
Exhaust manifolds
Intake manifold*

Fuel injection components
Emission control components
Spark plug wires and spark plugs
Ignition coils or coil assembly
Oil filter
Engine mounts and mount brackets
Flywheel (manual transmission)
Driveplate (automatic transmission)

* *The intake manifold will be installed after the engine is installed (V8 models).*

13 Initial start-up and break-in after overhaul

Warning: *Have a fire extinguisher handy when starting the engine for the first time.*

1 Once the engine has been installed in the vehicle, double-check the engine oil and coolant levels.

2 With the spark plugs out of the engine and the ignition system and fuel pump disabled (see Chapter 4, Section 2), crank the engine until oil pressure registers on the gauge or the light goes out.

3 Install the spark plugs, hook up the plug wires (V6 engine) or install the ignition coils (V8 engine) and restore the ignition and fuel pump functions.

4 Start the engine. It may take a few moments for the fuel system to build up pressure, but the engine should start without a great deal of effort.

5 After the engine starts, it should be allowed to warm up to normal operating temperature. While the engine is warming up, make a thorough check for fuel, oil and coolant leaks.

6 Shut the engine off and recheck the engine oil and coolant levels.

7 Drive the vehicle to an area with minimum traffic, accelerate from 30 to 50 mph, then allow the vehicle to slow to 30 mph with the throttle closed. Repeat the procedure 10 or 12 times. This will load the piston rings and cause them to seat properly against the cylinder walls. Check again for oil and coolant leaks.

8 Drive the vehicle gently for the first 500 miles (no sustained high speeds) and keep a constant check on the oil level. It is not unusual for an engine to use oil during the break-in period.

9 At approximately 500 to 600 miles, change the oil and filter.

10 For the next few hundred miles, drive the vehicle normally. Do not pamper it or abuse it.

11 After 2,000 miles, change the oil and filter again and consider the engine broken in.

COMMON ENGINE OVERHAUL TERMS

B

Backlash - The amount of play between two parts. Usually refers to how much one gear can be moved back and forth without moving gear with which it's meshed.

Bearing Caps - The caps held in place by nuts or bolts which, in turn, hold the bearing surface. This space is for lubricating oil to enter.

Bearing clearance - The amount of space left between shaft and bearing surface. This space is for lubricating oil to enter.

Bearing crush - The additional height which is purposely manufactured into each bearing half to ensure complete contact of the bearing back with the housing bore when the engine is assembled.

Bearing knock - The noise created by movement of a part in a loose or worn bearing.

Blueprinting - Dismantling an engine and reassembling it to EXACT specifications.

Bore - An engine cylinder, or any cylindrical hole; also used to describe the process of enlarging or accurately refinishing a hole with a cutting tool, as to bore an engine cylinder. The bore size is the diameter of the hole.

Boring - Renewing the cylinders by cutting them out to a specified size. A boring bar is used to make the cut.

Bottom end - A term which refers collectively to the engine block, crankshaft, main bearings and the big ends of the connecting rods.

Break-in - The period of operation between installation of new or rebuilt parts and time in which parts are worn to the correct fit. Driving at reduced and varying speed for a specified mileage to permit parts to wear to the correct fit.

Bushing - A one-piece sleeve placed in a bore to serve as a bearing surface for shaft, piston pin, etc. Usually replaceable.

C

Camshaft - The shaft in the engine, on which a series of lobes are located for operating the valve mechanisms. The camshaft is driven by gears or sprockets and a timing chain. Usually referred to simply as the cam.

Carbon - Hard, or soft, black deposits found in combustion chamber, on plugs, under rings, on and under valve heads.

Cast iron - An alloy of iron and more than two percent carbon, used for engine blocks and heads because it's relatively inexpensive and easy to mold into complex shapes.

Chamfer - To bevel across (or a bevel on) the sharp edge of an object.

Chase - To repair damaged threads with a tap or die.

Combustion chamber - The space between the piston and the cylinder head, with the piston at top dead center, in which air-fuel mixture is burned.

Compression ratio - The relationship between cylinder volume (clearance volume) when the piston is at top dead center and cylinder volume when the piston is at bottom dead center.

Connecting rod - The rod that connects the crank on the crankshaft with the piston. Sometimes called a con rod.

Connecting rod cap - The part of the connecting rod assembly that attaches the rod to the crankpin.

Core plug - Soft metal plug used to plug the casting holes for the coolant passages in the block.

Crankcase - The lower part of the engine in which the crankshaft rotates; includes the lower section of the cylinder block and the oil pan.

Crank kit - A reground or reconditioned crankshaft and new main and connecting rod bearings.

Crankpin - The part of a crankshaft to which a connecting rod is attached.

Crankshaft - The main rotating member, or shaft, running the length of the crankcase, with offset throws to which the connecting rods are attached; changes the reciprocating motion of the pistons into rotating motion.

Cylinder sleeve - A replaceable sleeve, or liner, pressed into the cylinder block to form the cylinder bore.

D

Deburring - Removing the burrs (rough edges or areas) from a bearing.

Deglazer - A tool, rotated by an electric motor, used to remove glaze from cylinder walls so a new set of rings will seat.

E

Endplay - The amount of lengthwise movement between two parts. As applied to a crankshaft, the distance that the crankshaft can move forward and back in the cylinder block.

F

Face - A machinist's term that refers to removing metal from the end of a shaft or the face of a larger part, such as a flywheel.

Fatigue - A breakdown of material through a large number of loading and unloading cycles. The first signs are cracks followed shortly by breaks.

Feeler gauge - A thin strip of hardened steel, ground to an exact thickness, used to check clearances between parts.

Free height - The unloaded length or height of a spring.

Freeplay - The looseness in a linkage, or an assembly of parts, between the initial application of force and actual movement. Usually perceived as slop or slight delay.

Freeze plug - See Core plug.

G

Gallery - A large passage in the block that forms a reservoir for engine oil pressure.

Glaze - The very smooth, glassy finish that develops on cylinder walls while an engine is in service.

H

Heli-Coil - A rethreading device used when threads are worn or damaged. The device is installed in a retapped hole to reduce the thread size to the original size.

I

Installed height - The spring's measured length or height, as installed on the cylinder head. Installed height is measured from the spring seat to the underside of the spring retainer.

J

Journal - The surface of a rotating shaft which turns in a bearing.

K

Keeper - The split lock that holds the valve spring retainer in position on the valve stem.

Key - A small piece of metal inserted into matching grooves machined into two parts fitted together - such as a gear pressed onto a shaft - which prevents slippage between the two parts.

Knock - The heavy metallic engine sound, produced in the combustion chamber as a result of abnormal combustion - usually detonation. Knock is usually caused by a loose or worn bearing. Also referred to as detonation, pinging and spark knock. Connecting rod or main bearing knocks are created by too much oil clearance or insufficient lubrication.

L

Lands - The portions of metal between the piston ring grooves.

Lapping the valves - Grinding a valve face and its seat together with lapping compound.

Lash - The amount of free motion in a gear train, between gears, or in a mechanical assembly, that occurs before movement can begin. Usually refers to the lash in a valve train.

Lifter - The part that rides against the cam to transfer motion to the rest of the valve train.

M

Machining - The process of using a machine to remove metal from a metal part.

Main bearings - The plain, or babbit, bearings that support the crankshaft.

Main bearing caps - The cast iron caps, bolted to the bottom of the block, that support the main bearings.

O

O.D. - Outside diameter.

Oil gallery - A pipe or drilled passageway in the engine used to carry engine oil from one area to another.

Oil ring - The lower ring, or rings, of a piston; designed to prevent excessive amounts of oil from working up the cylinder walls and into the combustion chamber. Also called an oil-control ring.

Oil seal - A seal which keeps oil from leaking out of a compartment. Usually refers to a dynamic seal around a rotating shaft or other moving part.

O-ring - A type of sealing ring made of a special rubberlike material; in use, the O-ring is compressed into a groove to provide the sealing action.

Overhaul - To completely disassemble a unit, clean and inspect all parts, reassemble it with the original or new parts and make all adjustments necessary for proper operation.

P

Pilot bearing - A small bearing installed in the center of the flywheel (or the rear end of the crankshaft) to support the front end of the input shaft of the transmission.

Pip mark - A little dot or indentation which indicates the top side of a compression ring.

Piston - The cylindrical part, attached to the connecting rod, that moves up and down in the cylinder as the crankshaft rotates. When the fuel charge is fired, the piston transfers the force of the explosion to the connecting rod, then to the crankshaft.

Piston pin (or wrist pin) - The cylindrical and usually hollow steel pin that passes through the piston. The piston pin fastens the piston to the upper end of the connecting rod.

Piston ring - The split ring fitted to the groove in a piston. The ring contacts the sides of the ring groove and also rubs against the cylinder wall, thus sealing space between piston and wall. There are two types of rings: Compression rings seal the compression pressure in the combustion chamber; oil rings scrape excessive oil off the cylinder wall.

Piston ring groove - The slots or grooves cut in piston heads to hold piston rings in position.

Piston skirt - The portion of the piston below the rings and the piston pin hole.

Plastigage - A thin strip of plastic thread, available in different sizes, used for measuring clearances. For example, a strip of plastigage is laid across a bearing journal and mashed as parts are assembled. Then parts are disassembled and the width of the strip is measured to determine clearance between journal and bearing. Commonly used to measure crankshaft main-bearing and connecting rod bearing clearances.

Press-fit - A tight fit between two parts that requires pressure to force the parts together. Also referred to as drive, or force, fit.

Prussian blue - A blue pigment; in solution, useful in determining the area of contact between two surfaces. Prussian blue is commonly used to determine the width and location of the contact area between the valve face and the valve seat.

R

Race (bearing) - The inner or outer ring that provides a contact surface for balls or rollers in bearing.

Ream - To size, enlarge or smooth a hole by using a round cutting tool with fluted edges.

Ring job - The process of reconditioning the cylinders and installing new rings.

Runout - Wobble. The amount a shaft rotates out-of-true.

S

Saddle - The upper main bearing seat.

Scored - Scratched or grooved, as a cylinder wall may be scored by abrasive particles moved up and down by the piston rings.

Scuffing - A type of wear in which there's a transfer of material between parts moving against each other; shows up as pits or grooves in the mating surfaces.

Seat - The surface upon which another part rests or seats. For example, the valve seat is the matched surface upon which the valve face rests. Also used to refer to wearing into a good fit; for example, piston rings seat after a few miles of driving.

Short block - An engine block complete with crankshaft and piston and, usually, camshaft assemblies.

Static balance - The balance of an object while it's stationary.

Step - The wear on the lower portion of a ring land caused by excessive side and back-clearance. The height of the step indicates the ring's extra side clearance and the length of the step projecting from the back wall of the groove represents the ring's back clearance.

Stroke - The distance the piston moves when traveling from top dead center to bottom dead center, or from bottom dead center to top dead center.

Stud - A metal rod with threads on both ends.

T

Tang - A lip on the end of a plain bearing used to align the bearing during assembly.

Tap - To cut threads in a hole. Also refers to the fluted tool used to cut threads.

Taper - A gradual reduction in the width of a shaft or hole; in an engine cylinder, taper usually takes the form of uneven wear, more pronounced at the top than at the bottom.

Throws - The offset portions of the crankshaft to which the connecting rods are affixed.

Thrust bearing - The main bearing that has thrust faces to prevent excessive endplay, or forward and backward movement of the crankshaft.

Thrust washer - A bronze or hardened steel washer placed between two moving parts. The washer prevents longitudinal movement and provides a bearing surface for thrust surfaces of parts.

Tolerance - The amount of variation permitted from an exact size of measurement. Actual amount from smallest acceptable dimension to largest acceptable dimension.

U

Umbrella - An oil deflector placed near the valve tip to throw oil from the valve stem area.

Undercut - A machined groove below the normal surface.

Undersize bearings - Smaller diameter bearings used with re-ground crankshaft journals.

V

Valve grinding - Refacing a valve in a valve-refacing machine.

Valve train - The valve-operating mechanism of an engine; includes all components from the camshaft to the valve.

Vibration damper - A cylindrical weight attached to the front of the crankshaft to minimize torsional vibration (the twist-untwist actions of the crankshaft caused by the cylinder firing impulses). Also called a harmonic balancer.

W

Water jacket - The spaces around the cylinders, between the inner and outer shells of the cylinder block or head, through which coolant circulates.

Web - A supporting structure across a cavity.

Woodruff key - A key with a radiused backside (viewed from the side).

Notes

Chapter 3
Cooling, heating and air conditioning systems

Contents

Specifications

General

Expansion tank cap pressure rating	16 psi
Thermostat rating (opening to fully open temperature range)	
V6 engine	194-201 to 223 degrees F
V8 engine	175-182 to 202 degrees F
Cooling system capacity	See Chapter 1
Refrigerant type	R-134a
Refrigerant capacity	Refer to HVAC specification tag

Torque specifications

Ft-lbs (unless otherwise indicated)

Air conditioning compressor manifold bolt	15 in-lbs
Air conditioning compressor mounting fasteners	
V6 engine	
Studs only	35 in-lbs
Nuts and bolts	18
V8 engine	18
Condenser inlet and outlet line fitting nuts	71 in-lbs
Condenser (power steering cooling line bracket) mounting nuts	71 in-lbs
Radiator mounting brackets bolts	89 in-lbs
Thermostat housing cover bolts	89 in-lbs
Water pump bolts	
V6 engine	89 in-lbs
V8 engine	18
Water pump pulley bolts	18

1 General information

Engine cooling system

The cooling system consists of a radiator, an expansion tank, a pressure cap (located on the expansion tank), a thermostat, a temperature-controlled electric cooling fan, and a pulley/belt-driven water pump.

The radiator cooling fan is mounted in a housing/shroud at the engine side of the radiator. It is designed to come on when the engine reaches a certain temperature, and shut off again when the engine cools down some, thereby keeping the engine in the desired operating-temperature range.

The expansion tank, referred to by the manufacturer as a "degas bottle," functions somewhat differently than a conventional recovery tank. Designed to separate any trapped air in the coolant, it is pressurized by the radiator and has a pressure cap on top. The radiator on these models does not have a pressure cap. When the thermostat is closed, no coolant flows in the expansion tank, but when the engine is fully warmed up, coolant flows from the top of the radiator through a small hose that enters the top of the expansion tank, where the air separates and the coolant falls into a coolant reservoir in the bottom of the tank, which is fed to the cooling system through a larger hose connected to the lower radiator hose. **Warning:** *Unlike a conventional coolant recovery tank, the pressure cap on the expansion tank should never be opened after the engine has warmed up, because of the danger of severe burns caused by steam or scalding coolant.*

Coolant in the left side of the radiator circulates through the lower radiator hose to the water pump, where it is forced through coolant passages in the cylinder block. The coolant then travels up into the cylinder head, circulates around the combustion chambers and valve seats, travels out of the cylinder head past the open thermostat into the upper radiator hose and back into the radiator.

When the engine is cold, the thermostat restricts the circulation of coolant to the engine. When the minimum operating temperature is reached, the thermostat begins to open, allowing coolant to return to the radiator.

Transmission cooling systems

Vehicles with an automatic transmission are equipped with a transmission oil cooler located near the radiator. Hot transmission fluid is directed to the cooler by lines and hoses. Once cooled, the fluid is directed back to the transmission. For more information on transmission oil coolers, refer to Chapter 7B.

Heating system

The heating system consists of the heater controls, the heater core, the heater blower assembly (which houses the blower motor and the blower motor resistor), and the hoses connecting the heater core to the engine cooling system. Hot engine coolant is circulated through the heater core. When the heater mode is activated, a flap door opens to expose the heater box to the passenger compartment. A fan switch on the heater controls activates the blower motor, which forces air through the core, heating the air.

Air conditioning system

The air conditioning system consists of the condenser, which is mounted in front of the radiator and the transaxle fluid cooler (if equipped), the evaporator case assembly under the dash, a compressor mounted on the engine, and the plumbing connecting all of the above components.

A blower fan forces the warmer air of the passenger compartment through the evaporator core (sort of a radiator-in-reverse), transferring the heat from the air to the refrigerant. The liquid refrigerant boils off into low pressure vapor, taking the heat with it when it leaves the evaporator.

2 Antifreeze - general information

Refer to illustration 2.5

Warning: *Do not allow antifreeze to come in contact with your skin or painted surfaces of the vehicle. Rinse off spills immediately with plenty of water. Antifreeze is highly toxic if ingested. Never leave antifreeze lying around in an open container or in puddles on the floor; children and pets are attracted by its sweet smell and may drink it. Check with local authorities about disposing of used antifreeze. Many communities have collection centers which will see that antifreeze is disposed of safely. Never dump used antifreeze on the ground or pour it into drains.*
Caution: *Do not mix coolants of different colors. Doing so might damage the cooling system and/or the engine. The manufacturer specifies either a green colored coolant or a yellow colored coolant to be used in these systems. Read the warning label in the engine compartment for additional information.*
Note: *Non-toxic antifreeze is now manufac-* tured and available at local auto parts stores, but even this type must be disposed of properly.

The cooling system should be filled with a water/ethylene glycol based antifreeze solution, which will prevent freezing down to at least -20-degrees F (even lower in cold climates). It also provides protection against corrosion and increases the coolant boiling point. The manufacturer recommends that the correct type of coolant be used and strongly urges that coolant types not be mixed (see Chapter 1).

Drain, flush and refill the cooling system at least every other year (see Chapter 1). The use of antifreeze solutions for periods of longer than two years is likely to cause damage and encourage the formation of rust and scale in the system.

Before adding antifreeze to the system, inspect all hose connections. Antifreeze can leak through very minute openings.

The exact mixture of antifreeze to water, which you should use, depends on the relative weather conditions. The mixture should contain at least 50-percent antifreeze, but should never contain more than 70-percent antifreeze. Consult the mixture ratio chart on the container before adding coolant.

Hydrometers are available at most auto parts stores to test the coolant **(see illustration)**. **Warning:** *Do not remove the expansion tank cap, drain the coolant or perform any service procedures on the cooling system until the engine has cooled completely.*

3 Thermostat - check and replacement

Check

1 Before assuming the thermostat is to blame for a cooling system problem, check the coolant level, drivebelt (see Chapter 1) and temperature gauge operation.
2 If the engine seems to be taking a long time to warm up, based on heater output or temperature gauge operation, the thermostat is probably stuck open. Replace the thermostat with a new one.

2.5 Use a hydrometer (available at auto parts stores) to test the condition of your coolant

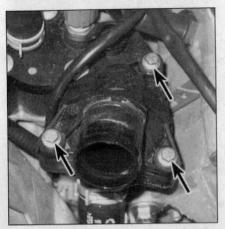

3.11 Thermostat housing cover mounting bolts (V6 engine)

3.12 Remove the O-ring seal and discard it (V6 engine)

3.16 Thermostat housing bolts (V8 engine)

3   If the engine runs hot, use your hand to check the temperature of the upper radiator hose. If the hose isn't hot, but the engine is, the thermostat is probably stuck closed, preventing the coolant inside the engine from escaping to the radiator. Replace the thermostat. **Caution:** *Don't drive the vehicle without a thermostat. The computer may stay in open loop and emissions and fuel economy will suffer.*

4   If the upper radiator hose is hot, it means that the coolant is flowing and the thermostat is open. Consult the *Troubleshooting* section at the front of this manual for cooling system diagnosis.

Replacement

Warning: *Wait until the engine is completely cool before beginning this procedure.*

5   Disconnect the cable from the negative battery terminal (see Chapter 5, Section 1).

6   Drain the cooling system (see Chapter 1). If the coolant is relatively new and still in good condition, save it and reuse it.

7   Remove the air duct between the air cleaner housing and the throttle body (see Chapter 4).

8   On V6 models, follow the upper radiator hose to the engine to locate the thermostat housing. On V8 models, the thermostat housing is located in between the engine and the cooling fan in front of the power steering pump.

9   On V6 models, remove the hose from the housing by loosening the hose clamp. Squeeze the ends of the clamp together with hose clamp pliers (regular pliers will work also). If the radiator hose is stuck, grasp it near the end with a pair of adjustable pliers and twist it to break the seal, then pull it off. If the hose is old or if it has deteriorated, cut it off and install a new one.

10   If the outer surface of the thermostat housing cover, which mates with the hose, is corroded, pitted, or otherwise deteriorated, it might be damaged even more by hose removal. If it is, replace the thermostat housing cover.

V6 models

Refer to illustrations 3.11 and 3.12

11   Remove the fasteners and detach the thermostat housing cover. If the cover is stuck, tap it lightly to jar it loose. Be prepared for some coolant to spill as the gasket seal is broken **(see illustration)**.

12   Remove the O-ring seal and discard it **(see illustration)**.

13   Note how the thermostat is installed and then remove the thermostat from the housing.

14   Install the replacement thermostat into the housing. Lubricate the new O-ring seal with fresh coolant and place it in position. Install the housing cover and tighten the bolts to the torque listed in this Chapter's Specifications.

15   Reattach the radiator hose to the thermostat housing cover. Make sure that the hose clamp still has tension when installed. If it doesn't, replace it.

V8 models

Refer to illustrations 3.16 and 3.17

16   Remove the fasteners and detach the thermostat housing cover **(see illustration)**. If the cover is stuck, tap it lightly to jar it loose. Be prepared for some coolant to spill as the seal is broken.

17   Note how the thermostat is installed and then remove it **(see illustration)**. Remove and discard the O-ring seal.

18   Lubricate the new O-ring seal with fresh coolant and place it in position. Install the replacement thermostat into the housing.

19   Install the thermostat housing cover and tighten the bolts to the torque listed in this Chapter's Specifications.

All models

20   Install the air duct between the air cleaner housing and the throttle body (see Chapter 4).

21   Refill the cooling system (see Chapter 1) and reconnect the battery (see Chapter 5, Section 1).

22   Start the engine and allow it to reach normal operating temperature, then check

3.17 Note the direction of the installed thermostat and which side the O-ring seal is positioned on the thermostat

for leaks and proper thermostat operation (as described in Steps 2 through 4).

---

4   Engine cooling fan - check and replacement

Warning: *Do not work with your hands near the fan any time the engine is running or with the key in the ON position. When the key is turned to the ON position, the fan can start at any time (even when the engine is not running).*

Note: *The fan and shroud may only be available as a complete assembly. Check with your local auto parts store or dealership for parts availability before separating the fan motor from the shroud.*

Check

Refer to illustration 4.2

1   The radiator cooling fan is controlled by the engine management system's Powertrain Control Module (PCM), acting on the information received from the cylinder head tem-

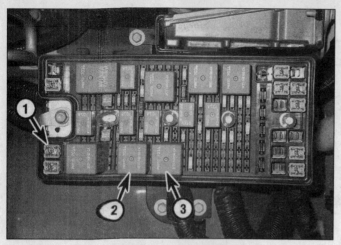

4.2 Cooling fan fuse and relay details:

1   *Cooling fan fuse*
2   *Cooling fan relay (high speed)*
3   *Cooling fan relay (low speed)*

4.8 Cooling fan shroud and motor details:

1   *Cooling fan electrical connectors*
2   *Cooling fan shroud mounting fastener locations*

perature (CHT) or engine coolant temperature (ECT) sensor (see Chapter 6).

2   Warm the engine up until the gauge on the instrument panel indicates the high side of NORMAL. The fan should come on. If it does not, check the cooling fan fuse and relays (see Chapter 12) **(see illustration)**. **Note:** *It's possible that the CHECK ENGINE light may come on indicating a stored diagnostic trouble code (DTC) if there is a problem with the cooling fan circuit. Refer to Chapter 6 regarding the retrieval and definition of any stored codes.*

3   To test the fan motor, unplug the fan electrical connector and use fused jumper wires to connect the fan directly to the battery. If the fan still does not work, replace the motor by removing the fan and its mounting fasteners.

4   If the motor works, the problem is with a temperature sensor or engine management system (see Section 6) or in the wiring (see Chapter 12).

Replacement

Refer to illustration 4.8

5   Disconnect the cable from the negative

battery terminal (see Chapter 5, Section 1).

6   Remove the air duct between the air cleaner housing and the throttle body (see Chapter 4).

7   Unbolt the coolant expansion tank (see Section 5) and the power steering fluid reservoir from the fan shroud and move them aside.

8   Disconnect the fan electrical connector on the fan shroud **(see illustration)**.

9   Remove the fan shroud mounting bolts and then carefully remove the assembly from the engine compartment **(see illustration 4.8)**.

10   Installation is the reverse of removal. Tighten the shroud fasteners securely and, if separated, tighten the fan mounting bolts securely.

5   Coolant expansion tank - removal and installation

Refer to illustration 5.2

Warning: *Wait until the engine is completely cool before beginning this procedure.*

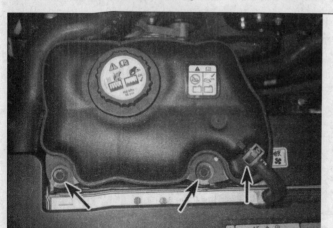

5.2 The expansion tank mounting bolts and upper hose

1   Drain the cooling system (see Chapter 1).

2   Disconnect the upper and lower expansion tank hoses **(see illustration)**. Plug the hoses to prevent leakage.

3   Remove the expansion tank mounting bolts. Lift the tank out of the engine compartment.

4   Clean out the tank with soapy water and a brush to remove any deposits inside. Inspect the reservoir carefully for cracks. If you find a crack, replace the reservoir.

5   Installation is the reverse of removal.

6   Radiator - removal and installation

Removal

Refer to illustrations 6.4, 6.6, 6.8 and 6.9

Warning: *Wait until the engine is completely cool before beginning this procedure.*

1   Disconnect the cable from the negative battery terminal (see Chapter 5, Section 1).

2   Drain the cooling system (see Chapter 1). If the coolant is relatively new and in good condition, save it and reuse it.

3   Remove the lower engine splash shield that is beneath the radiator and attached to the front bumper cover.

4   Remove the radiator cover **(see illustration)**.

5   Remove the cooling fan and fan shroud (see Section 4).

6   Disconnect the upper and lower radiator hoses and the expansion tank hose from the radiator **(see illustration)**. Loosen the hose clamps by squeezing the ends together. Hose clamp pliers work best, but regular pliers will work also. If any hose is stuck, grasp it near the end with a pair of adjustable pliers and twist it to break the seal, then pull it off. If any hose is old or deteriorated, cut it off and install a new one.

6.4 **Remove the plastic fasteners to remove the radiator cover**

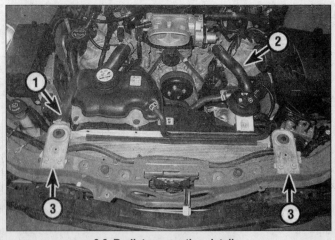

6.6 **Radiator mounting details:**

1	Upper radiator hose	3	Radiator mounting
2	Lower radiator hose		brackets and insulators

7   Remove the radiator mounting brackets and insulators **(see illustration 6.6)**.

8   Remove the brackets for the power steering lines **(see illustration)**.

9   Remove the fasteners that attach the air conditioning condenser brackets to the radiator **(see illustration)**.

10   On models with automatic transmissions, detach the transmission fluid cooler from the radiator and secure it aside (see Chapter 7B).

11   Carefully lift the radiator from the engine compartment.

12   Don't spill coolant on the vehicle or scratch the paint. The rubber insulators that help secure the bottom of the radiator may stick to the radiator when it's removed. They will need to be returned to their original positions during installation.

13   Remove bugs and dirt from the radiator with compressed air and a soft brush. Don't bend the cooling fins. Inspect the radiator for leaks and damage. If it needs repair, have a radiator shop do the work.

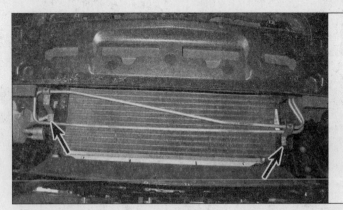

6.8 **Power steering line brackets and fasteners**

Installation

Refer to illustration 6.14

14   Inspect the rubber insulators (where the bottom of the radiator mounts in the engine compartment and in the top mounting brackets) for cracks and deterioration **(see illustration)**. Make sure that they're free of dirt and gravel. When installing the radiator, make

sure that it's correctly seated on the insulators before fastening the top brackets.

15   Installation is otherwise the reverse of the removal procedure. Tighten the radiator mounting bracket bolts to the torque listed in this Chapter's Specifications. After installation, fill the cooling system with the correct mixture of antifreeze and water (see Chap-

6.9 **Air conditioning condenser bracket mounting location (one side shown)**

6.14 **Radiator lower mounting insulator (one side shown)**

ter 1). Reconnect the battery (see Chapter 5, Section 1).

16   Start the engine and check for leaks. Allow the engine to reach normal operating temperature, indicated by the upper radiator hose becoming hot. Recheck the coolant level and add more if required.

7   Water pump - check

Refer to illustration 7.2

1   A failure in the water pump can cause serious engine damage due to overheating.
2   If a failure occurs at the pump seal, coolant will leak from the weep hole(s) on the water pump **(see illustration)**.
3   Using a flashlight, look for traces of coolant residue or dried coolant tracks around the weep hole(s). If the seal has leaked, it should be very apparent.
4   If the water pump shaft bearings fail, there may be a howling sound near the water pump while it's running. With the engine off, shaft wear can be felt if the water pump pulley is rocked up-and-down. Don't mistake

7.2 Water pump weep hole location on a V8 model - V6 models are similar

drivebelt slippage, which causes a squealing sound, for water pump bearing failure.
5   A quick water pump performance check is to turn the heater on. If the pump is failing, it won't be able to efficiently circulate hot water all the way to the heater core as it should.

8.4 Loosen (but don't remove) the water pump pulley mounting bolts with the drivebelt installed

8   Water pump - replacement

Refer to illustration 8.4

1   Disconnect the cable from the negative battery terminal (see Chapter 5, Section 1).
2   Drain the cooling system (see Chapter 1). If the coolant is relatively new and in good condition, save it and reuse it.
3   Remove the air duct between the air cleaner housing and the throttle body (see Chapter 4).
4   Loosen the water pump pulley bolts, but do not remove them yet **(see illustration)**.
5   Remove the drivebelt (see Chapter 1) and the water pump pulley.

V6 models

Refer to illustration 8.7

6   Remove the all hoses from the water pump.
7   Remove the water pump mounting bolts and then remove the pump from the engine **(see illustration)**. If the water pump is stuck, gently tap it with a soft-faced hammer to break the seal.
8   Clean the bolt threads and the threaded holes in the engine removing any corrosion and sealant. Remove all traces of old gasket material from the sealing surfaces.

V8 models

Refer to illustrations 8.9 and 8.10

9   Remove the water pump mounting bolts and then remove the pump **(see illustration)**.
10   Clean the O-ring surfaces on the pump and the housing **(see illustration)**.
11   If you're installing the old pump, install a new O-ring.
12   Clean the bolt threads and the threaded holes in the engine removing any corrosion or debris.

All models

13   Compare the new pump to the old one to make sure that they're identical.
14   On V6 models, apply a thin film of RTV

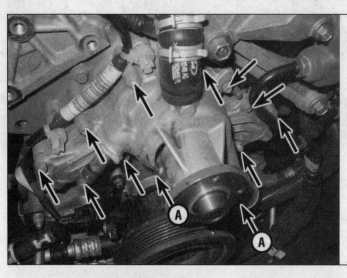

8.7 Water pump mounting bolt locations for the V6 engine (bolts A are hidden from view)

8.9 Water pump mounting bolts for the V8 engine (one hidden from view)

8.10 Inspect the sealing surface in the pump cavity for dirt or signs of pitting

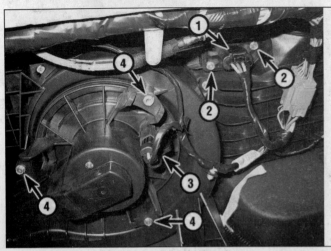

10.1 Blower motor resistor and blower motor details:

1 Blower motor resistor electrical connector
2 Blower motor resistor mounting fasteners
3 Blower motor electrical connector
4 Blower motor mounting fasteners

11.2 Mounting screws for the control assembly

sealant to hold the new gasket in place during installation. **Caution:** *Make sure that the gasket is correctly positioned on the water pump and the engine block surface is clean and free of old gasket material.*
15 Carefully mate the pump with the water pump housing.
16 Install the water pump bolts and tighten them to the torque listed in this Chapter's Specifications. Don't overtighten the water pump bolts; doing so will damage the pump.
17 The remainder of installation is the reverse of removal. Tighten the water pump pulley mounting bolts to the torque listed in this Chapter's Specifications. Refill the cooling system (see Chapter 1). Reconnect the battery (see Chapter 5, Section 1).
18 Operate the engine and thoroughly check for leaks.

9 Coolant temperature indicator - check

Warning: *Wait until the engine is completely cool before beginning this procedure.*
1 The coolant temperature indicator system consists of a temperature gauge on the dash and a sensor mounted on the engine. Depending on which model you're working with, a Cylinder Head Temperature (CHT) sensor (V8 engine) or Engine Coolant Temperature (ECT) sensor (V6 engine) (see Chapter 6), which are information sensors for the Powertrain Control Module (PCM), provide a signal to the PCM which controls and actuates the temperature gauge. **Note:** *Models equipped with a 'Message Center' may also advise that you check the instrument gauges to alert you of a possible malfunction.*
2 If an overheating indication has occurred, first check the coolant level in the system (see Chapter 1) and that the coolant mixture is correct (see Section 2). Also, refer to the *Troubleshooting* section at the beginning of this book before assuming that the temperature indicator is faulty.
3 Start the engine and warm it up for 10 minutes. If the temperature gauge has not

moved from the C position, check the wiring harness connections going to the instrument cluster.
4 If there is a problem with the temperature sensor, it is very likely that the CHECK ENGINE light will be illuminated and the sensor or circuit will need repair (see Chapter 6).

10 Blower motor resistor and blower motor - replacement

Warning: *The models covered by this manual are equipped with Supplemental Restraint systems (SRS), more commonly known as airbags. Always disarm the airbag system before working in the vicinity of any airbag system component to avoid the possibility of accidental deployment of the airbag, which could cause personal injury (see Chapter 12). Do not use a memory saving device to preserve the PCM's memory when working on or near the airbag system components.*

Blower motor resistor

Refer to illustration 10.1
1 Locate the blower motor resistor under the right-side of the instrument panel **(see illustration)**.
2 Disconnect the electrical connector from the blower motor resistor **(see illustration 10.1)**.
3 Remove the blower motor resistor mounting screws and remove the resistor from the housing.
4 Installation is the reverse of removal.

Blower motor

5 Locate the blower motor under the right-hand side of the instrument panel.
6 Disconnect the blower motor electrical connector **(see illustration 10.1)**.
7 Remove the blower motor mounting screws and remove the blower motor **(see illustration 10.1)**.
8 Remove the fan from the blower motor by removing the clip on the end of the shaft.
9 Installation is the reverse of removal.

11 Heater/air conditioner control assembly - removal and installation

Refer to illustration 11.2
Warning: *The models covered by this manual are equipped with Supplemental Restraint systems (SRS), more commonly known as airbags. Always disarm the airbag system before working in the vicinity of any airbag system component to avoid the possibility of accidental deployment of the airbag, which could cause personal injury (see Chapter 12). Do not use a memory saving device to preserve the PCM's memory when working on or near the airbag system components.*
1 Remove the instrument panel center trim bezel (see Chapter 11).
2 With the trim bezel removed, remove the mounting screws for the control assembly and separate it from the bezel **(see illustration)**.
3 Installation is the reverse of removal. Be sure to reconnect all electrical connectors before installing the center trim bezel.

12 Heater core - replacement

Refer to illustrations 12.3, 12.4a, 12.4b and 12.6
Warning 1: *The models covered by this manual are equipped with Supplemental Restraint systems (SRS), more commonly known as airbags. Always disarm the airbag system before working in the vicinity of any airbag system component to avoid the possibility of accidental deployment of the airbag, which could cause personal injury (see Chapter 12). Do not use a memory saving device to preserve the PCM's memory when working on or near the airbag system components.*
Warning 2: *The air conditioning system is under high pressure. DO NOT loosen any fittings or remove any components until after the system has been discharged. Air conditioning refrigerant must be properly discharged into an EPA-approved container at a dealer service department or an automotive air condi-*

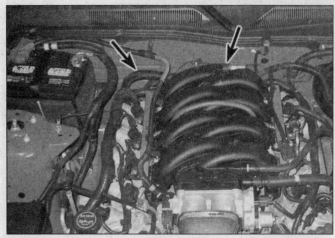

12.3 The heater hoses are routed to the rear of the engine and connect at the firewall

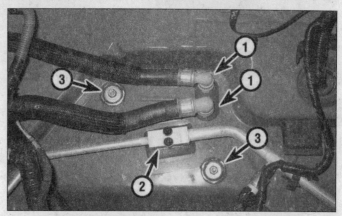

12.4a Evaporator/heater core housing mounting details:

1   *Heater hose fittings*
2   *Refrigerant line fitting to the evaporator (two fasteners)*
3   *Evaporator/heater core housing mounting fasteners*

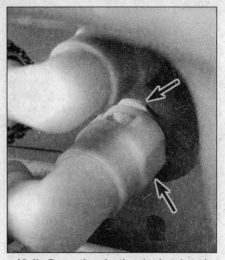

12.4b Press the plastic tabs (retainers) on the heater core pipes to release the hose fittings. Remove the plastic retainers from the pipes and discard them. Use new retainers for installation

12.6 Location of the antenna cable and interior mounting fastener

tioning repair facility. Always wear eye protection when disconnecting air conditioning system fittings.
Warning 3: *Wait until the engine is completely cool before beginning this procedure.*
Note: *At the time of this manual's writing, the manufacturer states that if the heater core is faulty, the evaporator/heater core housing assembly must be replaced; the heater core cannot be serviced individually. It is possible, however, that the aftermarket will manufacture replacement heater cores for installation into the housing.*
1   Have the air conditioning system discharged by a dealer service department or an automotive air conditioning shop before proceeding (see **Warning** above).
2   Remove the instrument panel from the vehicle (see Chapter 11).
3   With the appropriate tool, clamp-off the heater hoses in the engine compartment or

drain the cooling system (see Chapter 1) **(see illustration).**
4   Disconnect the heater hoses and evaporator lines from the evaporator/heater core housing at the firewall in the engine compartment **(see illustrations).** Plug all open lines and fittings to prevent contamination of heating, cooling and air conditioning systems.
5   Remove the evaporator/heater core housing mounting fasteners from the firewall in the engine compartment **(see illustration 12.4a).**
6   Detach the antenna cable and the interior mounting fastener from the housing **(see illustration).**
7   Disconnect any electrical connectors and anything else that may be attached to the housing and carefully pull the housing away from the firewall and then lift it out from the passenger compartment.
8   Installation is the reverse of removal. Use new O-rings for the refrigerant line fitting and lubricate them with refrigerant oil before installing them. Reconnect the battery (see Chapter 5, Section 1).
9   Refill the cooling system (see Chapter 1). Have the air conditioning system re-charged and leak-tested by the shop that discharged it.

13   Air conditioning and heating system - check and maintenance

Refer to illustration 13.1
Warning: *The air conditioning system is under high pressure. Do not loosen any hose fittings or remove any components until after the system has been discharged by a dealer service department or service station. Always wear eye protection when disconnecting air conditioning system fittings.*
1   The following maintenance checks should be performed on a regular basis to ensure the air conditioner continues to operate at peak efficiency.
a) *Check the compressor drivebelt. If it's worn or deteriorated, replace it (see Chapter 1).*
b) *Check the drivebelt tension and, if necessary, adjust it (see Chapter 1).*
c) *Check the system hoses. Look for cracks, bubbles, hard spots and deterioration. Inspect the hoses and all fittings for oil bubbles and seepage. If there's any evidence of wear, damage or leaks, replace the hose(s).*
d) *Inspect the condenser fins for leaves, bugs and other debris. Use a "fin comb"*

13.1 Look for the evaporator drain hose under the vehicle above the transmission on the right side

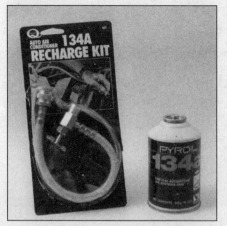

13.7 A basic R-134a charging kit is available at most auto parts stores - it must say R-134a (not R-12) on the kit and on the can of refrigerant

13.10 Always add refrigerant to the low side of the air conditioning system

13.13 Insert a thermometer in the center vent, turn on the air conditioning system and wait for it to cool down; depending on the humidity, the output air should be 30 to 40 degrees cooler than the ambient air temperature

or compressed air to clean the condenser.

e) *Make sure the system has the correct refrigerant charge.*

f) *Check the evaporator housing drain tube* (**see illustration**) *for blockage.*

2   It's a good idea to operate the system for about 10 minutes at least once a month, particularly during the winter. Long term non-use can cause hardening, and subsequent failure, of the seals.

3   Because of the complexity of the air conditioning system and the special equipment necessary to service it, in-depth troubleshooting and repairs are not included in this manual (refer to the *Haynes Automotive Heating and Air Conditioning Repair Manual*).

4   The most common cause of poor cooling is simply a low system refrigerant charge. If a noticeable drop in cool air output occurs, the addition of one can of refrigerant to the system may yield satisfactory results.

Checking the refrigerant charge

5   Warm the engine up to normal operating temperature.

6   Place a thermometer in the dashboard vent nearest the evaporator and operate the system until the indicated temperature is around 40 to 45 degrees F. If the ambient (outside) air temperature is very high, say 110 degrees F, the duct air temperature may be as high as 60 degrees F, but generally the air conditioning is 30-50 degrees F cooler than the ambient air. **Note:** *Humidity of the ambient air also affects the cooling capacity of the system. Higher ambient humidity lowers the effectiveness of the air conditioning system.*

Adding refrigerant

Refer to illustrations 13.7, 13.10 and 13.15

7   Buy an automotive air conditioning system charging kit at an auto parts store. A charging kit includes a can of refrigerant, a tap valve and a short section of hose that can

be attached between the tap valve and the system low side service valve (**see illustration**). Because one can of refrigerant may not be sufficient to bring the system charge up to the proper level, it's a good idea to buy an additional can. Make sure that one of the cans contains red refrigerant dye. If the system is leaking, the red dye will leak out with the refrigerant and help you pinpoint the location of the leak. **Caution:** *There are two types of refrigerant used in automotive systems; R-12 - which has been widely used on earlier models and the more environmentally-friendly R-134a used in all models covered by this manual. These two refrigerants (and their appropriate refrigerant oils) are not compatible and must never be mixed or components will be damaged. Use only R-134a refrigerant in the models covered by this manual.*

8   Hook up the charging kit by following the manufacturer's instructions. **Warning:** *DO NOT attempt to hook the charging kit hose to the system high side! The fittings on the charging kit are designed to fit only on the low side of the system.*

9   Back off the valve handle on the charging kit and screw the kit onto the refrigerant can, making sure first that the O-ring or rubber seal inside the threaded portion of the kit is in place. **Warning:** *Wear protective eyewear when dealing with pressurized refrigerant cans.*

10   Remove the dust cap from the low-side charging connection and attach the quick-connect fitting on the kit hose (**see illustration**).

11   Warm up the engine and turn on the air conditioner. Keep the charging kit hose away from the fan and other moving parts. **Note:** *The charging process requires the compressor to be running. Your compressor may cycle off if the pressure is low due to a low charge. If the clutch cycles off, you can pull the low-pressure cycling switch plug and attach a jumper wire. This will keep the compressor ON.*

12   Turn the valve handle on the kit until the stem pierces the can, then back the handle

out to release the refrigerant. You should be able to hear the rush of gas. Add refrigerant until the compressor discharge line (the small-diameter pipe) feels warm and the compressor inlet pipe (the large-diameter pipe) feels cool. Allow stabilization time between each addition.

13   If you have an accurate thermometer, place it in the center air conditioning vent (**see illustration**) and note the temperature of the air coming out of the vent. A fully-charged system which is working correctly should cool down to about 40 degrees F. Generally, an air conditioning system will put out air that is 30 to 40 degrees F cooler than the ambient air. For example, if the ambient (outside) air temperature is very high (over 100 degrees F), the temperature of air coming out of the registers should be 60 to 70 degrees F.

14   When the can is empty, turn the valve handle to the closed position and release the connection from the low-side port. Replace the dust cap. **Caution:** *Never add more than one can of refrigerant to the system. If more*

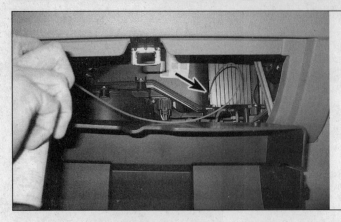

13.21 Remove the glove box (see Chapter 11) and insert the nozzle of the disinfectant through the air recirculation door

refrigerant than that is required, the system should be evacuated and leak tested.

15　Remove the charging kit from the can and store the kit for future use with the piercing valve in the UP position, to prevent inadvertently piercing the can on the next use.

Heating systems

16　If the carpet under the heater core is damp, or if antifreeze vapor or steam is coming through the vents, the heater core is leaking. Remove it (see Section 12) and install a new unit (most radiator shops will not repair a leaking heater core).

17　If the air coming out of the heater vents isn't hot, the problem could stem from any of the following causes:

a) *The thermostat is stuck open, preventing the engine coolant from warming up enough to carry heat to the heater core. Replace the thermostat (see Section 3).*

b) *There is a blockage in the system, preventing the flow of coolant through the heater core. Feel both heater hoses at the firewall. They should be hot. If one of them is cold, there is an obstruction in one of the hoses or in the heater core, or the heater control valve is shut. Detach the hoses and back flush the heater core with a water hose. If the heater core is clear but circulation is impeded, remove the two hoses and flush them out with a water hose.*

c) *If flushing fails to remove the blockage from the heater core, the core must be replaced (see Section 12).*

Eliminating air conditioning odors

Refer to illustration 13.21

18　Unpleasant odors that often develop in air conditioning systems are caused by the growth of a fungus, usually on the surface of the evaporator core. The warm, humid environment there is a perfect breeding ground for mildew to develop.

19　The evaporator core on most vehicles is difficult to access, and dealership service departments have a lengthy, expensive process for eliminating the fungus by opening up the evaporator case and using a powerful dis-

infectant and rinse on the core until the fungus is gone. You can service your own system at home, but it takes something much stronger than basic household germ-killers or deodorizers.

20　Aerosol disinfectants for automotive air conditioning systems are available in most auto parts stores, but remember when shopping for them that the most effective treatments are also the most expensive. The basic procedure for using these sprays is to start by running the system in the RECIRC mode for ten minutes with the blower on its highest speed. Use the highest heat mode to dry out the system and keep the compressor from engaging by disconnecting the wiring connector at the compressor (see Section 14).

21　Make sure that the disinfectant can comes with a long spray hose. Point the nozzle through the air recirculation door; just above the blower motor section of the housing **(see illustration)**, and spray according to the manufacturer's recommendations. Follow the manufacturer's recommendations for the length of spray and waiting time between applications. **Caution:** *Be careful not to let the spray hose get caught in the blower motor fan.*

22　Once the evaporator has been cleaned, the best way to prevent the mildew from coming back again is to make sure your evaporator housing drain tube is clear **(see illustration 13.1).**

14 Air conditioning compressor - removal and installation

Warning: *The air conditioning system is under high pressure. DO NOT loosen any fittings or remove any components until after the system has been discharged. Air conditioning refrigerant should be properly discharged into an EPA-approved container at a dealer service department or an automotive air conditioning repair facility. Always wear eye protection when disconnecting air conditioning system fittings.*

Note: *If you are replacing the compressor due to internal damage, you must also replace the accumulator (see Section 15) and the orifice tube (see Section 18).*

Removal

1　Have the air conditioning system discharged by a dealer service department or by an automotive air conditioning shop before proceeding (see **Warning** above).

2　Remove the drivebelt (see Chapter 1).

V6 models

3　Remove the air duct between the air cleaner housing and the throttle body (see Chapter 4).

4　Raise the front of the vehicle and secure it on jackstands.

5　Disconnect the electrical connector from the compressor clutch field coil.

6　Remove the small brackets mounted to the compressor mounting fasteners.

7　Disconnect the compressor inlet and outlet line manifold from the compressor **(see illustration 14.10)**. Remove and discard the old O-rings.

8　Remove the compressor mounting fasteners and carefully remove the compressor.

V8 models

Refer to illustrations 14.10 and 14.11

9　Raise the front of the vehicle and secure it on jackstands.

10　Disconnect the compressor inlet and outlet line manifold from the compressor **(see**

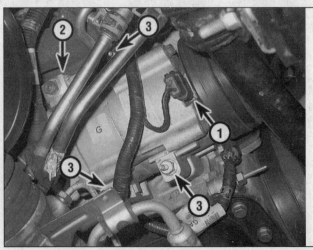

14.10 Air conditioning compressor details: (V8 engine shown, V6 similar)

1　Compressor clutch field coil electrical connector

2　Compressor inlet and outlet line manifold (the mounting bolt is out of view on the top - not the fastener on the side in this photo)

3　Compressor mounting bolt locations

14.11 Power steering line brackets at the steering gear and crossmember

illustration). Remove and discard the old O-rings.

11   Remove the two brackets that secure the power steering line in front of the steering gear **(see illustration)**.

12   Disconnect the crankshaft position sensor (CKP) electrical connector and position the harness aside (see Chapter 6).

13   Disconnect the electrical connector from the compressor clutch field coil **(see illustration 14.10)**.

14   Remove the wiring harness from the compressor mounting fasteners.

15   Remove the compressor mounting fasteners and the compressor **(see illustration 14.10)**.

Installation

16   If a new compressor is being installed, follow the directions with the compressor regarding the draining of excess oil prior to installation.

17   The clutch may have to be transferred from the original compressor to the replacement.

18   Before reconnecting the inlet and outlet lines to the compressor, replace all manifold O-rings and lubricate them with the appropriate refrigerant oil.

19   Installation is otherwise the reverse of

removal. Tighten the compressor line manifold and mounting fasteners to the torque listed in this Chapter's Specifications.

20   Replace the accumulator (see Section 15) and the orifice tube (see Section 18) if necessary (see **Note** above).

21   Have the system evacuated, recharged and leak tested by the shop that discharged it.

15 Air conditioning accumulator - removal and installation

Refer to illustrations 15.2a and 15.2b

Warning: *The air conditioning system is under high pressure. DO NOT loosen any fittings or remove any components until after the system has been discharged. Air conditioning refrigerant should be properly discharged into an EPA-approved container at a dealer service department or an automotive air conditioning repair facility. Always wear eye protection when disconnecting air conditioning system fittings.*

Note: *A spring lock coupling tool is necessary to remove the accumulator.*

1   Have the air conditioning system discharged by a dealer service department or

by an automotive air conditioning shop before proceeding (see **Warning** above).

2   Remove the clips from the refrigerant line couplings and install a spring lock tool to disconnect the lines from the accumulator **(see illustrations)**.

3   Remove and discard the O-rings from the lines. Plug all open lines and fittings to prevent contamination of the air conditioning system.

4   Unscrew the accumulator mounting fasteners, then remove the accumulator **(see illustration 15.2a)**.

5   Install new O-rings on all fittings and coat them with the appropriate refrigerant oil. If you are installing a replacement accumulator, drill a 1/2 inch hole in the accumulator and drain the oil into a measuring cup. Add the measured amount plus 2 ounces of refrigerant oil to the replacement accumulator.

6   Installation is otherwise the reverse of removal.

7   Take the vehicle to the shop that discharged it and have the system evacuated and recharged.

16 Air conditioning condenser - removal and installation

Refer to illustrations 16.4a, 16.4b, and 16.6

Warning: *The air conditioning system is under high pressure. DO NOT loosen any fittings or remove any components until after the system has been discharged. Air conditioning refrigerant should be properly discharged into an EPA-approved container at a dealer service department or an automotive air conditioning repair facility. Always wear eye protection when disconnecting air conditioning system fittings.*

1   Have the air conditioning system discharged by a dealer service department or by an automotive air conditioning shop before proceeding (see **Warning** above).

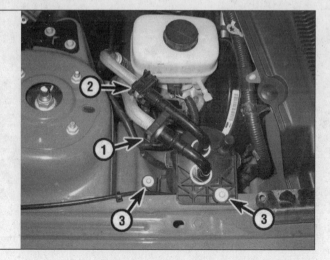

15.2a Air conditioning accumulator details:

1   *Inlet line coupling*
2   *Outlet line coupling*
3   *Mounting fasteners*

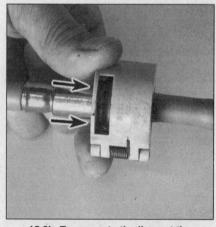

15.2b To separate the lines at the coupling, install the tool, push it towards the female fitting until you can see the coupling through the window in the tool, and then pull the lines apart

16.4a The condenser inlet line fitting is located at the upper left corner of the condenser. Remove the nut for the inlet line fitting to disconnect it

16.4b The condenser outlet line fitting is located at the lower right corner of the condenser. Remove the nut for the outlet line fitting to disconnect it

16.6 Condenser mounting details:

1   *Condenser mounting bracket and insulator (left side shown)*
2   *Air deflector (left side shown)*

2   Raise the front of the vehicle and secure it on jackstands.

3   Remove the lower engine splash shield that is beneath the radiator and attached to the front bumper cover.

4   Disconnect the refrigerant inlet and outlet lines from the condenser by removing the fasteners for the fittings **(see illustrations)**. Plug or cap all open line fittings to prevent dirt or contamination from entering the system.

5   Remove the fasteners for the power steering cooler line brackets and move the line aside **(see illustration 6.8)**.

6   Slide the lower part of the condenser off the mounting studs and pull it down, being careful not to damage the flexible air deflectors and seals mounted on the front and rear sides of the condenser. **Note:** *The upper part of the condenser is seated into a bracket and insulator that is mounted to the top part of the radiator.*

7   If you're going to install a new condenser, pour two ounces of the appropriate refrigerant oil directly into the condenser prior to installing it.

8   When placing the condenser into posi-tion, make certain that the condenser is seated into the upper mounts correctly and that the flexible air deflectors and seals are aligned properly **(see illustration 16.6)**.

9   Before reconnecting the refrigerant lines to the condenser, be sure to coat a pair of new O-rings with the appropriate refrigerant oil and install them in the refrigerant line fit-tings. Tighten the condenser line fittings and mounting fasteners to the torque listed in this Chapter's Specifications.

10   Installation is otherwise the reverse of removal.

11   Have the system evacuated, recharged and leak tested by the shop that discharged it.

---

17   Air conditioning cycling switch, dual function pressure switch and air conditioning pressure transducer - replacement

---

Refer to illustrations 17.1a and 17.1b

Note 1: *All switches and transducers are* threaded onto Schrader valves; therefore, it isn't necessary to discharge the air condition-ing system to remove them.

Note 2: *Transducers are used on V6 engine models only and resemble the dual function pressure switch found on V8 engine models. They are removed in the same way and are at the same location on both models.*

1   Unplug the electrical connector from the switch or transducer **(see illustrations)**.

2   Unscrew the switch or transducer from the Schrader valve with a wrench.

3   Lubricate the switch O-ring with clean refrigerant oil of the correct type.

4   Screw the new switch or transducer onto the threads until hand tight, then tighten it securely with a wrench.

5   Reconnect the electrical connector.

---

18   Air conditioning orifice tube - removal and installation

---

Refer to illustrations 18.2 and 18.3

Warning: *The air conditioning system is under*

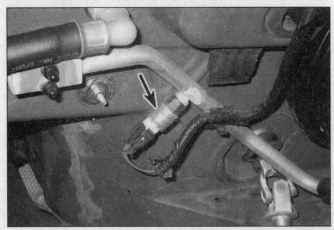

17.1a The air conditioning cycling switch is located on the evaporator outlet line against the firewall

17.1b The air conditioning dual function pressure switch (V8 models) or pressure transducer (V6 models) is located on a refrigerant line near the condenser

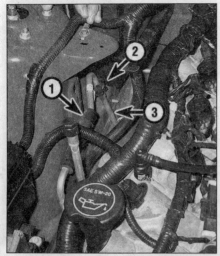

18.2 Orifice tube location details:

1   Condenser-to-evaporator
    refrigerant line
2   Line coupling
3   Evaporator outlet line where orifice
    tube is housed

high pressure. DO NOT loosen any hose fittings or remove any components until the system has been discharged. Air conditioning

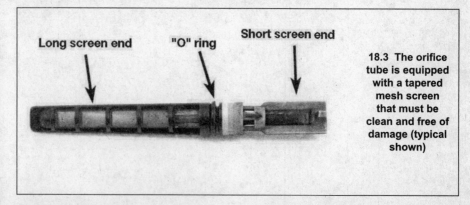

Long screen end   "O" ring   Short screen end

18.3 The orifice tube is equipped with a tapered mesh screen that must be clean and free of damage (typical shown)

refrigerant should be properly discharged into an EPA-approved recovery/recycling unit by a dealer service department or an automotive air conditioning repair facility. Always wear eye protection when disconnecting air conditioning system fittings.

Note 1: *A spring lock coupling tool is necessary to remove the orifice tube.*

Note 2: *The orifice tube changes the high-pressure liquid refrigerant into a low-pressure liquid.*

1   Have the air conditioning system discharged by a dealer service department or by an automotive air conditioning shop before proceeding (see **Warning** above).

2   Remove the line clip and disconnect the condenser to evaporator line with a spring lock tool **(see illustration 15.2b and the accompanying illustration)**.

3   Remove the orifice tube from the line with a pair of needle-nose pliers. If it's really stuck (or breaks off), special extractor tools are available at most auto parts stores **(see illustration)**.

4   Installation is the reverse of removal. Be sure to remove and discard the old O-rings and replace them with new ones.

5   Take the vehicle back to the shop that discharged it. Have the system evacuated, recharged and leak tested.

Notes

Chapter 4
Fuel and exhaust systems

Contents

Specifications

Fuel pressure

Fuel system pressure (at idle)
2005 models	
4.0L V6	35 to 55 psi
4.6L V8	40 to 70 psi
2006 and later models (both engines)	35 to 70 psi
Fuel system hold pressure (after 5 minutes)	Less than 5-psi loss from indicated operating pressure

Torque specifications

Throttle body mounting bolts/nuts (both engines)	89 in-lbs

1 General information

Sequential Multiport Fuel Injection (SFI) system

The fuel system consists of the Powertrain Control Module (PCM), the fuel pump relay, the Inertia Fuel Shutoff (IFS) switch, the Fuel Pump Driver Module (FPDM), the fuel tank, the electric in-tank fuel pump, the fuel level sensor, the fuel rail pressure and temperature sensor, the fuel rail and fuel injectors, the air filter housing and the electronic throttle body. For a more detailed description of the SFI system, refer to Section 10.

Fuel pump circuit
Fuel pump relay

The fuel pump relay is equipped with a primary and secondary voltage circuit. The primary circuit is controlled by the PCM and the secondary circuit is linked directly to battery voltage from the ignition switch. With the ignition switch ON (engine not running), the PCM will ground the relay for one second. During cranking, the PCM grounds the fuel pump relay as long as the Camshaft Position (CMP) sensor sends its position signal. If there are no reference pulses, the fuel pump will shut off after two or three seconds.

Inertia Fuel Shutoff (IFS) switch

The Inertia Fuel Shutoff (IFS) switch disables the fuel pump circuit in the event of a collision. The IFS switch is located behind the left (driver's side) kick panel. Here's how it works: A cylindrical magnet inside the switch has a steel ball sitting on top of it. Under normal driving conditions, the magnetic attraction between the magnet and the ball holds the ball in position on top of the magnet. When a collision occurs, the steel ball breaks away from the magnet, rolls up a conical ramp and strikes a target plate, which opens the switch electrical contacts and the fuel pump circuit. Once the IFS switch is open, you must manually reset it (see Section 3).

Fuel pump

Fuel is circulated from the fuel tank to the fuel injection system through a metal line running along the underside of the vehicle. An electric fuel pump is located inside the fuel tank. The fuel pump assembly consists of the pump, an inlet filter (sometimes referred to as a *sock* or *strainer*), a check valve to maintain pressure after the pump is shut off and a pressure relief valve to protect the pump from overpressurization in the event of a blocked fuel line. But what sets this pump apart from conventional in-tank pumps is its variable speed capability. The PCM controls fuel pressure by controlling the speed (rpm) of the pump. The PCM alters the fuel pressure by controlling the duty cycle of the Fuel Pump Driver Module (FPDM), which in turn controls the speed of the fuel pump by modulating the voltage to the fuel pump.

Exhaust system

The exhaust system includes the exhaust manifolds, the catalytic converters (or catalysts), the mufflers and the exhaust pipes connecting all of these components together.

The catalytic converters are emission-control devices that reduce the three principal tailpipe pollutants: hydrocarbons (HC), carbon monoxide (CO) and oxides of nitrogen (NOx). For more information about how catalytic converters work, refer to Chapter 6.

There are two catalytic converters on all models. The inlet and outlet exhaust pipes are welded onto each catalyst assembly. Each inlet pipe has a flange welded onto its upper end; the inlet pipe flanges are bolted to the exhaust manifolds. The catalyst outlet pipes are welded together into a single assembly. So to remove or replace either catalyst you must remove or replace the entire assembly (Ford calls this assembly the Y-pipe). The Y-pipe is connected to a single pipe (which Ford calls the intermediate pipe). The rest of the exhaust system consists of a muffler and tailpipe welded into one single assembly.

2   Fuel pressure relief procedure

Refer to illustration 2.2

Warning: Gasoline is extremely flammable, so take extra precautions when you work on any part of the fuel system. Don't smoke or allow open flames or bare light bulbs near the work area, and don't work in a garage where a gas-type appliance (such as a water heater or a clothes dryer) is present. Since gasoline is carcinogenic, wear fuel-resistant gloves when there's a possibility of being exposed to fuel, and, if you spill any fuel on your skin, rinse it off immediately with soap and water. Mop up any spills immediately and do not store fuel-soaked rags where they could ignite. The fuel system is under constant pressure, so, if any fuel lines are to be disconnected, the fuel pressure in the system must be relieved first. When you perform any kind of work on the fuel system, wear safety glasses and have a Class B type fire extinguisher on hand.
Note: After the fuel pressure has been relieved, it's a good idea to lay a shop towel over any fuel connection to be disassembled, to absorb the residual fuel that may leak out when servicing the fuel system.

1   Remove the cover from the engine compartment fuse and relay box.
2   Remove the fuel pump relay **(see illustration)** from the fuse and relay box.

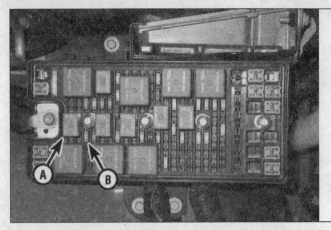

2.2 To relieve fuel pressure, remove the fuel pump relay (A), start the engine and wait for it to stall (it might not even start at all). If you're troubleshooting a non-operational fuel pump, always check the fuel pump relay and the fuel pump fuse (B) before checking for a wiring problem

3   Start the engine and allow it to idle until it stalls.
4   After the engine stalls, crank the engine for about 5 seconds to make sure fuel pressure in the fuel rail is released.
5   Turn the ignition switch to OFF, then disconnect the cable from the negative terminal of the battery (see Chapter 5, Section 1) before performing any work on the fuel system.

3   Fuel pump/fuel pressure - check

Warning: Gasoline is extremely flammable, so take extra precautions when you work on any part of the fuel system. See the **Warning** in Section 2.

Preliminary checks

1   If you suspect insufficient fuel delivery check the following items first:
  a) Check the battery and make sure that it's fully charged (see Chapter 5).
  b) Check the fuel filter for obstructions (see Chapter 1).
  c) Inspect the fuel line and quick-connect fittings (see Section 4). Verify that the problem is not simply a leak in a line.
2   Verify that the fuel pump actually runs. Remove the fuel filler cap and have an assistant turn the ignition switch to ON while you listen carefully for the sound of the fuel pump operating. You should hear a brief whirring

noise (for about one second) as the pump comes on and pressurizes the system. If the fuel pump makes no sound, check the fuel pump relay fuse, the Inertia Fuel Shutoff (IFS) switch (see Step 5) and the fuel pump relay.

Fuel pump electrical circuit check

3   If the pump does not turn on, check the fuel pump relay fuse and the fuel pump relay **(see illustration 2.2)**, both of which are located in the engine compartment fuse and relay box.
4   If the fuse and relay are good, but the fuel pump still doesn't operate, inspect the fuel pump circuit.

Resetting the Inertia Fuel Shutoff (IFS) switch

Refer to illustration 3.7

Note: The IFS switch is located behind the left kick panel. In the event that the fuel pump doesn't operate, the fuel pump fuse and the IFS switch should always be checked first.

5   To access the IFS switch, remove the left kick panel (see Chapter 11).
6   Make sure the ignition switch is turned to OFF.
7   Reset the IFS switch by pushing the large button on top of the switch **(see illustration)**.
8   Run the test in Step 2 again. If the pump still doesn't run, check the fuel pump relay, the fuel pump wiring harness for a loose connector or damaged wiring, or have the fuel pump

3.7 To reset the Inertia Fuel Shutoff (IFS) switch, depress the large button on top

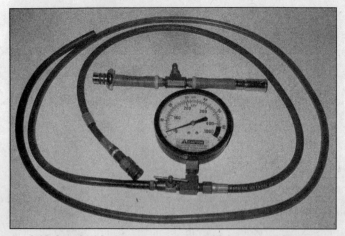

3.11 A typical fuel pressure gauge, with hoses and fittings suitable for tee-ing into the fuel system between the fuel delivery line and the fuel rail

3.13 Tee into the fuel system at the spring-lock coupling between the fuel supply line and the fuel rail (if you're unfamiliar with this type of coupling, refer to Section 4)

circuit diagnosed by an automotive service technician.

9   If the fuel pump runs, but a fuel system problem persists, proceed to the fuel pump pressure check.

Fuel pump pressure check

Refer to illustrations 3.11 and 3.13

Note: *Before proceeding, obtain a fuel pressure gauge capable of measuring fuel pressure well above the specified operating range of the fuel system that you're going to test, and you'll also need fittings suitable for tee-ing the gauge into the fuel system between the fuel delivery line and the fuel rail.*

10   Relieve the fuel system pressure (see Section 2).

11   In addition to a fuel pressure gauge capable of reading fuel pressure up to 50 psi, you'll need a hose and an adapter suitable for tee-ing into the fuel system at the quick-connect fitting between the fuel delivery hose and the fuel rail **(see illustration)**.

12   Disconnect the quick-connect fitting at the connection between the fuel delivery hose and the fuel rail (if you're unfamiliar with quick-connect fittings, refer to Section 4).

13   Tee in the fuel pressure gauge between the fuel delivery hose and the fuel rail **(see illustration)**.

14   Turn off all the accessories, then start the engine and let it idle. The fuel pressure should be within the operating range listed in this Chapter's Specifications. If the pressure reading is within the specified range, the system is operating correctly.

15   If the fuel pressure is higher than specified, the fuel rail pressure and temperature sensor, the Fuel Pump Driver Module (FPDM), the Powertrain Control Module (PCM) or the circuit connecting these components is probably defective. But checking this circuit is beyond the scope of the home mechanic, so have the circuit checked by a professional.

16   If the fuel pressure is lower than specified, inspect the fuel delivery lines and hoses for an obstruction or a kink. Also inspect all

fuel delivery line and hose quick-connect fittings for leaks. Replace the fuel filter (see Chapter 1) and re-check the pressure. If the lines, hoses, connections and the fuel filter are all in good shape, remove the fuel pump module (see Section 5) and inspect the fuel pump inlet strainer for restrictions. If everything else is okay, replace the fuel pump.

17   Turn the ignition switch to OFF, wait five minutes and recheck the pressure on the gauge. Compare the reading with the hold pressure listed in this Chapter's Specifications. If the hold pressure is less than specified:

a) *The fuel delivery line or a quick-connect fitting might be leaking.*

b) *A fuel injector (or injectors) may be leaking.*

c) *The fuel pump might be defective.*

18   After the testing is complete, relieve the fuel pressure (see Section 2), remove the fuel pressure gauge and reconnect the fuel delivery line to the fuel rail (see Section 4 if you're unfamiliar with quick-connect fittings).

---

4   Fuel lines and fittings - general information

Refer to illustration 4.3

Warning: *Gasoline is extremely flammable, so take extra precautions when you work on any part of the fuel system. See the* **Warning** *in Section 2.*

1   Always relieve the system fuel pressure before servicing fuel lines or fittings (see Section 2).

2   The fuel supply line extends from the fuel tank to the engine compartment. The EVAP purge line extends from the EVAP canister, which is located behind the fuel tank, up to the purge valve, which is located on the firewall in the engine compartment. Anytime you raise the vehicle for underbody service, inspect the lines underneath the vehicle for leaks, kinks and dents.

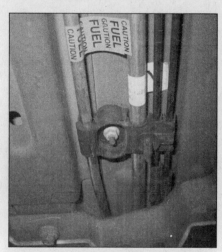

4.3 A typical plastic clip for the fuel delivery and EVAP lines

3   The fuel and EVAP lines are secured to the underbody with plastic and metal clips **(see illustration)**. Whenever the vehicle is raised, inspect thees clips for damage, replacing them as necessary.

Steel tubing

4   If it's necessary to replace a fuel line or EVAP line, use steel tubing that complies with the manufacturer's specifications, or its equivalent.

5   Don't use copper or aluminum tubing to replace steel tubing. These materials cannot withstand normal vehicle vibration.

6   Because steel fuel lines are under high pressure when the engine is running, they require special consideration:

a) *Inspect all O-rings for cuts, cracks and deterioration. If an O-ring is torn, cracked, hardened or otherwise damaged, replace it.*

b) *If the lines are replaced, always use original equipment parts, or parts that meet the original equipment standards specified in this Section.*

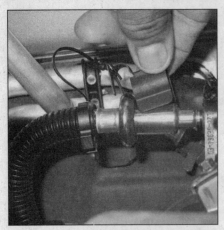

4.17 Remove the safety clamp from the spring lock coupling

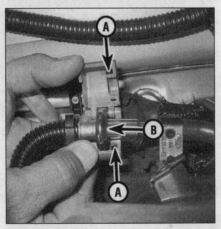

4.19a Place the spring-lock coupling tool around the coupling and close it so the lip (A) of the tool is flush against the garter spring housing (B) . . .

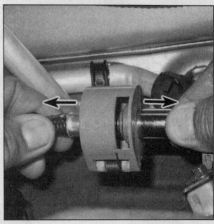

4.19b . . . push the spring-lock coupling tool toward the garter housing until the garter spring is disengaged, then pull the fuel lines apart

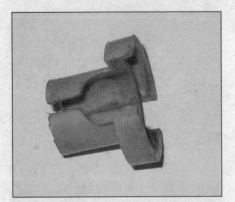

4.19c This type of spring-lock coupling tool can be used in places where access to the fitting is limited. To use it . . .

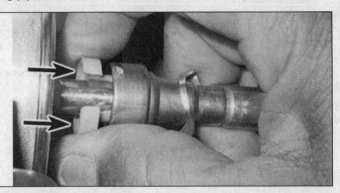

4.19d . . . slip it over the line, press it into the coupling and pull the lines apart

c)   *Never allow metal lines to chafe against the frame. Maintain a minimum of 1/4-inch clearance around a line to prevent contact with the frame.*

7   If you find dirt in the system during disassembly, disconnect the fuel supply line and then blow it out with compressed air. And be sure to inspect the fuel filter (see Chapter 1) and the fuel pump inlet strainer for contamination (see Section 5).

Flexible hoses

Caution: *Use only original equipment replacement hoses or their equivalent. Unapproved hose material might fail when subjected to the fuel pressure at which this system operates.*

8   Don't route fuel hose within four inches of any part of the exhaust system or within ten inches of the catalytic converter. Never allow rubber hoses to chafe against the frame. Maintain a minimum of 1/4-inch clearance around a hose to prevent contact with the frame.

9   If a hose is equipped with quick-connect fittings, the quick-connect fittings cannot be serviced separately. If the fitting or hose is damaged, replace the entire fuel hose assembly. Do not attempt to repair fuel hoses.

Replacing fuel lines and hoses and EVAP lines

10   If a fuel line or hose or an EVAP line is damaged, replace it with factory replacement parts. Do not substitute fuel lines or hoses or EVAP lines of inferior quality. They might not be suitable for, and might fail when subjected to, the operating pressure of this system.

11   Always relieve the fuel system pressure before replacing fuel or EVAP lines (see Section 2).

12   Always disconnect the cable from the negative terminal of the battery (see Chapter 5, Section 1) before replacing fuel or EVAP lines.

13   Remove all clips that secure the fuel or EVAP line to the vehicle body. Pay close attention to all clips; they not only secure the fuel line and hoses, they also route them correctly. The hoses and line must be reattached to their respective clips when reassembled.

14   Be sure to use the correct tool and the correct procedure when disconnecting any fuel or EVAP couplings or fittings.

Disconnecting and connecting fuel line and EVAP line fittings

Spring-lock couplings

Refer to illustrations 4.17, 4.19a, 4.19b, 4.19c and 4.19d

15   The fuel supply line utilizes spring lock

couplings at some connections such as the connection between the fuel supply line and the fuel rail. The male side of the coupling, which is sealed by two O-rings, is simply the end of a fuel line with a flared end. The male side of the coupling is inserted into the female side of the coupling, which is secured by a garter spring that prevents unintentional disconnection by gripping the flared end of the male side of the coupling. On some of these fittings on some models, a safety clamp provides additional security. These clamps are often tethered to the female side of the coupling so that you don't lose them while the coupling is disconnected.

16   BEFORE DISCONNECTING SPRING LOCK COUPLINGS, ALWAYS RELIEVE SYSTEM FUEL PRESSURE (see Section 2), then DISCONNECT THE CABLE FROM THE NEGATIVE BATTERY TERMINAL (see Chapter 5, Section 1).

17   Remove the safety clamp **(see illustration)**.

18   If you're using a clamshell-type tool, install the tool over the coupling. Other types of release tools simply fit over the fuel line. **Note:** *These tools are available at most auto parts stores. They come in a variety of sizes, so it's a good idea to purchase a set of them to be sure you'll have the right size.*

19   Push the coupling tool firmly toward the garter spring housing to disengage the garter spring from the flared end of the female side

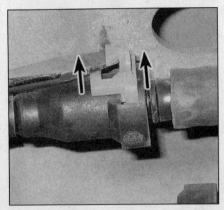

4.30 Using a screwdriver, carefully lever the locking tab up to its released position

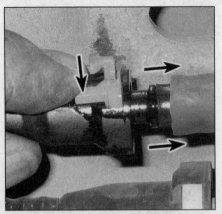

4.31 Press down on the corners of the locking tab to release it, push the fuel line into the coupling to release it from the retainer, then pull the fuel line out of the coupling

4.32 Inspect the condition of the O-ring inside the coupling. If the O-ring is cracked, torn or deteriorated, replace the coupling

of the connection **(see illustrations)**. Then pull the two fuel lines apart to disengage the tool from the garter spring, open and remove the tool and disconnect the lines.

20   Before reconnecting the coupling, wipe off the ends of both fuel lines with a clean cloth. Inspect the condition of the O-rings and the garter spring. If either O-ring or the garter spring is damaged or worn, replace it. Also inspect the inside of the female side of the fitting and make sure it's clean.

21   Lubricate the O-rings with some clean engine oil, then press the two sides of the connection together until the flared end of the male side of the fitting is locked into place by the garter spring. Pull on the coupling to verify that it's fully engaged.

22   When you have verified that the coupling is reconnected, install the safety clamp.

23   Reconnect the cable to the negative battery terminal (see Chapter 5, Section 1).

24   Turn the ignition key to ON (*not* START) and re-pressurize the fuel system, which will take a moment, then check for fuel leakage around the coupling. If there are no signs of leaks, start the engine and check again.

Quick-connect couplings

25   Besides spring-lock couplings, the vehicles covered by this manual also use various types of quick-connect couplings to connect fuel lines, EVAP lines and PCV lines. After you locate the coupling, determine which type it is, then use whichever of the following procedures that applies to that particular style of coupling.

26   There are several types of quick-connect couplings in use on these vehicles. If you're in need of a replacement, take it with you to the auto parts store or dealer parts department. If you show the counterperson the coupling, retainer and/or O-ring and tell them the location of the coupling, they will be able to determine which part you need.

27   Also, at the time of publication, the O-rings inside the following quick-connect couplings were not available separately from the coupling themselves. So if an O-ring is damaged, you must *replace the coupling* (which also includes the fuel hose or line to which it's permanently attached). And if one

of these couplings is damaged, you cannot repair it. So, whether you're replacing a coupling because the O-ring is damaged or because the coupling itself is damaged, you must replace the coupling, and the fuel line to which it's permanently connected, as a single assembly.

28   Always relieve the system fuel pressure before disconnecting fuel line couplings (see Section 2).

29   Always disconnect the cable from the negative battery terminal before disconnecting fuel lines (see Chapter 5, Section 1).

Type 1

Refer to illustrations 4.30. 4.31, 4.32 and 4.33

Note: *The retainer clip for a Type 1 quick-connect coupling can be replaced separately, but the rest of the assembly (the coupling, O-ring and fuel hose or line to which the coupling is attached, must be replaced as a single assembly).*

30   Pull up the locking tab to the RELEASE position **(see illustration)**.

31   Press down the corners of the locking tab to release it, push the fuel or EVAP line into the coupling to disengage it from the retainer, then pull the fuel line out of the coupling **(see illustration)**.

32   Inspect the condition of the coupling and the O-ring inside **(see illustration)**. If the

coupling is damaged or if the O-ring inside the coupling is cracked, torn or deteriorated, replace the coupling.

33   Inspect the condition of the retainer **(see illustration)**. If the retainer is damaged, remove it from the coupling and install a new retainer. (It's not a bad idea to replace the retainer anytime that you disconnect this type of coupling because retainers are inexpensive but critical parts.)

34   Before reconnecting the coupling to the fuel or EVAP line, apply a film of clean engine oil to the end of the fuel line.

35   To connect the coupling, insert the fuel or EVAP line into the coupling until the fuel line clicks into place.

36   Verify that the coupling is fully connected by trying to pull the fuel or EVAP line and the coupling apart. If the coupling is fully connected, press down the locking tab to its LOCKED position.

Type II

Refer to illustration 4.37

Note: *This type of coupling has a brightly-colored, slender release button running along one side of the coupling body. You'll find this type of coupling used on both fuel and EVAP line connections anywhere from the fuel tank to the fuel and EVAP lines running under the vehicle. It's mainly used to connect EVAP lines together and to connect them to com-*

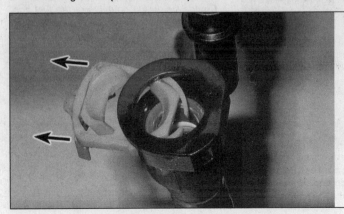

4.33 Inspect the condition of the retainer. If it's damaged, pull it out of the coupling, discard it and install a new retainer

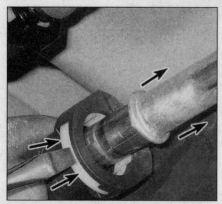

4.37 To disconnect a fuel line from this type of quick-connect coupling, depress the release button, then pull the fuel line out of the coupling

4.42a To disconnect this type of coupling, locate the locking tab on the side of the coupling . . .

4.42b . . . press it down firmly with the tip of a screwdriver and pull off the coupling

ponents such as the EVAP canister. If the coupling, the O-rings inside the coupling or the fuel hose or line to which the coupling is attached is damaged, they must be replaced as a single assembly.

37   Press the button on the coupling and pull the fuel or EVAP line out of the coupling **(see illustration)**.

38   Inspect the condition of the quick-connect coupling and the O-rings inside the coupling. If the coupling is damaged or if the O-rings are cracked, torn or deteriorated, replace the coupling and the EVAP line to which it's attached.

39   Apply a film of clean engine oil to the fuel or EVAP line and to the O-rings inside the coupling.

40   To reconnect this type of coupling, simply insert the fuel or EVAP line into the coupling until it clicks into place.

41   Verify that the coupling is properly connected by trying to pull the coupling and the fuel or EVAP line apart.

Type III

Refer to illustrations 4.42a and 4.42b

Note: *This type of coupling has a locking tab*

on the side of the coupling body. It's used on some models to connect the PCV fresh air inlet and crankcase ventilation lines to the valve covers, the PCV valve and the intake manifold. If the coupling, the O-rings inside the coupling or the PCV line to which the coupling is attached is damaged, they must be replaced as a single assembly.

42   Depress the locking tab on the side of the coupling body and disconnect the coupling **(see illustrations)**.

43   Inspect the condition of the quick-connect coupling and the O-rings inside the coupling. If the coupling is damaged or if the O-rings are cracked, torn or deteriorated, replace the coupling and the EVAP line to which it's attached.

44   Apply a film of clean engine oil to the O-rings inside the coupling.

45   To reconnect this type of coupling, release the locking tab by pushing it down again, then push the coupling onto the PCV pipe until the coupling clicks into place.

46   Verify that the coupling is properly connected by trying to pull the coupling and the EVAP line apart.

Type IV

Refer to illustration 4.47

Note: *This is another type of EVAP line quick-connect coupling that you'll find on some models. It's used to connect the EVAP purge line to the EVAP canister purge valve and to some other EVAP components. If the coupling, the O-rings inside the coupling or the fuel hose or line to which the coupling is attached is damaged, they must be replaced as a single assembly.*

47   To disconnect this type of quick-connect coupling, depress the two release tabs on the top and bottom of the fitting and pull it off the purge valve pipe (shown) or EVAP line **(see illustration)**.

48   Inspect the condition of the quick-connect coupling and the O-rings inside the coupling. If the coupling is damaged or if the O-rings are cracked, torn or deteriorated, replace the coupling and the EVAP line to which it's attached.

49   Apply a film of clean engine oil to the

EVAP line and to the O-rings inside the coupling.

50   To reconnect this type of coupling, simply insert the EVAP line into the coupling, or push the coupling onto the EVAP line, until the line clicks into place.

51   Verify that the coupling is properly connected by trying to pull the coupling and the EVAP line apart.

5   Fuel pump module - removal and installation

Refer to illustrations 5.4, 5.5, 5.6, 5.7, 5.8 and 5.9

Warning: *Gasoline is extremely flammable, so take extra precautions when you work on any part of the fuel system. See the* **Warning** *in Section 2.*

1   Relieve the system fuel pressure (see Section 2).

2   Disconnect the cable from the negative battery terminal (see Chapter 5, Section 1).

3   Remove the rear seat cushion (see Chapter 11).

4   Remove the fuel pump module access cover, which is the one on the left **(see illustration)**.

5.4 To remove the access plate for the fuel pump module (or for the fuel level sensor), carefully pry it loose from the floorpan

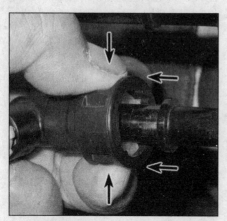

4.47 To disconnect this type of quick-connect coupling, depress the two release tabs on the top and bottom of the coupling and pull it off the pipe

5.5 Disconnect the fuel pump module
electrical connector

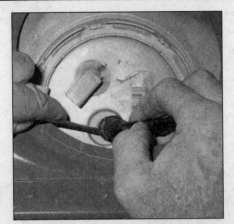

5.6 Disconnect the (Type II) fuel pump
module quick-connect coupling (see
illustration 4.37 for more information)

5.7 Using a hammer and a *brass* punch,
loosen the fuel pump lock ring by turning
it counterclockwise

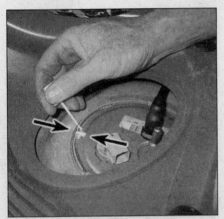

5.8 Mark the position of the fuel
pump module assembly in relation to
the fuel tank

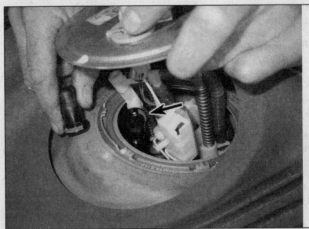

5.9 Before you can lift
the fuel pump module
out of the fuel tank, you
must disconnect the
(Type II) quick-connect
coupling that connects
the crossover tube to
the pump assembly
(see illustration 4.37 for
more information)

5   Disconnect the fuel pump module electrical connector **(see illustration)**.
6   Disconnect the fuel supply line quick-connect fitting **(see illustration)**.
7   Using a hammer and a *brass* punch (do NOT use a steel punch!), loosen the fuel pump module lock ring **(see illustration)**. When the lock ring is loose enough, fully unscrew it.
8   Before removing the fuel pump module from the fuel tank, mark the relationship of the module to the tank **(see illustration)**.
9   Lift up the fuel pump module assembly far enough to access the quick-connect coupling for the crossover tube connected to the underside of the fuel pump assembly **(see illustration)**. Disconnect the crossover tube's quick-connect fitting from the pump. If you're unfamiliar with this type of fitting, **see illustration 4.37**.
10   After disconnecting the crossover tube, carefully lift the fuel pump module assembly out of the fuel tank.
11   Remove the old O-ring type seal **(see illustration 6.10)** and discard it.
12   If you're planning to reinstall the original fuel pump unit, inspect the condition of the fuel inlet strainer. The strainer is a permanent part of the pump and cannot be removed. If

it's only slightly dirty, try scrubbing it gently with a small brush, then rinse it off with clean solvent. But if the strainer is so dirty that it's obstructed - and you're unable to clean it - you must replace the pump.
13   Clean the fuel pump mounting flange and the tank mounting surface, particularly the area where the O-ring type seal is installed. Be sure to use a new O-ring type seal and apply a thin coat of heavy grease to the new seal ring to hold it in place while installing the fuel pump module.
14   After installing the fuel pump assembly in the fuel tank, make sure the alignment arrows on the pump mounting flange and on the top of the fuel tank are aligned **(see illustration 5.8)**.
15   Installation is otherwise the reverse of removal.

6   **Fuel level sensor - removal and installation**

Refer to illustrations 6.5, 6.8 and 6.10
Warning: *Gasoline is extremely flammable, so take extra precautions when you work on any part of the fuel system. See the* **Warning** *in Section 2.*

1   Relieve the system fuel pressure (see Section 2).
2   Disconnect the cable from the negative battery terminal (see Chapter 5, Section 1).
3   Remove the rear seat cushion (see Chapter 11).
4   Remove the fuel level sensor access cover, which is the cover on the right **(see illustration 5.4)**.
5   Disconnect the fuel level sensor electrical connector **(see illustration)**.

6.5 Disconnect the electrical connector
from the fuel level sensor

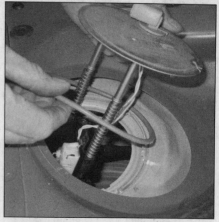

6.8 Before you can lift the fuel level sensor out of the fuel tank, you must disconnect the (Type II) quick-connect coupling that connects the crossover tube to the sensor assembly (see illustration 4.37 for more information)

6.10 Remove and discard the old O-ring type seal for the fuel level sensor

6   Using a hammer and a *brass* punch (do NOT use a steel punch!), loosen the fuel level sensor lock ring **(see illustration 5.7)**. When the lock ring is loose enough, fully unscrew it.

7   Before removing the fuel level sensor from the fuel tank, mark the relationship of the sensor to the tank **(see illustration 5.8)**.

8   Lift up the fuel level sensor far enough to access the quick-connect coupling for the crossover tube connected to the underside of the sensor assembly **(see illustration)**. Disconnect the crossover tube's quick-connect fitting from the pump. If you're unfamiliar with this type of fitting, **see illustration 4.37**.

9   After disconnecting the crossover tube, carefully lift the fuel level sensor out of the fuel tank. Angle the sensor assembly as necessary to protect the float and the float arm from damage.

10   Remove the old O-ring type seal **(see illustration)** and discard it.

11   Clean the fuel sensor mounting flange and the tank mounting surface, particularly the area where the O-ring type seal is installed. Be sure to use a new O-ring type seal and apply a thin coat of heavy grease to the new seal ring to hold it in place while installing the fuel sensor.

12   After installing the fuel sensor in the fuel tank, make sure the alignment arrows on the sensor mounting flange and on the top of the fuel tank are aligned **(see illustration 5.8)**.

13   Installation is otherwise the reverse of removal.

7   Fuel tank - removal and installation

Refer to illustrations 7.12, 7.13, 7.15 and 7.16

Warning 1: *Gasoline is extremely flammable, so take extra precautions when you work on any part of the fuel system. See the* **Warning** *in Section 2.*

Warning 2: *Don't begin this procedure until the gauge indicates that the tank is empty or nearly empty. If the tank must be removed when it's full (for example, if the fuel pump malfunctions), siphon any remaining fuel from the tank prior to removal.*

1   Relieve the system fuel pressure (see Section 2).

2   Disconnect the cable from the negative terminal of the battery (see Chapter 5, Section 1).

3   Unless the vehicle has been driven far enough to completely empty the tank, it's critical to siphon the residual fuel out before removing the tank from the vehicle because the tank is quite heavy when filled with fuel. You can siphon some, but not all, of the fuel from the tank through the fuel filler neck hose (see next Step). **Warning:** *Do NOT start the siphoning action by mouth! Use a siphoning kit (available at most auto parts stores).*

4   Remove the fuel tank filler neck cap and insert a siphon hose into the filler neck. Push the hose down all the way until it bottoms out inside the fuel tank, then siphon out as much fuel as possible into an approved fuel container. Because of the shape of the fuel tank, you probably won't be able to siphon all the fuel from the tank through the fuel filler neck. You might encounter some resistance when the siphoning hose reaches the fuel tank because it must go through the filler pipe check valve, but the hose will push past the check valve as long as the hose diameter isn't too big. When you're done siphoning the fuel out of the tank, leave the siphon hose in place. Don't attempt to remove it until you have removed the fuel tank, because when you pull up on the siphoning hose it will snag at the spring-loaded check valve and could damage the valve if you use force to pull out

the hose. Siphon out the rest of the fuel when you get to Step 9.

5   Remove the rear seat cushion (see Chapter 11).

6   Remove the fuel pump module and fuel level sensor access covers (see Sections 5 and 6).

7   Disconnect the electrical connector and the fuel tube quick-connect coupling from the fuel pump module **(see illustrations 5.5 and 5.6)**.

8   Disconnect the electrical connector from the fuel level sensor **(see illustration 6.5)**.

9   Remove the fuel pump module (see Section 5) and the fuel level sensor (see Section 6) and insert a siphon hose through each opening and siphon out as much fuel as possible. **Warning:** *If the fuel inside the tank were to slosh around while lowering the tank and destabilize it, you could drop the tank and damage it, and you could incur serious injuries to yourself.* When you're done, install the fuel pump module and the fuel level sensor in the tank (see Sections 5 and 6, respectively).

10   Raise the rear of the vehicle and support it securely on jackstands.

11   Remove the intermediate exhaust pipe (see Section 14). Remove the driveshaft (see Chapter 8). If you're working on a convertible model, remove the rear support braces (see Chapter 10).

12   Detach the pin-type taillight harness retainers from the outer edge of the fuel tank

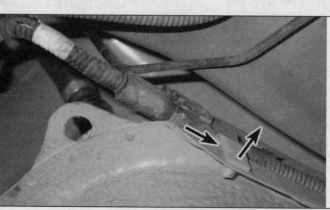

7.12 To detach these pin-type taillight harness retainers from the edge of the fuel tank, insert a screwdriver between the tank flange and the clip and pry the clip loose from the flange

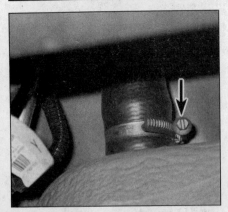

7.13 To disconnect the fuel filler neck hose from the filler neck pipe, loosen this hose clamp and pull off the hose

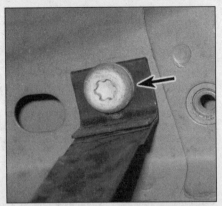

7.15 To detach the fuel tank straps, remove the tank strap bolts

7.16 Lower the tank just enough to disconnect this EVAP tube quick-connect coupling (if you're unfamiliar with quick-connect couplings, refer to Section 4)

(see illustration).

13   Loosen the hose clamp screw **(see illustration)** and disconnect the fuel filler neck hose from the fuel tank.

14   Support the fuel tank securely. Have at least one assistant standing by to help out until the tank is on the floor. **Warning:** *Do NOT attempt to lower the fuel tank by yourself. You could be seriously injured in the event that the tank gets out of control and falls on you.*

15   Remove the two fuel tank strap retaining bolts **(see illustration)**.

16   Lower the tank just far enough to access the fuel tank EVAP tube quick-connect fitting **(see illustration)**. Disconnect this fitting. If you're unfamiliar with quick-connect fittings, see Section 4.

17   Remove the fuel tank strap retainer pins and straps.

18   With the help of at least one assistant to help you steady the tank, carefully lower the tank to the floor.

19   If you're replacing the tank, or having it cleaned or repaired, refer to Section 8.

20   Installation is the reverse of removal.

8   Fuel tank - cleaning and repair

1   All repairs to the fuel tank or filler neck should be carried out by a professional who has experience in this critical and ptentially dangerous work. Even after cleaning and flushing of the fuel tank, explosive fumes can remain and ignite during repair of the tank.

2   If the fuel tank is removed from the vehicle, it should not be placed in an area where sparks or open flames could ignite the fumes coming out of the tank. Be especially careful inside a garage where a gas-type appliance is located, because it could cause an explosion.

9   Air intake duct and air filter housing - removal and installation

Air intake duct

Refer to illustrations 9.1 and 9.2

1   Disconnect the PCV fresh air inlet hose (crankcase breather hose) from the air intake

duct **(see illustration)**. If you're unfamiliar with a Type III quick-connect coupling, **see illustrations 4.42a and 4.42b**. Also loosen the hose clamp at the throttle body end of the air intake duct and disconnect the duct from the throttle body.

2   Loosen the hose clamp at the air filter housing end of the air intake duct **(see illustration)**, disconnect the intake duct from the filter housing and remove it.

3   Installation is the reverse of removal.

Air filter housing

4   Disconnect the air intake duct from the air filter housing **(see illustration 9.2)**.

5   Disconnect the electrical connector from the MAF sensor.

6   Remove the air filter housing retaining bolt and remove the filter housing. When you pull up on the air filter housing you will feel a slight resistance; that's the locator pin and grommet on the underside of the filter housing. Grasp the filter housing firmly and lift it straight up and the grommet will disengage.

9.1 To disconnect the air intake duct from the throttle body, disconnect the PCV fresh air inlet hose (crankcase breather hose) Type III quick-connect coupling (A), then loosen the hose clamp screw (B) and pull the duct off the throttle body (V8 model shown, V6 models similar)

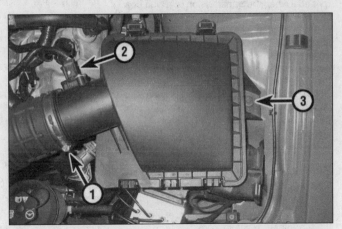

9.2 To disconnect the air intake duct from the air filter housing, loosen the hose clamp screw (1) and pull the duct off the air filter housing. To remove the air filter housing, disconnect the air intake duct from the housing, then disconnect the electrical connector (2) from the MAF sensor and remove the filter housing retaining bolt (3)

7   Inspect the condition of the rubber grommet for the filter housing locator pin. If the grommet is cracked, torn or deteriorated, replace it.

8   If you want to disassemble the air filter housing to replace or inspect the air filter element, refer to Chapter 1. If you want to disassemble the air filter housing to replace the Mass Air Flow (MAF) sensor, refer to Chapter 6.

9   Installation is the reverse of removal.

10   Sequential Multiport Fuel Injection (SFI) system - general information

Sequential Multiport Fuel Injection (SFI) system

All models are equipped with a Sequential Multiport Fuel Injection (SFI) system. The Powertrain Control Module (PCM) controls the fuel injectors, which inject fuel directly into the intake port of each cylinder in the engine firing order. When the engine is running, the PCM constantly monitors an array of engine operating conditions (cylinder head temperature, intake air temperature, mass of air moving through the intake system, engine rpm, load, etc.) and alters the pulse width of the injectors accordingly, delivering the right amount of fuel into the intake ports where it mixes with incoming air.

Electronic Throttle Control (ETC) system

In a conventional induction system with an accelerator cable, the position of the throttle plate inside the throttle body is determined by the position of the accelerator pedal, which is determined by your foot, isn't always appropriate to the prevailing operating conditions. For example, if you mash the accelerator pedal when the transmission is in high gear and the vehicle is cruising down the freeway under no load, it takes a moment for the PCM to downshift the transmission and spin up the engine speed so that it can respond to your new demand. On vehicles with an automatic transmission, the Electronic Throttle Control (ETC) system produces the ideal transmission output shaft torque because the position of the throttle plate inside the throttle body is no longer based solely on driver demand (the position of the accelerator pedal).

Electronic throttle body and Throttle Position (TP) sensors

Instead, the PCM-controlled electronic throttle body regulates the amount of air entering the intake manifold in response to driver demand *and* in response to the operating conditions. There is no accelerator cable or cruise control cable connected to the throttle body. Both of these functions are handled by the PCM. There is also no Idle Air Control (IAC) motor on the electronic throttle body. This

function is also handled by the PCM, which opens the throttle plate slightly in response to any load imposed on the engine during idle or low-speed maneuvers.

The electronic throttle body uses two Throttle Position (TP) sensors (TP1 and TP2) because the monitor for this system requires a redundant TP sensor. TP1 has a negative slope (increasing angle, decreasing voltage) and TP2 has a positive slope (increasing angle, increasing voltage). When the engine is running, the negatively-sloped TP1 is used by the ETC system as the actual TP sensor and TP2 is used as a reference sensor by the monitor. The replacement procedure for the electronic throttle body is in this Chapter.

Electronic Throttle Control (ETC) module and Accelerator Pedal Position Sensors (APPS)

The accelerator pedal is equipped with three Accelerator Pedal Position Sensors (APPS), all three of which are housed inside a small plastic housing, known as the Electronic Throttle Control (ETC) module, at the top of the pedal. APP1 has a negative slope (increasing angle, decreasing voltage) and APP2 and APP3 have a positive slope (increasing angle, decreasing voltage). When the engine is running, APP1 is the actual APP sensor. The PCM uses APP2 and APP3 as reference sensors to calculate where a signal *should* be so that it can infer whether a sensor signal makes sense or not. If any *one* signal is irrational (doesn't match the other two signals), the PCM is still able to compute the correct outcome. If *two* of the three input signals are bad, the PCM substitutes a default value and turns on the Malfunction Indicator Lamp (MIL). The ETC module and the accelerator pedal are integrated into a single assembly. Neither component can be serviced separately. If one of the APP sensors is defective, you must replace the "pedal and sensor assembly," which is the manufacturer's term for this assembly (see Chapter 6 for the replacement procedure).

Besides eliminating the accelerator and cruise control cables and the IAC motor, the torque based ETC system also results in, according to the manufacturer, an improved airflow range, a more responsive powertrain at altitude and improved shift quality.

Electronically controlled returnless fuel system

In a conventional fuel system with a return line, a fuel pressure regulator maintains the pressure within the correct operating range. When the vehicle decelerates, intake manifold goes up and a vacuum hose between the intake manifold and the pressure regulator lifts the spring-loaded diaphragm inside the regulator, allowing excess fuel pressure to bleed off and the unused fuel to return to the fuel tank. When the vehicle accelerates again, intake manifold vacuum goes down and the spring inside the regulator closes the diaphragm, shutting off the return line and allowing fuel pressure to rise again.

But all of the models covered by this manual use a *returnless* fuel system, i.e. there is no fuel pressure regulator and no fuel return line.

Fuel Pump Driver Module (FPDM)

In the type of returnless fuel system used by the vehicles covered by this manual, the PCM controls the fuel pressure by controlling the duty cycle of the Fuel Pump Driver Module (FPDM), which in turn controls the speed of the fuel pump by modulating the voltage to the fuel pump. The FPDM is located in the spare tire compartment. To replace the FPDM, refer to Chapter 6.

Fuel Rail Pressure and Temperature (FRPT) sensor

The FRPT sensor, which is used on all models, measures the pressure *and the temperature* of the fuel in the fuel rail. The FRPT sensor uses intake manifold vacuum as a reference to determine the pressure difference between the fuel rail and the intake manifold. The relationship between fuel pressure and fuel temperature is used to determine the likelihood of the presence of fuel vapor in the fuel rail. Both the pressure and temperature signals are used to control the speed of the fuel pump. The speed of the fuel pump controls the pressure inside the fuel rail in order to keep the fuel in a liquid state. Keeping the fuel in a liquid state increases the efficiency of the injectors because the higher fuel rail pressure allows a decrease in the injector pulse width (the interval of time during which the injector is open). The FRPT sensor is located on the left fuel rail.

11   Fuel injection system - check

Warning: *Gasoline is extremely flammable, so take extra precautions when you work on any part of the fuel system. See the* **Warning** *in Section 2.*
Note: *The following procedure is based on the assumption that the fuel pump is working and the fuel pressure is adequate (see Section 3).*

Preliminary checks

1   Inspect all electrical connectors that are part of the SFI system. Loose electrical connectors and poor grounds can cause many problems that resemble more serious malfunctions.

2   Verify that the battery is fully charged. The Powertrain Control Module (PCM), the information sensors and the output actuators must receive adequate and stable voltage to function correctly.

3   Inspect the condition of the air filter element (see Chapter 1). A dirty or partially blocked filter will severely degrade performance and fuel economy.

4   Check the fuses that are related to the SFI system. If you find a blown fuse, replace it and see if it blows again. If it does, look for a grounded or shorted wire in the harness for the relevant circuit (fuel pump, fuel injectors, etc.). (For more information on fuses, refer

11.7 Use an automotive stethoscope to listen to each injector. If an injector is working correctly, it should make a steady clicking sound that rises and falls in response to engine rpm

12.2a To disconnect the electrical connector from the electronic throttle body throttle motor, slide out the lock (toward the harness), then depress the release tab and pull off the connector (4.6L V8 shown, 4.0L V6 similar)

to Chapter 12. For a complete guide to the fuses on your vehicle, refer to your owner's manual.)

System checks

Refer to illustration 11.7

5   Inspect the condition of any vacuum hoses connected to the intake manifold or to the throttle body.

6   Disconnect the air intake duct from the throttle body (see Section 9) and inspect the bore of the throttle body for dirt, carbon or other residue build-up, particularly around the throttle plate. **Caution:** *The bore inside the throttle bodies used on all vehicles covered in this manual is coated with a special film designed to protect the bore and throttle plate and to resist the accumulation of sludge. Do not attempt to clean the interior of the throttle body with carburetor or other spray cleaners. Cleaning the throttle body bore with carb cleaner can damage the throttle body bore and impair performance.*

7   With the engine running, place an automotive stethoscope against each injector, one at a time, and listen for a clicking sound, indicating operation **(see illustration)**. If you

don't have a stethoscope, you can place the tip of a long screwdriver against the injector and listen through the handle.

8   If an injector isn't operating (not clicking), inspect the condition of the injector wiring harness. Make sure that the wiring is in good shape and that the injector electrical connector is correctly connected.

9   If the wiring harness seems to be OK, purchase a special injector test light (sometimes called a "noid" light) and install it into the injector wiring harness connector. Start the engine and see if the noid light flashes. If it does, the injector is receiving proper voltage. If it doesn't flash, further diagnosis is necessary. You might want to have it checked by a dealership service department or other qualified repair shop.

10   If the light flashes, turn off the engine and disconnect the electrical connectors from the fuel injectors, then measure the resistance of each injector with an ohmmeter. If the resistance of the injector that didn't make the clicking sound is significantly different than the other injectors, it most likely is faulty.

11   Any further diagnosis of the fuel injection system should be left to a professional service technician.

12 Throttle body - removal and installation

Refer to illustrations 12.2a, 12.2b, 12.4 and 12.5

Note: *The photos accompanying this section depict the electronic two-barrel throttle body used on 4.6L V8 models. The electronic one-barrel throttle body used on 4.0L V6 models is similar. For the purposes of this procedure, the only real difference between the two units are the means by which they're secured to the intake manifold: the 4.0L V6 throttle body uses four mounting bolts; the 4.6L V8 unit uses two nuts and two bolts.*

1   Disconnect the cable from the negative battery terminal (see Chapter 5, Section 1).

2   Disconnect the electrical connectors from the electronic throttle control motor and from the Throttle Position (TP) sensor **(see illustrations)**.

3   Remove the air intake duct (see Section 9).

4   Remove the four throttle body mount-

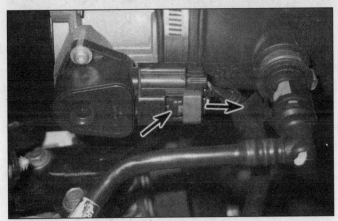

12.2b To disconnect the electrical connector from the Throttle Position (TP) sensor, slide out the lock (toward the harness), then depress the release tab and pull off the connector (4.6L V8 shown, 4.0L V6 similar)

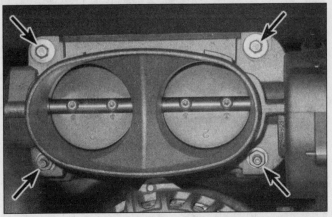

12.4 To detach the throttle body from the intake manifold on a 4.6L V8 engine, remove these two mounting bolts and two nuts (the throttle body on 4.0L V6 engines uses four bolts)

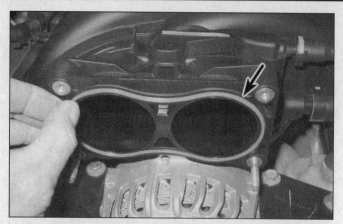

12.5 Remove the old throttle body O-ring seal from the intake manifold and discard it. Always use a new O-ring seal when installing the throttle body (4.6L V8 shown, 4.0L V6 similar)

13.16 To disconnect the electrical connectors from the fuel injectors, depress the release tab and pull off each connector

ing bolts (4.0L V6) or bolts and nuts (4.6L V8) **(see illustration)** and remove the throttle body.

5   Remove the old throttle body gasket **(see illustration)**. Inspect the condition of the gasket. If it's damaged, replace it. If it's undamaged, you can reuse it.

6   Installation is the reverse of removal. Be sure to use a new gasket, if necessary, and tighten the throttle body mounting fasteners to the torque listed in this Chapter's Specifications.

13   Fuel rail and injectors - removal and installation

1   Relieve the system fuel pressure (see Section 2).

2   Disconnect the cable from the negative battery terminal (see Chapter 5, Section 1).

4.0L V6

3   Disconnect the fuel supply line spring-lock coupling from the fuel rail supply tube. If you're unfamiliar with spring-lock couplings, see Section 4.

4   Remove the upper and lower fuel rail supply tube mounting bracket bolts.

5   Remove the four bolts (two per flange) that secure the fuel rail supply tube flanges to the two fuel rail flanges, then remove the fuel rail supply tube.

6   Remove the fuel rail flange O-rings and discard them. (Each O-ring might be stuck on the fuel rail supply tube or it might be in the fuel rail flange.)

7   Disconnect the vacuum hose and the electrical connector from the Fuel Rail Pressure and Temperature (FRPT) sensor.

8   Disconnect the electrical connectors from all six fuel injectors, detach the wiring harness retainer clip from the right valve cover and set the injector wiring harness aside.

9   Remove the four fuel rail bolts, then remove the fuel rail and injectors as a single assembly.

10   Grasp each fuel injector firmly and pull it out of the fuel rail.

11   Remove the two old O-rings from each injector **(see illustration 13.20)**.

12   Installation is the reverse of removal. Be sure to install new O-rings on the fuel injectors and at the mounting flanges between the fuel rails and the fuel rail supply tube.

4.6L V8

Refer to illustrations 13.16, 13.17, 13.18, 13.19 and 13.20

13   Disconnect the fuel supply line spring-lock coupling from the fuel rail **(see illustration 3.13)**. If you're unfamiliar with spring-lock couplings, refer to Section 4.

14   Disconnect the electrical connector and the vacuum hose from the Fuel Rail Pressure and Temperature (FRPT) sensor (see *Fuel Rail Pressure and Temperature (FRPT) sensor - replacement* in Chapter 6).

15   Detach the wiring harness retainer clips from the fuel rails.

16   Disconnect the electrical connectors from all eight fuel injectors **(see illustration)**.

17   Remove the four fuel rail mounting bolts **(see illustration)**.

18   Remove the fuel rail and injectors as a single assembly **(see illustration)**.

13.17 To detach the fuel rail from the intake manifold, remove these four bolts

13.18 Remove the fuel rail and injectors as a single assembly. First, work the injectors on one side out of their bores, then work the injectors on the other fuel rail free

13.19 To remove the injectors from the fuel rail, remove the retainer clips

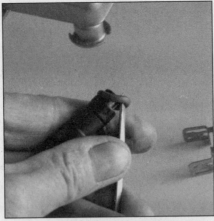

13.20 Remove the two old O-rings from each injector and discard them. Always use new O-rings when installing the injectors

14.1a A typical exhaust system rubber hanger insulator. This one's at the transmission mount

19   Remove the injector retainer clips **(see illustration)** and remove the injectors from the fuel rail. If an injector sticks in its bore, work it out of the fuel rail by wiggling it from side-to-side while simultaneously pulling on it.

20   Remove the old upper and lower O-rings from each fuel injector **(see illustration)** and discard them.

21   Install new upper and lower O-rings on each injector.

22   Coat each injector O-ring with a small amount of clean engine oil, then install the injectors in the fuel rail.

23   Install the injector retainer clips.

24   Coat each lower injector O-ring with a small amount of clean engine oil, then install the fuel rail and injector assembly.

25   Installation is otherwise the reverse of removal.

14   Exhaust system servicing - general information

Refer to illustrations 14.1a and 14.1b

Warning: *Inspect and repair exhaust system components only after sufficient time has elapsed after driving the vehicle to allow the system components to cool completely. Also, when working under the vehicle, make sure it is securely supported on jackstands.*

1   The exhaust system consists of the exhaust manifolds, the catalytic converters, the muffler, the tailpipe and all connecting pipes, brackets, hangers and clamps. The exhaust system is suspended from the underside of the vehicle by a series of rubber hanger insulators **(see illustrations)**. Inspect these insulators periodically for cracks or other signs of deterioration, and replace them as necessary.

2   To keep the exhaust system safe and

quiet, conduct regular inspections of the exhaust system anytime that you're servicing anything underneath the vehicle. Look for any damaged or bent parts, open seams, holes, loose connections, excessive corrosion or other defects which could allow exhaust fumes to enter the vehicle. Deteriorated exhaust system components should not be repaired; they should be replaced with new parts.

3   If the exhaust system components are extremely corroded or rusted together, you'll probably need welding equipment to remove them. The convenient way to accomplish this is to have a muffler repair shop remove the corroded sections with a cutting torch. If, however, you want to save money by doing it yourself (and you don't have a welding outfit with a cutting torch), simply cut off the old components with a hacksaw. If you have compressed air, special pneumatic cutting chisels can also be used. If you do decide to tackle the job at home, be sure to wear safety goggles to protect your

eyes from metal chips and work gloves to protect your hands.

4   Here are some simple guidelines to follow when repairing the exhaust system:

a) *Work from the back to the front when removing exhaust system components.*

b) *Apply penetrating oil to the exhaust system component fasteners to make them easier to remove.*

c) *Use new gaskets, hangers and clamps when installing exhaust systems components.*

d) *Apply anti-seize compound to the threads of all exhaust system fasteners during reassembly.*

e) *Be sure to allow sufficient clearance between newly installed parts and all points on the underbody to avoid overheating the floor pan and possibly damaging the interior carpet and insulation. Pay particularly close attention to the catalytic converter and heat shield.*

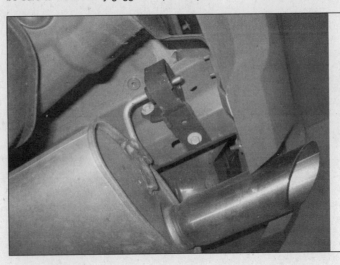

14.1b Another type of rubber exhaust hanger. This one's at the rear of the vehicle and mounted with two bolts

Notes

Chapter 5
Engine electrical systems

Contents

Specifications

Charging system
Charging voltage .... 13.5 to 15 volts

Torque specifications   **Ft-lbs**
Starter motor mounting bolts (all engines) 18

---

1 General information, precautions and battery disconnection

The engine electrical systems include all ignition, charging and starting components. Because of their engine-related functions, these components are discussed separately from body electrical devices such as the lights, the instruments, etc. (all of which are included in Chapter 12).

Precautions

Always observe the following precautions when working on the electrical system:

a) *Be extremely careful when servicing engine electrical components. They are easily damaged if checked, connected or handled improperly.*

b) *Never leave the ignition switched on for long periods of time when the engine is not running.*

c) *Never disconnect the battery cables while the engine is running.*

d) *Maintain correct polarity when connect-*ing battery cables from another vehicle during jump starting - see the "Booster battery (jump) starting" section at the front of this manual.*

e) *Always disconnect the negative battery cable from the battery before working on the electrical system, but read the following battery disconnection procedure first.*

It's also a good idea to review the safety-related information regarding the engine electrical systems located in the "Safety first!" section at the front of this manual, before beginning any operation included in this Chapter.

Battery disconnection

Several systems need battery power all the time, either to maintain continued operation (radio, alarm system, power windows and door locks, etc.), or to preserve memory (Powertrain Control Module and other modules) which is lost if the battery is disconnected. Therefore, whenever the battery is to be disconnected, first note the following to ensure that there are no unforeseen consequences of this action:

a) *The engine management system's PCM might lose some of the information stored in its memory when the battery is disconnected. This includes idling and operating values, any fault codes detected and system monitors required for emissions testing. Whenever the battery is disconnected, the computer might require a certain period of time to relearn these operating values (see Chapter 6 for more information about the engine management system and the PCM).*

b) *On any vehicle with power door locks, it is a wise precaution to remove the key from the ignition and to keep it with you, so that it does not get locked inside if the power door locks should engage accidentally when the battery is reconnected!*

Devices known as "memory-savers" can be used to avoid some of the above problems. Precise details vary according to the device used. Typically, you plug it into the cigarette lighter and connect it to a spare battery. Then you disconnect the vehicle battery from the electrical system. The memory-saver passes sufficient current to maintain audio unit security codes, PCM memory values, etc. and it also maintains always-hot circuits such as the clock and radio memory. **Warning 1:** *Some of these devices allow a considerable amount of current to pass, which can mean that many of the vehicle's systems are still operational when the main battery is disconnected. If a "memory-saver" is used, ensure that the circuit concerned is actually "dead" before carrying out any work on it!* **Warning 2:** *If work is to be performed around any of the airbag system components, the battery must be disconnected and a memory-saver device must NOT be used. If a memory-saver device is used, power will be supplied to the airbag and personal injury may result if the airbag is accidentally deployed.*

To disconnect the battery for service procedures requiring power to be cut from the vehicle, peel back the insulator (if equipped), loosen the negative cable clamp nut and detach the cable from the negative battery terminal (see Section 3). Isolate the cable end to prevent it from coming into accidental contact with the battery post.

2   Battery - emergency jump starting

Refer to the *Booster battery (jump) starting* procedure at the front of this manual.

3   Battery - check and replacement

Warning: *Hydrogen gas is produced by the battery, so keep open flames and lighted cigarettes away from it at all times. Always wear eye protection when working around a battery. Rinse off spilled electrolyte immediately with large amounts of water.*

Check

Refer to illustrations 3.1a, 3.1b and 3.1c

1   A battery cannot be accurately tested until it is at or near a fully charged state. Disconnect the negative battery cable from the battery and perform the following tests:

a)   **Battery state of charge test** - *Visually inspect the indicator eye (if equipped) on the top of the battery. If the indicator eye is dark in color, charge the battery as described in Chapter 1. If the battery is equipped with removable caps, check the battery electrolyte. The electrolyte level should be above the upper edge of the plates. If the level is low, add distilled water. DO NOT OVERFILL. The excess electrolyte may spill over during periods of heavy charging. Test the specific gravity of the electrolyte using a hydrometer* **(see illustration)**. *Remove the caps and extract a sample of the electrolyte and observe the float inside the barrel of the hydrometer. Follow the instructions from the tool manufacturer and determine the specific gravity of the electrolyte for each cell. A fully charged battery will indicate approximately 1.270 (green zone) at 68-degrees F (20-degrees C). If the specific gravity of the electrolyte is low (red zone), charge the battery as described in Chapter 1.*

b)   **Open circuit voltage test** - *Using a digital voltmeter, perform an open circuit voltage test* **(see illustration)**. *Connect the negative probe of the voltmeter to the negative battery post and the positive probe to the positive battery post. The battery voltage should be greater than 12.5 volts. If the battery is less than the specified voltage, charge the battery before proceeding to the next test. Do not proceed with the battery load test until the battery is fully charged.*

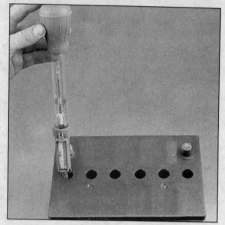

3.1a   Use a battery hydrometer to draw electrolyte from the battery cell (this hydrometer is also equipped with a thermometer to make temperature corrections)

c)   **Battery load test** - *An accurate check of the battery condition can only be performed with a load tester (available at most auto parts stores). This test evaluates the ability of the battery to operate the starter and other accessories during periods of heavy amperage draw (load). Connect a load-testing tool to the battery terminals* **(see illustration)**. *Load test the battery according to the tool manufacturer's instructions. This tool increases the load demand (amperage draw) on the battery. Maintain the load on the battery for 15 seconds and observe that the battery voltage does not drop below 9.6 volts. If the battery condition is weak or defective, the tool will indicate this condition immediately.* **Note:** *Cold temperatures will cause the minimum voltage reading to drop slightly. Follow the chart given in the tool manu-*

3.1b   To test the open circuit voltage of the battery, connect the black probe of the voltmeter to the negative terminal and the red probe to the positive terminal of the battery. A fully charged battery should indicate about 12.5 volts

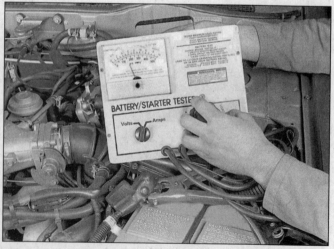

3.1c   Some battery load testers are equipped with an ammeter (shown) which enables the battery load to be precisely dialed in; less expensive testers only have a load switch and a voltmeter

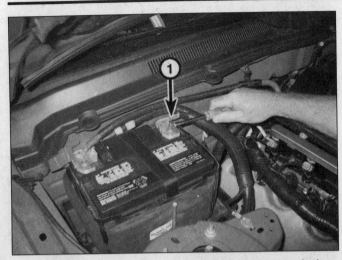

3.2 When disconnecting the cable from the battery terminals, ALWAYS loosen this nut (1) and disconnect the ground cable first

3.3 When disconnecting the cables from the positive terminal, loosen this nut (2) to disconnect the alternator cable, then loosen this nut (3) to loosen the clamp and disconnect the starter cable

facturer's instructions to compensate for cold climates. Minimum load voltage for freezing temperatures (32 degrees F/0-degrees C) should be approximately 9.1 volts.

d) *Battery drain test* - This test will indicate whether there's a constant drain on the vehicle's electrical system that can cause the battery to discharge. Make sure all accessories are turned off. If the vehicle has an underhood light, verify that it's working correctly, then disconnect it. Connect one lead of a digital ammeter to the disconnected negative battery cable clamp and the other lead to the negative battery post. A drain of approximately 100 milliamps or less is considered normal (due to the engine control computer, digital clock, digital radios and other components that normally cause a key-off battery drain). An excessive drain (approximately 500 milliamps or more) will cause the battery to discharge. The problem circuit or com-

ponent can be located by removing the fuses, one at a time, until the excessive drain stops and normal drain is indicated on the meter.

Replacement

Refer to illustrations 3.2, 3.3, 3.4, 3.5 and 3.7

Caution: *Always disconnect the ground cable first and hook it up last or you might accidentally short the battery with the tool that you're using to loosen the cable clamps.*

2   Loosen the cable clamp nut and disconnect the battery ground cable from the negative battery post **(see illustration)**. Isolate the cable end to prevent it from accidentally coming into contact with the battery post.

3   Loosen the cable clamp nut and disconnect the battery cable from the positive battery post **(see illustration)**.

4   Remove the battery heat shield **(see illustration)**.

5   Remove the battery hold-down clamp bolt **(see illustration)**.

6   Lift out the battery. Be careful - it's heavy. Battery lifting straps are available at most auto parts stores for a reasonable price. They make removing and installing the battery both easier and safer.

7   While the battery is out, inspect the battery tray for corrosion. If corrosion exists, clean the deposits with a mixture of baking soda and water to prevent further corrosion. Flush the area with plenty of clean water and dry thoroughly. If you need to remove the battery tray for any reason - to clean it more thoroughly, to replace it, etc. - simply remove the mounting bolts **(see illustration)** and lift out the tray.

8   If you are replacing the battery, make sure you replace it with a battery with identical dimensions, amperage rating, cold cranking rating, etc.

9   Installation is the reverse of removal.

10   After connecting the cables to the battery apply a light coating of petroleum jelly or grease to the connections to help prevent corrosion.

3.4 Remove the battery heat shield

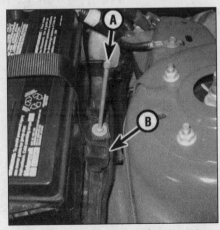

3.5 Remove the battery hold-down bolt (A), then remove the hold-down clamp (B)

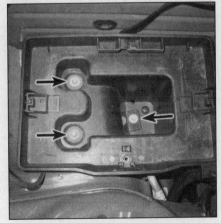

3.7 To detach the battery tray, remove these bolts

4.4a One battery cable is connected to the B+ terminal on the engine compartment fuse and relay box and another battery cable is connected to the B+ terminal of the alternator (see illustration 10.11)

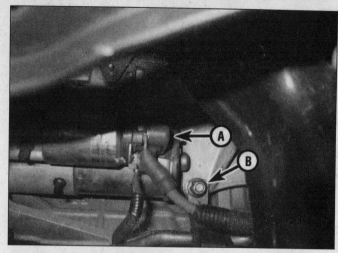

4.4b The third battery cable (A) is connected to the B+ terminal on the starter motor solenoid. The engine ground cable bolt (B) is also located near the starter motor

4   Battery cables - check and replacement

Refer to illustrations 4.4a and 4.4b

1   Periodically inspect the entire length of each battery cable for damage, cracked or burned insulation and corrosion. Poor battery cable connections can cause starting problems and decreased engine performance.

2   Check the cable-to-terminal connections at the ends of the cables for cracks, loose wire strands and corrosion. The presence of white, fluffy deposits under the insulation at the cable terminal connection is a sign that the cable is corroded and should be replaced. Check the terminals for distortion, missing mounting bolts and corrosion.

3   When removing the cables always disconnect the negative cable from the negative battery post first and hook it up last or you might accidentally short the battery with the tool that you're using to loosen the cable clamps. Even if only the positive cable is being replaced, be sure to disconnect the negative cable from the negative battery post first (see Chapter 1 for further information regarding battery cable maintenance).

4   First disconnect the ground cable from the negative terminal, then disconnect the cables from the positive terminal (see Section 3). Then disconnect the other end of each cable. Disconnect the positive cable from the B+ terminal inside the engine compartment fuse and relay box **(see illustration)** and disconnect the battery cable from the B+ terminal on the alternator **(see illustration 10.11)**. A third battery cable is connected to the B+ terminal on the starter motor solenoid **(see illustration)**. The ground cable is bolted to the engine near the starter motor. Note the routing of each cable to ensure correct installation, then detach any cable retaining clips or clamps and remove the cables.

5   If you are replacing either or both of the battery cables, take them with you when buying new cables. It is vitally important that you replace the cables with identical parts. Cables have characteristics that make them easy to identify. Positive cables are usually red and larger in cross-section; ground cables are usually black and smaller in cross-section.

6   Clean the threads of the starter solenoid or ground connection with a wire brush to remove rust and corrosion. Apply a light coat of battery terminal corrosion inhibitor or petroleum jelly to the threads to prevent future corrosion.

7   Attach the cable to the terminal and tighten the mounting nut/bolt securely.

8   Before connecting a new cable to the battery make sure that it reaches the battery post without having to be stretched.

9   After installing the cables connect the negative cable to the negative battery post.

5   Ignition system - general information and precautions

General information

4.0L V6 models

1   The Electronic Distributorless Ignition System (EDIS) is controlled by the Powertrain Control Module (PCM). The EDIS consists of the PCM, the Crankshaft Position (CKP) sensor, the ignition coil, the spark plug wires, and the spark plugs. The ignition coil is located on the left valve cover. The PCM-controlled coil, which is really three coils integrated into a single unit, fires two spark plugs simultaneously. For example, when it fires the plug in cylinder No. 1 (the right front cylinder) at Top Dead Center (TDC) on its *compression* stroke, it also fires the plug in cylinder No. 5, which is at TDC on its *exhaust* stroke. When it fires the plug in cylinder No. 4 at TDC on its compression stroke, it also fires the plug in cylinder No. 3 at TDC on its exhaust stroke. And when it fires the plug in cylinder No. 2 at TDC on its compression stroke, it also fires the plug in cylinder No. 6 at TDC on its exhaust stroke. Then when it fires the other three cylinders (Nos. 5, 3 and 6, respectively), it fires the plugs in cylinder Nos. 1, 4 and 2, respectively. So cylinder Nos. 1 and 5, 4 and 3, and 2 and 6 are referred to as *companion cylinders*. The simultaneous spark that goes to each companion cylinder is wasted, i.e. it doesn't do anything because there is no compression and no air/fuel mixture to ignite. This system is therefore generally referred to as a "waste spark" system. The CKP sensor provides base timing and crankshaft speed (rpm) signals to the PCM. For more information about the CKP sensor and the PCM, refer to Chapter 6. For more information about the spark plug wires and the spark plugs, refer to Chapter 1.

4.6L V8 models

2   The ignition system consists of the PCM, the Crankshaft Position (CKP) sensor, eight ignition coils and the spark plugs. The ignition coils are located on the valve covers, right above each spark plug. This setup is generally referred to as a "coil-over-plug" system. Each coil is connected directly to the spark plug that it fires. There are no spark plug wires. The CKP sensor provides base timing and crankshaft speed (rpm) signals to the Powertrain Control Module (PCM). For more information about the CKP sensor and the PCM, refer to Chapter 6. For more information about the spark plugs, refer to Chapter 1.

Precautions

3   When working on the ignition system take the following precautions:

a) *Do not keep the ignition switch on for more than 10 seconds if the engine will not start.*

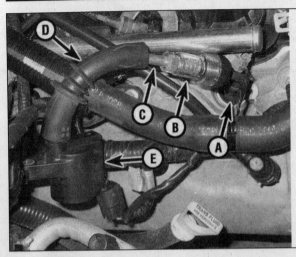

6.4 Here's the setup used for checking to see if the ignition coil is sending power to the spark plug. If the coil is delivering power, the tester will flash or spark, depending on the type of tester being used (coil-over-plug type ignition shown)

A   *Tester spark plug boot*
B   *Clear plastic tester body*
C   *Insert this end of tester into the spark plug boot*
D   *Ignition coil spark plug boot*
E   *Ignition coil*

b) *Always connect a tachometer in accordance with the manufacturer's instructions. Some tachometers may be incompatible with this ignition system. Consult an auto parts counterperson before buying a tachometer for use with this vehicle.*

c) *Never allow the ignition coil terminals to touch ground. Grounding the coil could result in damage to the PCM and/or the ignition coil.*

d) *Do not disconnect the battery when the engine is running.*

6 Ignition system - check

Refer to illustration 6.4

Warning: *Because of the very high voltage generated by the ignition system, use extreme care when performing a procedure involving ignition components. This not only includes the coil and spark plugs, but related items connected to the system as well, such as the electrical connectors, tachometer and any test equipment.*

Note: *The ignition system components on these models are expensive and difficult to diagnose. In the event of ignition system failure, if the checks do not clearly indicate the source of the ignition system problem, have the vehicle tested by a dealer service department or other qualified repair facility.*

1   If a malfunction occurs and the vehicle won't start, do not immediately assume that the ignition system is causing the problem. First, check the following items:

a) *Make sure the battery cable clamps, where they connect to the battery, are clean and tight.*

b) *Test the condition of the battery (see Section 3). If it does not pass all the tests, replace it with a new battery.*

c) *Check the ignition coil wiring and connections.*

d) *Check the appropriate fuses in the fuse and relay box, which is located behind the right kick panel inside the vehicle (see Chapter 12). If they're burned, determine the cause and repair the circuit.*

2   If the engine turns over but won't start, make sure there is sufficient secondary ignition voltage to fire the spark plug. Obtain a calibrated ignition tester (such testers are available at most auto parts stores.)

3   Disable the fuel injection system (see the fuel pressure relief procedure in Chapter 4). The fuel system must be disabled while checking the ignition system.

4   To test the spark on a 4.6L V8, remove an ignition coil (see Section 7), insert the tester into the boot on the coil, then clip the other end to a good ground or, depending on the type of tester being used, connect the other end to the spark plug **(see illustration)**.

5   To test the spark on a 4.0L V6, disconnect a spark plug wire from a spark plug, insert the tester into the spark plug wire boot and clip the other end to a good ground or, depending on the type of tester being used, connect the other end to the spark plug.

6   Crank the engine and watch the tester.

7   If the tester flashes (or sparks, depending on type), sufficient voltage is reaching the spark plug to fire it. Repeat the check on the remaining cylinders. Keep in mind that even if the coil is functioning normally and is firing the tester, one or more of the plugs themselves might be fouled, so remove, inspect and, if necessary, clean or replace the plugs (see Chapter 1)

8   If no sparks occur or if the spark is weak or intermittent, check for battery voltage to the primary terminal of the ignition coil. If there is no battery voltage at the coil primary terminal, have the ignition system checked out by a dealer service department or other qualified repair shop.

7 Ignition coil(s) - replacement

1   Disconnect the cable from the negative battery terminal (see Section 1).

4.0L V6 models

2   Disconnect the electrical connector from the ignition coil.

3   Disconnect the spark plug wires from the ignition coil (see Chapter 1).

4   Remove the ignition coil mounting bolts and remove the coil from the valve cover.

5   Installation is the reverse of removal.

4.6L V8 models

Refer to illustrations 7.6, 7.7, 7.8 and 7.9

Note: *This procedure applies to all eight ignition coils.*

6   Disconnect the electrical connector from the ignition coil **(see illustration)**. **Note:** *If you're going to remove more than one ignition coil, mark each electrical connector with its corresponding coil to prevent mix-ups during reassembly.*

7   Remove the ignition coil hold-down bolt **(see illustration)**.

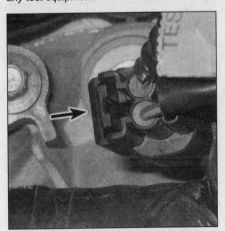

7.6 To disconnect the electrical connector from an ignition coil on a 4.6L V8, depress this release tab, then pull off the connector

7.7 To detach the ignition coil on a 4.6L V8, remove the retaining bolt . . .

7.8 . . . then grasp the coil firmly and pull straight up

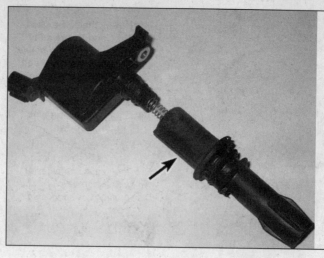

7.9 Inspect the condition of the rubber boot that insulates the connection between the ignition coil and the spark plug. If it's cracked, torn or otherwise deteriorated, replace it. To disconnect the boot, simply pull it off the coil (the spring, which carries the current to the plug, is a permanent part of the coil, so don't try to pull it off!)

8   Grasp the ignition coil firmly and pull it straight up **(see illustration)**.
9   Inspect the condition of the rubber boot that insulates the connection between the coil high-tension terminal and the spark plug **(see illustration)**. If it's cracked, torn or otherwise deteriorated, replace it. Prior to installing the coil, coat the interior of the rubber boot with silicone dielectric compound.
10   Installation is otherwise the reverse of removal.

8   Charging system - general information and precautions

The charging system includes the alternator, a voltage regulator, a charge indicator or warning light, the battery, three fusible links and the wiring between all the components. The charging system supplies electrical power for the ignition system, the lights, the radio, etc. The alternator is driven by a drivebelt at the front of the engine.

The voltage regulator is located inside the alternator and is not separately serviceable. The voltage regulator limits the alternator's voltage to a preset value. This prevents power surges, circuit overloads, etc., during peak voltage output.

The charging system doesn't require much maintenance. However, you should inspect the drivebelt and battery at the intervals outlined in Chapter 1.

The charging system is protected by a fusible link, located in the wiring harness to the alternator. In the event of charging system problems, check for battery voltage to the alternator at the B+ terminal (large red wire). If no voltage is present, check the fusible link, replacing it if it's burned.

To protect the alternator and the charging system circuit, be very careful when making electrical connections:

a) When reconnecting wires to the alternator from the battery, be sure to note the polarity.

b) Before using arc-welding equipment to repair any part of the vehicle, disconnect the wiring from the alternator and the cables from the battery.
c) Never start the engine with a battery charger connected.
d) Always disconnect both battery cables before using a battery charger.
e) The alternator is turned by an engine drivebelt that could cause serious injury if your hands, hair or clothes become entangled in it with the engine running.
f) Because the alternator is connected directly to the battery, it could arc or cause a fire if overloaded or shorted out.

9   Charging system - check

1   If a malfunction occurs in the charging circuit, do not immediately assume that the alternator is causing the problem. First, check the following items:

a) Make sure the battery cable clamps, where they connect to the battery, are clean and tight.
b) Test the condition of the battery (see Section 3). If it does not pass all the tests, replace it with a new battery.
c) Inspect the external alternator wiring and connections.
d) Check the drivebelt tension and inspect the condition of the drivebelt (see Chapter 1).
e) Make sure that the alternator mounting bolts are tight.
f) Run the engine and verify that the alternator isn't making any abnormal noises.
g) Check the fusible links in the charging system (see Chapter 12). If they're burned, determine the cause and repair the circuit.
h) Check the charge light on the dash. It should illuminate when the ignition key is turned ON (engine not running). If it does not, check the circuit from the alternator to the charge light on the dash.

2   With the ignition key off, check the battery voltage with no accessories operating

(see illustration 3.1b). It should be about 12.5 volts. (It might be slightly higher if the engine has been operating within the last hour.)
3   Start the engine and check the battery voltage again. It should now be greater than the voltage indicated in Step 2, but not more than 15 volts. Turn on all the vehicle accessories (air conditioning, rear window defogger, blower motor, etc.) and increase the engine speed to 2,000 rpm - the voltage should not drop below the voltage indicated in Step 2.
4   If the indicated voltage is greater than the specified charging voltage, replace the alternator.
5   If the indicated voltage reading is less than the specified charging voltage, the alternator is probably defective. Have the charging system checked at a dealer service department or other properly equipped repair facility. **Note:** *Many auto parts stores will bench test an alternator off the vehicle. Refer to your local auto parts store regarding their policy, many will perform this service free of charge.*

10   Alternator - removal and installation

4.0L V6 models

1   Disconnect the cable from the negative terminal of the battery (see Section 1).
2   Remove the alternator drivebelt (see Chapter 1).
3   Disconnect the electrical connector from the alternator.
4   Remove the rubber weather boot from the alternator B+ terminal nut, remove the nut and disconnect the battery cable from the B+ terminal.
5   Remove the nut from the alternator mounting stud, remove the two alternator mounting bolts and remove the alternator.
6   Installation is the reverse of removal.

4.6L V8 models

Refer to illustrations 10.8, 10.10 and 10.11

7   Disconnect the cable from the nega-

10.8 On a 4.6L V8 model, remove these four bolts and remove the alternator support bracket (throttle body removed for clarity)

10.10 To detach the alternator from the engine, remove the nuts from these two lower mounting studs (early-build models, shown) or remove the two lower mounting bolts (later models)

tive terminal of the battery (see Section 1). Remove the air intake duct (see Chapter 4).

8   Remove the four alternator mounting bracket bolts **(see illustration)**.

9   Remove the drivebelt (see Chapter 1).

10   Remove the two lower alternator mounting nuts (early models) from the mounting studs **(see illustration)** or remove the two lower bolts (later models).

11   Tilt the alternator forward to access the electrical connectors on the backside of the alternator **(see illustration)** B+ terminal nut and disconnect the battery cable from the B+ terminal, then disconnect the electrical connector from the alternator.

12   Installation is the reverse of removal. Be sure to tighten the alternator mounting nuts or bolts and the support bracket bolts securely.

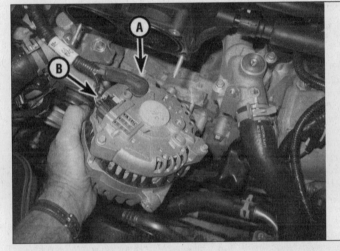

10.11 Pull back the rubber weather boot (A) that protects the B+ terminal, remove the nut from the terminal and disconnect the battery cable, then disconnect the electrical connector (B) from the alternator (4.6L V8 models)

11 Starting system - general information and precautions

The starting system consists of the battery, the ignition switch fuse, the ignition switch, the starter relay, the clutch start switch (manual transmissions), the Digital Transmission Range (DTR) sensor (automatic transmissions), the starter motor/solenoid assembly and the wiring connecting all of these components. The ignition switch fuse and starter relay are located in the fuse and relay box located behind the right kick panel inside the vehicle.

When the ignition key is turned to the START position, battery voltage is directed through the ignition switch fuse to the starter relay. The starter relay closes the starter control circuit, which runs through either the clutch start switch (manual transmission) or the DTR sensor (automatic transmission). If the clutch pedal is depressed or the transmission is in Neutral (manual transmission), or if the shift lever is in PARK or NEUTRAL (automatic transmission), battery voltage is sent to

the starter solenoid, which engages the starter motor pinion gear with the flywheel/driveplate and the starter motor cranks the engine.

The starter motor on a vehicle equipped with a manual transmission can be operated only when the clutch pedal is depressed. The starter on a vehicle equipped with an automatic transmission can be operated only when the transmission selector lever is in PARK or NEUTRAL.

Always observe the following precautions when working on the starting system:

a) Excessive cranking of the starter motor can overheat it and cause serious damage. Never operate the starter motor for more than 15 seconds at a time without pausing for at least two minutes to allow it to cool.

b) The starter is connected directly to the battery and could arc or cause a fire if mishandled, overloaded or short-circuited.

c) Always detach the cable from the negative battery terminal before working on the starting system.

12 Starter motor and circuit - in-vehicle check

Refer to illustrations 12.5 and 12.6

1   If a malfunction occurs in the starting circuit, do not immediately assume that the starter is causing the problem. First, check the following items:

a) Make sure that the battery cable clamps, where they connect to the battery terminals, are clean and tight.

b) Inspect the condition of the battery cables (see Section 4). Always replace defective battery cables with new ones.

c) Test the condition of the battery (see Section 3). If it does not pass all the tests, replace it with a new battery.

d) Inspect the condition of the starter solenoid wiring and connections. Refer to the wiring diagrams at the end of Chapter 12.

e) Make sure that the starter mounting bolts are tight.

12.5 To measure starter current draw with an inductive ammeter, simply hold the ammeter over the positive or negative cable (whichever cable has better clearance)

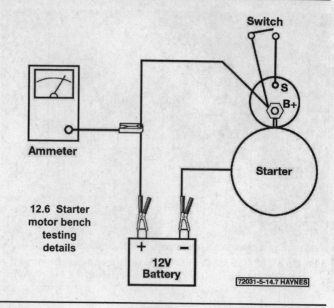

12.6 Starter motor bench testing details

f) *Make sure the starter is receiving voltage on the S terminal (the small wire) of the starter solenoid when the ignition key is turned to Start.*

g) *Check the operation of the Digital Transmission Range (DTR) sensor (automatic transmission) or clutch start switch (manual transmission). Make sure that the shift lever is in PARK or NEUTRAL (automatic transmission) or the clutch pedal is pressed (manual transmission). The DTR sensor or clutch start switch must operate correctly to provide battery voltage to the ignition switch. To replace and/or adjust the DTR sensor, refer to Chapter 6. To replace and/or adjust the clutch start switch, refer to Chapter 8.*

h) *Check the operation of the starter relay. The starter relay is located in the fuse and relay box behind the right kick panel inside the vehicle. Refer to Chapter 12 for the relay testing procedure.*

2　If the starter does not actuate when the ignition switch is turned to the START position, check for battery voltage to the solenoid. Connect a test light or voltmeter to the starter solenoid positive terminal and have an assistant turn the ignition switch to the START position.

3　If there's no voltage at the solenoid, check the ignition switch fuse and the starter relay.

4　If there is voltage at the solenoid, but the starter motor does not operate, remove the starter (see Section 13) and bench test it (see Step 6).

5　If the starter turns over slowly, check the starter cranking voltage and the current draw from the battery. This test must be performed with the starter assembly on the engine. Crank the engine over (for 10 seconds or less) and observe the battery voltage. It should not drop below 8.0 volts on manual transmission models or 8.5 volts on automatic transmission models. Also, observe the current draw using

an ammeter **(see illustration)**. It should not exceed 400 amps or drop below 250 amps. **Caution:** *The battery cables may be excessively heated because of the large amount of amperage being drawn from the battery. Discontinue the testing until the starting system has cooled down.* If the starter motor cranking amp values are not within the correct range, replace it with a new unit. There are several conditions that may affect the starter cranking potential. The battery must be in good condition and the battery cold-cranking rating must not be under-rated for the particular application. Be sure to check the battery specifications carefully. The battery terminals and cables must be clean and not corroded. Also, in cases of extreme cold temperatures, make sure the battery and/or engine block is warmed before performing the tests.

6　If the starter is getting voltage but doesn't operate, remove the starter/solenoid assembly (see Section 13) and test it on the bench **(see illustration)**. Most likely the solenoid is defective. In some rare cases, the engine might be seized, so be sure to try to rotate the crankshaft pulley (see Chapter 2A or 2B) before proceeding. With the starter/solenoid assembly mounted in a vise on the bench, install one

jumper cable from the negative battery terminal to the body of the starter. Install the other jumper cable from the positive battery terminal to the B+ terminal on the starter. Install a starter switch and apply battery voltage to the solenoid S terminal (for 10 seconds or less) and see if the solenoid plunger, shift lever and overrunning clutch extends and rotates the pinion drive. If the pinion drive extends but does not rotate, the solenoid is operating but the starter motor is defective. If there is no movement but the solenoid clicks, the solenoid and/or the starter motor is defective. If the solenoid plunger extends and rotates the pinion drive, the starter/solenoid assembly is working properly.

13 Starter motor - removal and installation

Refer to illustrations 13.3 and 13.4

1　Disconnect the cable from the negative battery terminal (see Section 1).

2　Raise the vehicle and support it securely on jackstands.

3　Disconnect the wiring from the terminals

13.3 Remove the protective cover (1) from the B+ terminal, remove the nut (2) and disconnect the battery cable from the B+ terminal on the solenoid, then remove the other nut (3) and disconnect the starter cable from the S terminal on the solenoid

on the starter motor solenoid **(see illustration)**.

4   Remove the starter motor mounting bolts **(see illustration)** and detach the starter from the engine. The starter on 4.0L V6 engines is retained by two bolts. The starter on 4.6L V8 engines is secured by three bolts.

5   Installation is the reverse of removal. Be sure to tighten the starter mounting bolts to the torque listed in this Chapter's Specifications

13.4 To detach the starter motor from the engine, remove the starter mounting bolts (V8 model shown, V6 models only have two bolts)

Notes

Chapter 6
Emissions and engine control systems

Contents

Specifications

Torque specifications
	Ft-lbs (unless otherwise indicated)
Cylinder Head Temperature (CHT) sensor	89 in-lbs
Exhaust Gas Recirculation (EGR) valve (4.0L V6)	
EGR valve mounting bolts	18
Exhaust manifold-to-EGR valve tube fittings (both ends)	29
Fuel rail pressure sensor mounting bolts (both engines)	53 in-lbs
Knock sensor retaining bolt (both engines)...	180 in-lbs
Oxygen sensors (all engines, all sensors)...	35
Variable Camshaft Timing (VCT) oil control solenoid	
mounting bolt (4.6L V8)...	44 in-lbs

1   General information

Refer to illustrations 1.6

To minimize atmospheric pollution caused by evaporative emissions, unburned hydrocarbons and other exhaust gas pollutants, and to maintain good driveability and fuel economy, a number of emission control systems are utilized on the vehicles covered in this manual. They include the following systems:

Catalytic converters
Charge Motion Control Valve (CMCV) system (4.6L V8 models)

Electronic Fuel Injection (EFI) system
Evaporative Emissions Control (EVAP) system
Exhaust Gas Recirculation (EGR) system (4.0L V6 models)
Idle Air Control (IAC) system
Intake Manifold Runner Control (IMRC) system (4.2L V6 models)
Intake Manifold Tuning (IMT) system (4.6L V8 models)
On-Board Diagnostics (OBD-II) system
Positive Crankcase Ventilation (PCV) system
Variable Camshaft Timing (VCT) system (4.6L V8 models)

The Sections in this Chapter include general descriptions, checking procedures within the scope of the home mechanic and component replacement procedures (when possible) for each of the systems listed above.

Before assuming that an emissions control system is malfunctioning, check the fuel and ignition systems carefully. The diagnosis of some emission control devices requires specialized tools, equipment and training. If checking and servicing become too difficult or if a procedure is beyond your ability, consult a dealer service department or other repair shop. Remember, the most frequent cause of emissions problems is simply a loose or bro-

1.6 The Vehicle Emission Control Information (VECI) label contains such essential information as the types of emission control systems installed on the engine, and a vacuum diagram

2.1 Simple code readers are an economical way to extract trouble codes when the CHECK ENGINE light comes on

ken wire or vacuum hose, so always check the hose and wiring connections first.

This doesn't mean, however, that emissions control systems are particularly difficult to maintain and repair. You can quickly and easily perform many checks and do most of the regular maintenance at home with common tune-up and hand tools. **Note:** *Because of a Federally mandated warranty which covers the emissions control system components, check with your dealer about warranty coverage before working on any emissions-related systems. Once the warranty has expired, you may wish to perform some of the component checks and/or replacement procedures in this Chapter to save money.*

Pay close attention to any special precautions outlined in this Chapter. It should be noted that the illustrations of the various systems might not exactly match the system installed on your vehicle because of changes made by the manufacturer during production or from year-to-year.

2.2 Scanners like these from Actron and AutoXray are powerful diagnostic aids - they can tell you just about anything that you want to know about your engine management system

A Vehicle Emissions Control Information (VECI) label **(see illustration)** is attached to the underside of the hood. This label contains important emissions specifications and adjustment information. Part of this label, the vacuum hose routing diagram, provides a vacuum hose schematic with emissions components identified. When servicing the engine or emissions systems, always refer to the VECI label and the vacuum hose routing diagram on your vehicle for up-to-date information.

2   On-Board Diagnostic (OBD) system and trouble codes

Scan tool information

Refer to illustrations 2.1 and 2.2

1   Hand-held scanners are handy for analyzing the engine management systems used on late-model vehicles. Because extracting the Diagnostic Trouble Codes (DTCs) from an engine management system is now the first step in troubleshooting many computer-controlled systems and components, even the most basic generic code readers are capable of accessing a computer's DTCs **(see illustration)**. More powerful scan tools can also perform many of the diagnostics once associated with expensive factory scan tools. If you're planning to obtain a generic scan tool for your vehicle, make sure that it's compatible with OBD-II systems. If you don't plan to purchase a code reader or scan tool and don't have access to one, you can have the codes extracted by a dealer service department or by an independent repair shop.

2   With the advent of the Federally mandated emission control system known as On-Board Diagnostics-II (OBD-II), specially designed scanners were developed. Several tool manufacturers have released OBD-II scan tools for the home mechanic **(see illustration)**.

OBD system general description

3   All models are equipped with the sec-

ond-generation on-board diagnostic (OBD-II) system. This system consists of an on-board computer known as the Powertrain Control Module (PCM), and information sensors, which monitor various functions of the engine and send data to the PCM. This system incorporates a series of diagnostic monitors that detect and identify fuel injection and emissions control systems faults and store the information in the computer memory. This updated system also tests sensors and output actuators, diagnoses drive cycles, freezes data and clears codes.

4   This powerful diagnostic computer must be accessed using an OBD-II code reader or scan tool and the 16 pin Data Link Connector (DLC) located under the driver's dash area. The PCM is the brain of the electronically controlled fuel and emissions system. It receives data from a number of sensors and other electronic components (switches, relays, etc.). Based on the information it receives, the PCM generates output signals to control various relays, solenoids (i.e. fuel injectors) and other actuators. The PCM is specifically calibrated to optimize the emissions, fuel economy and driveability of the vehicle.

5   It isn't a good idea to attempt diagnosis or replacement of the PCM or emission control components at home while the vehicle is under warranty. Because of a Federally mandated warranty which covers the emissions system components and because any owner-induced damage to the PCM, the sensors and/or the control devices may void this warranty, take the vehicle to a dealer service department if the PCM or a system component malfunctions.

Information sensors

Note: *The following list provides a brief description of the function, operation and location of each of the important information sensors.*

6   **Accelerator Pedal Position (APP) sensor** - The Electronic Throttle Control (ETC) system uses an electronic throttle body instead of a conventional cable-operated throttle body. The Powertrain Control Module (PCM) controls the position of the

throttle plate with a solenoid that's located in the throttle body. The PCM's commands are based on the inputs that it receives from the APP sensor, which is located inside a small plastic module at the top of the accelerator pedal assembly. Its signal is converted into a rotary angle (degrees of pedal travel) by the PCM, and then it's converted into "counts," which is the input used by the ETC system. The APP sensor is an integral component of the accelerator pedal assembly. If it fails, you must replace the pedal and sensor as a single assembly.

7   Camshaft Position (CMP) sensor - The CMP sensor monitors the position of the camshaft(s) and tells the Powertrain Control Module (PCM) when the piston in the No. 1 cylinder is on its compression stroke. The PCM uses the CMP sensor signal to synchronize ignition timing and the sequential firing of the fuel injectors. On 4.0L V6 models, the CMP sensor is located on top of the forward end of the left valve cover. On 4.6L V8 engines, the CMP sensor is located on the front end of the right cylinder head.

8   Crankshaft Position (CKP) sensor - The CKP sensor is a magnetic transducer mounted on the engine block near the front end of the crankshaft. The CKP sensor is the primary sensor that provides ignition information to the PCM. The PCM uses the CKP sensor to determine crankshaft position (which piston will be at TDC next) and crank speed (rpm), both of which it needs to synchronize the ignition system. A pulse wheel or tone wheel, which is mounted on the front of the crankshaft, has 35 teeth about 10 degrees apart, with one empty space for a missing tooth. Basic timing is set by the position of this missing tooth. The PCM uses the input generated by this missing tooth to synchronize the ignition system and to track the rotation of the crankshaft.

On 4.0L V6 engines, the CKP sensor is located on the front of the timing chain cover, at about 11 o'clock in relation to the crankshaft. The pulse wheel is mounted on the front end of the crankshaft, right behind the damper pulley. On 4.6L V8 engines, the CKP sensor is mounted on the lower right side of the timing cover.

9   Cylinder Head Temperature (CHT) sensor - The CHT sensor, which is used on 4.6L V8 engines, is located on the inner wall of the right cylinder head, facing in, i.e. toward the valley between the two cylinder heads. Unlike a conventional Engine Coolant Temperature (ECT) sensor, the CHT sensor measures the temperature of the aluminum cylinder head, not the temperature of the engine coolant. But the PCM is able to infer the temperature of the coolant from the CHT sensor signal nonetheless. If the CHT indicates a cylinder head temperature of about 250 degrees F. the PCM initiates a fail-safe cooling strategy that allows you to drive home in limp-home mode. Basically, the PCM disables half of the fuel injectors. It alternates which injectors are disabled every 32 engine cycles. The cylinders that are

not injected with fuel act as air pumps to help cool down the engine. If the CHT sensor indicates a temperature of about 330 degrees F. or higher, the PCM shuts down all of the injectors until the temperature goes below about 310 degrees F.

The CHT sensor is a thermistor, i.e. its resistance decreases as the temperature increases, and its resistance increases as the temperature decreases. This type of thermistor is also referred to as a Negative Temperature Coefficient (NTC) thermistor. This variable resistance produces an analogous voltage drop across the sensor terminals, thus providing an electrical signal to the PCM that accurately reflects the cylinder head temperature. The CHT sensor is a passive sensor; i.e. it's connected to a voltage divider network in which varying the resistance of the sensor produces a variation in total current flow. Voltage is dropped across a fixed resistor that's installed in series with the resistor inside the CHT sensor, and together both resistors determine the voltage signal that's provided to the PCM. This voltage signal is equal to the reference voltage minus the voltage drop across the fixed resistor.

10   Digital Transmission Range (DTR) sensor - Like the Park/Neutral Position (PNP) switch that it replaces, the DTR sensor prevents you from starting the engine unless the automatic transaxle is in Park or Neutral, and it activates the back-up lights when you put the shift lever in Reverse. Unlike the PNP switch, however, the DTR sensor also tells the PCM and/or the Transmission Control Module (TCM) what gear the transaxle is in. The PCM uses this information to determine what gear the transaxle should be in and to determine when to upshift and downshift. If the DTR sensor fails, the PCM uses pressure switch data to calculate the correct shift lever position. The DTR sensor is mounted on the left side of the transmission.

11   Engine Coolant Temperature (ECT) sensor - The ECT sensor, which is used on 4.0L V6 engines, measures the temperature of the engine coolant. It's located on top of the engine, near the upper radiator hose connection. The PCM uses the ECT sensor's input for fuel and cooling fan control.

The ECT sensor is a thermistor, i.e. its resistance decreases as the coolant temperature increases, and its resistance increases as the coolant temperature decreases. This type of thermistor is also referred to as a Negative Temperature Coefficient (NTC) thermistor. This variable resistance produces an analogous voltage drop across the sensor terminals, thus providing an electrical signal to the PCM that accurately reflects the coolant temperature. The oil temperature sensor is a passive sensor; i.e. it's connected to a voltage divider network in which varying the resistance of the sensor produces a variation in total current flow. Voltage is dropped across a fixed resistor that's installed in series with the resistor inside the oil temperature sensor, and together both resistors determine the voltage

signal that's provided to the PCM. This voltage signal is equal to the reference voltage minus the voltage drop across the fixed resistor.

12   Fuel Rail Pressure and Temperature (FRPT) sensor - The FRPT sensor measures the pressure and the temperature of the fuel in the fuel rail. The FRPT sensor uses intake manifold vacuum as a reference to determine the pressure difference between the fuel rail and the intake manifold. The relationship between fuel pressure and fuel temperature is used to determine the likelihood of the presence of fuel vapor in the fuel rail. Both the pressure and temperature signals are used to control the speed of the fuel pump. The speed of the fuel pump controls the pressure inside the fuel rail in order to keep the fuel in a liquid state. Keeping the fuel in a liquid state increases the efficiency of the injectors because the higher fuel rail pressure allows a decrease in the injector pulse width (the interval of time during which the injector is open). The FRPT sensor is located on the left fuel rail.

13   Intake Air Temperature (IAT) sensor - The IAT sensor is used by the PCM to calculate air density, which is one of the variables it must know in order to calculate injector pulse width and adjust ignition timing (to prevent spark knock when air intake temperature is high). Like the ECT sensor, the IAT sensor is a Negative Temperature Coefficient (NTC) type thermistor, whose resistance decreases as the temperature increases. The IAT sensor is an integral component of the Mass Air Flow (MAF) sensor on all models. It cannot be serviced separately from the MAF sensor. For more information about the MAF sensor, refer to Step 15.

14   Knock sensor - The knock sensor monitors engine vibration caused by detonation. Basically, a knock sensor converts abnormal engine vibration to an electrical signal. When the knock sensor detects a knock in one of the cylinders, it signals the PCM so that the PCM can retard ignition timing accordingly. The knock sensor contains a piezoelectric material, a certain type of piezoresistive crystal, that has the ability to produce a voltage when subjected to a mechanical stress. The piezoelectric crystal in the knock sensor vibrates constantly and produces an output signal that's proportional to the intensity of the vibration. As the intensity of the vibration increases, so does the voltage of the output signal. When the intensity of the crystal's vibration reaches a specified threshold, the PCM stores that value in its memory and retards ignition timing in all cylinders (the PCM does not selectively retard timing only at the affected cylinder). The PCM doesn't respond to the knock sensor's input when the engine is idling; it only responds when the engine reaches a specified speed.

On 4.0L V6 engines, the knock sensor is located in the valley between the two cylinder heads. On 4.6L V8 engines, the knock sensors are located in the valley between the cylinder heads. You must remove the intake manifold to access the knock sensor(s) on either engine.

15 Mass Air Flow (MAF) sensor - The MAF sensor is the principal means by which the PCM monitors intake airflow. It uses a hot-wire sensing element to measure the amount of air entering the engine. Air passing over the hot wire causes it to cool down. The hot wire's temperature is maintained at 392 degrees F. above the ambient temperature by electrical current supplied to the wire and controlled by the PCM. A constantly "cold" wire located right next to the hot wire measures the ambient air temperature.) As intake air passes through the MAF sensor and over the hot wire, it cools the wire, and the control system immediately corrects the temperature back to its constant value. The current required to maintain the specified constant temperature value is used by the PCM as an indicator of airflow. The MAF sensor is mounted on the air filter housing.

16 Oxygen sensors - Oxygen sensors generate a voltage signal that varies in accordance with the amount of oxygen in the exhaust stream. The PCM uses the data from the upstream oxygen sensor to calculate the injector pulse width. The downstream oxygen sensor monitors the oxygen content of the exhaust gases as they exit the catalytic converter. This information is used by the PCM to predict catalyst deterioration and/or failure. One job of the catalytic converter is to store excess oxygen. As long as the catalyst is functioning correctly, the downstream sensor should show little activity because there should be little oxygen exiting the catalyst. But as the catalyst deteriorates, its ability to store oxygen is compromised. When the output signal from the downstream sensor starts to look like the output signal from the upstream sensor, the PCM stores a DTC and turns on the MIL to let you know that it's time to replace the catalyst.

There are four oxygen sensors on all models. The upstream oxygen sensors are located directly below the exhaust manifold flanges. The downstream sensors are located on the catalysts.

17 Power Steering Pressure (PSP) sensor - The PSP sensor monitors the hydraulic pressure of the power steering fluid in the power steering system. The PSP sensor provides a voltage input to the PCM that varies in accordance with changes in the hydraulic pressure. The PCM uses the input signal from the PSP sensor to elevate the idle speed when the engine is already under some other load, such as the air conditioning compressor, while maneuvering the vehicle at low speed, such as parking or stop-and-go driving. The PSP sensor also signals the PCM to adjust the Electronic Pressure Control (EPC) pressure during high-load situations such as parking. The PSP sensor is located on the power steering system's high-pressure line, between the power steering pump and the power steering gear.

18 Throttle Position (TP) Sensor - The TP sensor, which is located on the throttle body, is a rotary potentiometer, which is a type of variable resistor, that produces a variable voltage signal in proportion to the opening angle of the throttle plate. The PCM sends 5 volts to the TP sensor. As the plate opens and closes, the resistance of the TP sensor changes with it, altering the signal back to the PCM. The output voltage of the TP sensor is about 0.6 volt at idle (closed throttle plate) to 4.5 volts at wide-open throttle. This variable signal enables the PCM to calculate the position (opening angle) of the throttle plate. The PCM uses the TP sensor input, along with other sensor inputs, to adjust fuel injector pulse-width and ignition timing. The TP sensor is an integral part of the electronic throttle body and is not serviceable separately from the throttle body.

19 Transmission speed sensors - On vehicles with an automatic transmission, there are two speed sensors: the Turbine Shaft Speed (TSS) sensor and the Output Shaft Speed (OSS) sensor. Both sensors are located on the right side of the transmission. The TSS sensor provides the PCM with turbine shaft speed information, which the PCM uses to determine. The OSS sensor provides the PCM with output shaft speed information, which the PCM uses to determine transmission shift scheduling, Torque Converter Clutch (TCC) engagement scheduling and Electronic Pressure Control (EPC) pressure.

20 Vehicle Speed Sensor (VSS) - On vehicles with a manual transmission, the VSS provides information to the PCM to indicate vehicle speed. The VSS is a variable reluctance or Hall-Effect sensor that generates a waveform with a frequency that's proportional to the speed of the vehicle. If the vehicle is moving at a relatively low speed, the VSS generates a signal with a lower frequency. As the vehicle speed increases, the VSS generates a signal with a higher frequency. The PCM uses this information to determine when acceleration or deceleration occurs, so that it can alter parameters such as fuel injector pulse-width and ignition advance or retard. The VSS is located on the right side of the transmission.

Output actuators

Note: *Based on the information it receives from the information sensors described above, the PCM adjusts fuel injector pulse width, idle speed, ignition spark advance, ignition coil dwell and EVAP canister purge operation. It does so by controlling the output actuators. The following list provides a brief description of the function, location and operation of each of the important output actuators.*

21 Charge Motion Control Valve (CMCV) - The CMCV, which is used on 4.6L V8 engines, provides more intake airflow to improve torque, lower emissions and increase performance. The CMCV system consists of the PCM-controlled CMCV on the intake manifold, a crank arm, a pair of actuating rods, and butterfly valves located inside the lower ends of the intake manifold runners. Below the specified rpm (1500 to 3000 rpm), the CMCV is not energized, the actuating rods are fully extended and the butterfly valves are closed. When the butterflies are closed, airflow through the intake runners is restricted, which decreases the amount of air that can move through the runners but increases the speed of the airflow, which improves torque, lowers emissions and increases performance. When engine speed reaches the threshold rpm level, the PCM energizes the CMCV and the crank arm on the actuator moves the actuating rods, which open the butterfly valves inside the manifold. When the butterfly valves are opened, the intake airflow through the intake runners increases. The CMCV is located on the back (firewall) side of the intake manifold, so you must remove the intake manifold to replace the CMCV.

22 Electronic Throttle Body - All vehicles covered by this manual use an electronic throttle body. There is no accelerator cable, no cruise control cable and no Idle Air Control (IAC) motor. All of these functions are handled electronically by the Powertrain Control Module (PCM). The electronic throttle body is part of Ford's Generation II (Gen II) Torque Based Electronic Throttle Control (ETC), which is a hardware and software strategy designed to deliver output shaft torque based on driver demand (the position of the accelerator pedal). Torque based ETC allows earlier upshifts and later downshifts, which improves shift quality, performance and high-altitude responsiveness, reduces emissions and saves fuel. For more information about the electronic throttle body, see Chapter 4.

23 Evaporative Emission Control (EVAP) canister purge valve - When the engine is cold or still warming up, no captive fuel vapors are allowed to escape from the EVAP canister. After the engine is warmed up, the PCM energizes the canister purge valve, which regulates the flow of these vapors from the canister to the intake manifold. The rate of the flow of vapors is regulated by the purge valve in response to commands from the PCM, which controls the duty cycle of the valve. The EVAP canister purge valve is located in the engine compartment, on a bracket bolted to the left strut tower. For more information about the EVAP system, see Section 21.

24 Exhaust Gas Recirculation (EGR) valve - When the engine is under load, the temperature inside the combustion chambers heats up dramatically. When the temperature inside the combustion chambers reaches 2500 degrees F., the engine begins to produce oxides of nitrogen (NOx), which is an odorless, colorless and toxic gas. The PCM-controlled EGR valve reduces NOx by introducing a controlled amount of spent exhaust gases into the intake manifold, which dilutes the air/fuel mixture, lowers combustion chamber temperatures and reduces the creation of NOx.

The typical Ford Differential Pressure Feedback EGR (DPFE EGR) system consists of the EGR sensor, EGR vacuum regulator solenoid, EGR valve, orifice tube assembly, the Powertrain Control Module (PCM) and the electrical wiring and vacuum hoses connecting these components. A DPFE type EGR system works like this: Based upon inputs that it receives from various information sensors, the PCM determines the right amount of EGR flow for the conditions and calculates the correct pressure drop across the metering orifice that's necessary to achieve the desired flow rate. Then it energizes the EGR vacuum regulator solenoid. The solenoid receives a variable duty cycle between zero and 100 percent. The higher the duty cycle the greater the vacuum diverted by the solenoid to the EGR valve. When vacuum is applied to the EGR valve diaphragm, it overcomes the pressure of a spring and begins to lift the valve off its seat, allowing spent exhaust gases to flow into the exhaust manifold. The spent exhaust gases must flow through a metering orifice before they flow through the EGR valve. One side of the metering orifice is exposed to the backpressure created by hot exhaust gases rushing out through the exhaust manifold and the other side of the orifice is exposed to the intake manifold. This setup creates a pressure drop across the orifice when exhaust gases are flowing through the EGR valve into the intake manifold. The PCM targets an appropriate pressure drop across the metering orifice to achieve the right EGR flow. The differential pressure feedback EGR sensor measures the pressure drop across the metering orifice and sends an analog voltage feedback signal, between 0 and 5 volts, to the PCM that's proportional to the pressure drop. The PCM uses this feedback signal to constantly trim the pressure drop across the orifice in order to maintain the correct flow rate.

Ford refers to the newest version of the DPFE EGR system, which is used only on 4.0L V6 models, as the EGR System Module (ESM), or simply the EGR module, because everything except the EGR pipe itself has been combined into a single module. The EGR module is located on the left side of the intake manifold, between the throttle body and the plenum. For more information about the EGR system, see Section 22.

25 Fuel injectors - The PCM opens the fuel injectors sequentially (in firing order sequence). The PCM also controls the pulse width, which is the interval of time during which each injector is open. The pulse width of an injector (measured in milliseconds) determines the amount of fuel delivered. For more information on the fuel delivery system and the fuel injectors, including injector replacement, refer to Chapter 4.

26 Fuel Pump Driver Module (FPDM) - The fuel systems used on all vehicles covered by this manual are returnless, i.e. there is no intake-manifold-vacuum-actuated fuel pressure regulator and there is no fuel return line between the fuel rail and the fuel tank. Instead, system fuel pressure is controlled by the speed of the fuel pump. The PCM controls the duty cycle to the FPDM, which modulates

the voltage to the fuel pump to maintain the correct fuel pressure. The FPDM is located inside the spare tire cavity in the trunk.

27 Fuel pump relay - When grounded by the PCM, the fuel pump relay connects battery voltage to the fuel pump. The fuel pump relay provides battery voltage, through the Inertia Fuel Shutoff (IFS) switch, to the fuel pump. The fuel pump relay is located inside the engine compartment fuse and relay box **(see illustration 2.2 in Chapter 4)**.

28 Ignition coils - The ignition coils are triggered by the PCM. The ignition coils are mounted on the valve cover. Refer to Chapter 5 for more information on the ignition coils.

29 Positive Crankcase Ventilation (PCV) system heater - On 4.0L V6 engines, the PCV valve is equipped with a PCM-controlled heating element. When the intake temperature is below 32 degrees F. the PCM grounds the PCV valve's heating element control circuit and turns on the heater. When the intake air reaches 48 degrees F. the PCM opens the circuit and turns off the heater. To avoid an unnecessary current drain, the PCV valve heater is always off when the engine is turned off. The PCV valve is located on the left valve cover, and looks like a conventional PCV valve except for the electrical connector plugged into it.

30 Variable Camshaft Timing (VCT) system - The VCT system, which is used on the 4.6L V8 engine, produces lower emissions, more power, better fuel efficiency and improved idle quality. There are several slightly different versions of the VCT system used on various Ford models. On the Mustang's 4.6L V8, the VCT system consists of:

- Two Camshaft Position (CMP) sensors, one on the front end of each cylinder head
- Two trigger wheels, one on the front of each camshaft timing-chain sprocket
- Two PCM-controlled hydraulic positioning control solenoids (we call them oil control solenoids), one on each head

Here's how the VCT system works: Each CMP sensor's trigger wheel has four evenly spaced teeth (equal to the number of cylinders in each head) and one extra tooth. The extra tooth between the evenly-spaced teeth represents the CMP sensor signal for that cylinder bank. The PCM also uses the signal from the Crankshaft Position (CKP) sensor for information on the crankshaft position and as a reference signal for the positioning signal from the CMP sensor.

Besides the signals from the CMP and CKP sensors, the PCM also uses signals from the Intake Air Temperature (IAT), Cylinder Heat Temperature (CHT), Engine Oil Temperature (EOT), Throttle Position (TP) and Mass Air Flow (MAF) sensors to determine the operating conditions of the engine. At idle, the PCM controls the position of the camshafts in relation to ambient air temperature and cylinder head temperature. During part and wide-open throttle, the PCM controls cam position in response to engine rpm, load and throttle

position. The PCM doesn't enable the VCT system until the engine is warmed up and the operating conditions are right.

The PCM-controlled VCT oil control solenoid controls the flow of engine oil in the VCT actuator assembly, which is an integral component of the camshaft timing chain sprocket. When the oil control solenoid is energized by the PCM, engine oil flows into the actuator assembly, which advances or retards the camshaft timing. One half of the actuator is connected to the timing chain and the other half is coupled to the camshaft. Oil chambers between the two halves couple the timing chain to the cam. When the flow of oil is shifted from one side of the chamber to the other, the differential change in oil pressure forces the camshaft to rotate in either an advance or retard position, depending on which way the oil flows. As the PCM alters the duty cycle of the solenoid valve, oil pressure/flow advances or retards the camshaft timing. When a fixed camshaft phase is called for, the PCM achieves a steady cam phase by *dithering* (oscillating) the duty cycle of the oil control solenoid valve.

OBD-II Diagnostic Trouble Codes (DTCs) and the Malfunction Indicator Light (MIL)

31 To test the critical emission control components, circuit and systems on an OBD-II vehicle, the PCM runs a series of *monitors* during each vehicle *trip*. The monitors are a series of testing protocols used by the PCM to determine whether each monitored component, circuit or system is functioning satisfactorily. The monitors must be run in a certain order. For example, the oxygen sensor monitor cannot run until the engine, the catalytic converter and the oxygen sensors are all warmed up. Another example, the misfire monitor cannot run until the engine is in closed-loop operation. And so on. (For a good overview of the OBD-II monitors, see the Haynes *OBD-II and Electronic Engine Management Systems* Techbook.) An OBD-II *trip* consists of operating the vehicle (after an engine-off period) and driving it in such a manner that the PCM's monitors test all of the monitored components, circuits and systems at least once.

32 If the PCM recognizes a fault in some component, circuit or system while it's running the monitors, it stores a Diagnostic Trouble Code (DTC) and turns on the Malfunction Indicator Light (MIL) on the instrument cluster. A DTC can self-erase, but only after the MIL has been extinguished. For example, the MIL might be extinguished for a misfire or fuel system malfunction if the fault doesn't recur when monitored during the next three subsequent sequential driving cycles in which the conditions are similar to those under which the malfunction was first identified. (For other types of malfunctions, the criteria for extinguishing the MIL can vary.)

33 Once the MIL has been extinguished,

the PCM must pass the diagnostic test for the most recent DTC for 40 *warm-up cycles* (80 warm-up cycles for the fuel system monitor and the misfire monitor). A warm-up cycle consists of the following chain of events:

The engine has been started and is running

The engine temperature rises by at least 40-degrees above its temperature when it was started

The engine coolant temperature crosses the 160-degree F mark

The engine is turned off after meeting the above criteria

Obtaining DTCs

Refer to illustration 2.34

34   Of course, if the MIL does NOT go out after several driving cycles, it's probably an indication that something must be repaired or replaced before the DTC can be erased and the MIL extinguished. This means that you will need to extract the DTC(s) from the

PCM, make the necessary repair or replace a component, then erase the DTC yourself. You can extract the DTCs from the PCM by plugging a code reader or generic OBD-II scan tool **(see illustrations 2.1 and 2.2)** into the PCM's data link connector **(see illustration)**, which is located under the left side of the dash. Plug the tool into the 16-pin data link connector (DLC), then follow the instructions included with the tool to extract all the diagnostic codes.

Erase the DTC(s), turn off the MIL and verify the repair

35   Once you've completed the repair or replaced the component, use your code reader or scan tool to erase the DTC(s) and turn off the MIL. On most tools, you simply press a button to erase DTCs and turn off the MIL, but on some tools you'll have to locate this function by using the menu on the tool's display. If it isn't obvious, follow the instructions that come with your tool.

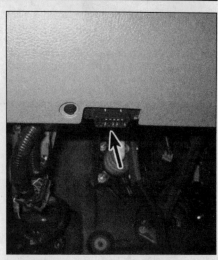

2.34 The 16-pin Data Link Connector (DLC) is located under the left side of the dash

OBD-II trouble codes

Note: *Not all trouble codes apply to all models.*

Code	Probable cause
P0010	Intake camshaft position actuator circuit open (Bank 1)
P0011	Intake camshaft position timing over-advanced (Bank 1)
P0012	Variable camshaft position timing over-retarded (Bank 1)
P0016	Crankshaft position-camshaft position correlation, sensor A (Bank 1)
P0018	Crankshaft position-camshaft position correlation, sensor A (Bank 2)
P0020	Intake camshaft position actuator circuit open (Bank 2)
P0021	Intake camshaft position timing over-advanced (Bank 2)
P0022	Intake camshaft position timing over-retarded (Bank 2)
P0040	Upstream oxygen sensors swapped from bank to bank (crossed wiring harnesses)
P0041	Downstream oxygen sensors swapped from bank to bank (crossed wiring harnesses)
P0053	Upstream oxygen sensor 1 heater resistance, heater current requirements too low or too high (Bank 1)
P0054	Downstream oxygen sensor 2 heater resistance, heater current requirements too low or too high (Bank 1)
P0055	Downstream oxygen sensor 3 heater resistance, heater current requirements too low or too high (Bank 1)
P0059	Upstream oxygen sensor 1 heater resistance, heater current requirements too low or too high (Bank 2)
P0060	Downstream oxygen sensor 2 heater resistance, heater current requirements too or too high (Bank 2)
P0061	Downstream oxygen sensor 3 heater resistance, heater current requirements too low or too high (Bank 2)
P0068	Throttle Position (TP) sensor inconsistent with Mass Air Flow sensor
P0097	Intake Air Temperature (IAT) sensor 2 circuit, low voltage
P0098	Intake Air Temperature (IAT) sensor 2 circuit, high voltage
P0102	Mass Air Flow (MAF) sensor A circuit, low input

Code	Probable cause
P0103	Mass Air Flow (MAF) sensor A circuit, high input
P0104	Mass Air Flow (MAF) sensor A circuit, intermittent or erratic
P0106	Manifold Absolute Pressure (MAP)/Barometric pressure (BARO) sensor circuit, range/performance problem
P0107	Manifold Absolute Pressure (MAP)/Barometric pressure (BARO) sensor circuit, low voltage detected
P0108	Manifold Absolute Pressure (MAP)/Barometric pressure (BARO) sensor circuit, high voltage detected
P0109	Manifold Absolute Pressure (MAP)/Barometric pressure (BARO) sensor circuit, intermittent
P0111	Intake Air Temperature (IAT) sensor 1 circuit, range or performance problem
P0112	Intake Air Temperature (IAT) sensor 1 circuit, low input
P0113	Intake Air Temperature (IAT) sensor 1 circuit, high input
P0114	Intake Air Temperature (IAT) sensor circuit, intermittent or erratic
P0116	Engine Coolant Temperature (ECT) sensor 1 circuit, range or performance problem
P0117	Engine Coolant Temperature (ECT) sensor 1 circuit, low input
P0118	Engine Coolant Temperature (ECT) sensor 1 circuit, high input
P0119	Engine Coolant Temperature (ECT) sensor 1 circuit, intermittent or erratic
P0121	Throttle Position (TP1) sensor circuit, range or performance problem
P0121	Throttle Position (TP)/Accelerator Pedal Position (APP) sensor circuit, range or performance problem
P0122	Throttle Position (TP1) sensor circuit, low input
P0122	Throttle Position (TP)/Accelerator Pedal Position (APP) sensor circuit, low input
P0123	Throttle Position (TP1) sensor circuit, high input
P0123	Throttle Position (TP)/Accelerator Pedal Position (APP) sensor circuit, high input
P0125	Insufficient coolant temperature for closed loop fuel control
P0127	Intake Air Temperature (IAT) sensor, temperature too high
P0128	Coolant thermostat, temperature below thermostat regulating temperature
P0131	Upstream oxygen sensor 1 circuit, out of range or low voltage (Bank 1)
P0132	Upstream oxygen sensor 1 circuit, high voltage (Bank 1)
P0133	Upstream oxygen sensor 1 circuit, slow response (Bank 1)
P0135	Upstream oxygen sensor 1 circuit, excessive current draw, open circuit or short circuit (Bank 1)
P0136	Downstream oxygen sensor 2 circuit, circuit malfunction (Bank 1)
P0138	Downstream oxygen sensor 2 circuit, high voltage (Bank 1)
P0139	Downstream oxygen sensor 2 circuit, slow response (Bank 1)
P0141	Downstream oxygen sensor circuit 2, excessive current draw, open circuit or short circuit (Bank 1)
P0144	Downstream oxygen sensor circuit 3, high voltage (Bank 1)
P0147	Downstream oxygen sensor circuit 3, excessive current draw, open circuit or short circuit (Bank 1)
P0148	Fuel delivery error
P0151	Upstream oxygen sensor 1 circuit out of range or low voltage (Bank 2)

OBD-II trouble codes (continued)

Note: *Not all trouble codes apply to all models.*

Code	Probable cause
P0152	Upstream oxygen sensor 1 circuit, high voltage (Bank 2)
P0153	Upstream oxygen sensor 1 circuit, slow response (Bank 2)
P0155	Upstream oxygen sensor 1 circuit, excessive current draw, open circuit or short circuit (Bank 2)
P0156	Downstream oxygen sensor 2 circuit, circuit malfunction (Bank 2)
P0158	Downstream oxygen sensor 2 circuit, high voltage (Bank 2)
P0159	Downstream oxygen sensor 2 circuit, slow response (Bank 2)
P0161	Downstream oxygen sensor 2 circuit, excessive current draw, open circuit or short circuit (Bank 2)
P0167	Downstream oxygen sensor 3 circuit, excessive current draw, open circuit or short circuit (Bank 2)
P0171	System too lean (Bank 1)
P0172	System too rich (Bank 1)
P0174	System too lean (Bank 2)
P0175	System too rich (Bank 2)
P0180	Fuel temperature sensor A circuit malfunction, high or low voltage
P0181	Fuel temperature sensor A circuit, range or performance problem
P0182	Fuel temperature sensor A circuit, low input
P0183	Fuel temperature sensor A circuit, high input
P0190	Fuel Rail Pressure (FRP) sensor circuit malfunction
P0191	Fuel Rail Pressure (FRP) sensor circuit, range or performance problem
P0192	Fuel Rail Pressure (FRP) sensor circuit, low input
P0193	Fuel Rail Pressure (FRP) sensor circuit, high input
P0196	Engine Oil Temperature (EOT) sensor circuit, range or performance problem
P0197	Engine Oil Temperature (EOT) sensor circuit, low input
P0198	Engine Oil Temperature (EOT) sensor circuit, high input
P0201	Injector No. 1 circuit malfunction
P0202	Injector No. 2 circuit malfunction
P0203	Injector No. 3 circuit malfunction
P0204	Injector No. 4 circuit malfunction
P0205	Injector No. 5 circuit malfunction
P0206	Injector No. 6 circuit malfunction
P0207	Injector No. 7, circuit malfunction
P0208	Injector No. 8, circuit malfunction
P0217	Engine coolant over-temperature condition
P0218	Transmission Fluid Temperature (TFT), over-temperature condition

Code	Probable cause
P0219	Engine over-speed condition
P0221	Throttle Position (TP) sensor 2 circuit, range or performance problem
P0221	Throttle Position (TP)/Accelerator Pedal Position (APP) sensor B circuit, range or performance problem
P0222	Throttle Position (TP) sensor 2 circuit, low input
P0222	Throttle Position (TP)/Accelerator Pedal Position (APP) sensor B circuit, low voltage
P0223	Throttle Position (TP) sensor 2 circuit, high input
P0223	Throttle Position (TP)/Accelerator Pedal Position (APP) sensor B circuit, high voltage
P0230	Fuel pump primary circuit malfunction
P0231	Fuel pump secondary circuit, low voltage
P0232	Fuel pump secondary circuit, high voltage
P0261	Cylinder 1 injector circuit, low voltage
P0262	Cylinder 1 injector circuit, high voltage
P0264	Cylinder 2 injector circuit, low voltage
P0265	Cylinder 2 injector circuit, high voltage
P0267	Cylinder 3 injector circuit, low voltage
P0268	Cylinder 3 injector circuit, high voltage
P0270	Cylinder 4 injector circuit, low voltage
P0271	Cylinder 4 injector circuit, high voltage
P0273	Cylinder 5 injector circuit, low voltage
P0274	Cylinder 5 injector circuit, high voltage
P0276	Cylinder 6 injector circuit, low voltage
P0277	Cylinder 6 injector circuit, high voltage
P0279	Cylinder 7 injector circuit, low voltage
P0280	Cylinder 7 injector circuit, high voltage
P0282	Cylinder 8 injector circuit, low voltage
P0283	Cylinder 8 injector circuit, high voltage
P0297	Vehicle over-speed condition
P0298	Engine oil high-temperature condition
P0300	Random misfire detected
P0301	Cylinder No. 1, misfire detected
P0302	Cylinder No. 2, misfire detected
P0303	Cylinder No. 3, misfire detected
P0304	Cylinder No. 4, misfire detected
P0305	Cylinder No. 5, misfire detected
P0306	Cylinder No. 6, misfire detected

OBD-II trouble codes (continued)

Note: *Not all trouble codes apply to all models.*

Code	Probable cause
P0307	Cylinder No. 7, misfire detected
P0308	Cylinder No. 8, misfire detected
P0310	Misfire detection monitor
P0315	PCM unable to learn crankshaft pulse wheel tooth spacing (exceeds allowable correction tolerances)
P0316	Misfire detected during first 1000 engine revolutions
P0320	Ignition engine speed input, circuit malfunction
P0325	Knock sensor 1, circuit malfunction (Bank 1)
P0326	Knock sensor No. 1 circuit, range or performance problem (Bank 1)
P0330	Knock sensor 2, circuit malfunction (Bank 2)
P0331	Knock sensor 2 circuit, range or performance problem (Bank 2)
P0340	Camshaft Position (CMP) sensor A, circuit malfunction (Bank 1, or single sensor)
P0344	Camshaft Position (CMP) sensor A circuit, intermittent (Bank 1, or single sensor)
P0345	Camshaft Position (CMP) sensor A, circuit malfunction (Bank 2)
P0349	Camshaft Position (CMP) sensor A circuit, intermittent (Bank 2)
P0350	Ignition coil (undetermined), primary or secondary circuit malfunction
P0351	Ignition coil A, primary or secondary circuit malfunction
P0352	Ignition coil B, primary or secondary circuit malfunction
P0353	Ignition coil C, primary or secondary circuit malfunction
P0354	Ignition coil D, primary or secondary circuit malfunction
P0355	Ignition coil E, primary or secondary circuit malfunction
P0356	Ignition coil F, primary or secondary circuit malfunction
P0357	Ignition coil G, primary or secondary circuit malfunction
P0358	Ignition coil H, primary or secondary circuit malfunction
P0400	Exhaust Gas Recirculation (EGR) flow failure (outside minimum or maximum limits)
P0401	Exhaust Gas Recirculation (EGR) flow, insufficient flow detected
P0402	Exhaust Gas Recirculation (EGR) flow, excessive flow detected
P0403	Exhaust Gas Recirculation (EGR) control circuit
P0405	Differential Pressure Feedback Exhaust Gas Recirculation (DPFEGR) sensor circuit, low voltage detected
P0406	Differential Pressure Feedback Exhaust Gas Recirculation (DPFEGR) sensor circuit, high voltage detected
P0410	Secondary Air Injection (AIR) system, lack of air flow detected with secondary AIR pump operating
P0411	Secondary Air Injection (AIR) system, incorrect upstream flow detected
P0412	Secondary Air Injection (AIR) system, switching valve A circuit malfunction
P0420	Catalyst system efficiency below threshold (Bank 1)

Code	Probable cause
P0430	Catalyst system efficiency below threshold (Bank 2)
P0442	Evaporative Emission (EVAP) control system, small leak detected
P0443	Evaporative Emission (EVAP) control system, canister purge control valve circuit malfunction
P0446	Evaporative Emission (EVAP) control system, canister vent solenoid or control circuit malfunction
P0451	Evaporative Emission (EVAP) system, Fuel Tank Pressure (FTP) sensor circuit, out of range or intermittent
P0452	Fuel Tank Pressure (FTP) sensor circuit, low voltage detected
P0453	Fuel Tank Pressure (FTP) sensor circuit, high voltage detected
P0454	Fuel Tank Pressure (FTP) sensor, circuit noisy
P0454	Evaporative Emission (EVAP) system pressure sensor, intermittent circuit
P0455	Evaporative Emission (EVAP) control system, leak detected (no purge flow or large leak)
P0456	Evaporative Emission (EVAP) control system, very small leak detected
P0457	Evaporative Emission (EVAP) control system, leak detected (fuel filler neck cap loose or off)
P0460	Fuel level sensor, circuit malfunction
P0461	Fuel level sensor circuit, range or performance problem
P0462	Fuel level sensor circuit, low input
P0463	Fuel level sensor circuit, high input
P0480	Low Fan Control (LFC)/Fan Control 1 (FC1), primary circuit malfunction
P0481	High Fan Control (HFC)/Fan Control 3 (FC3), primary circuit malfunction
P0481	Fan 2 control circuit
P0482	Medium Fan Control (MFC), primary circuit failure
P0482	Fan 3 control circuit
P0483	Fan performance
P0491	Secondary Air Injection (AIR) system, insufficient flow (Bank 1)
P0500	Vehicle Speed Sensor (VSS) malfunction
P0503	Vehicle Speed Sensor (VSS) circuit, intermittent, erratic or high voltage
P0505	Idle Air Control (IAC) system malfunction
P0506	Idle Air Control (IAC) system, rpm lower than expected
P0507	Idle Air Control (IAC) system, rpm higher than expected
P050A	Cold start idle air control performance
P050B	Cold start ignition timing performance
P050E	Cold start engine exhaust temperature out of range
P0511	Idle Air Control (IAC) system, circuit malfunction
P0512	Starter request circuit, short circuit
P0528	Fan speed sensor circuit, no signal
P0532	Air Conditioning Pressure (ACP) sensor, low voltage detected

OBD-II trouble codes (continued)

Note: *Not all trouble codes apply to all models.*

Code	Probable cause
P0533	Air Conditioning Pressure (ACP) sensor, high voltage detected
P0534	Low air conditioning cycling period or refrigerant charge loss (frequent A/C compressor clutch cycling)
P0537	Air Conditioning Evaporator Temperature (ACET) circuit, low input
P0538	Air Conditioning Evaporator Temperature (ACET) circuit, high input
P053A	Positive Crankcase Ventilation (PCV) heater control circuit, open circuit
P0552	Power Steering Pressure (PSP) sensor circuit, low input
P0553	Power Steering Pressure (PSP) sensor circuit, high input
P0579	Cruise control multifunction input circuit, range or performance problem
P0581	Cruise control multifunction input circuit, high voltage
P0600	Serial communication link
P0602	Powertrain Control Module (PCM) programming error
P0603	Powertrain Control Module (PCM), internal Keep-Alive-Memory (KAM) error
P0604	Powertrain Control Module (PCM), internal control module Random Access Memory (RAM) error
P0605	Powertrain Control Module (PCM), Read Only Memory (ROM) error
P0606	Powertrain Control Module (PCM), internal communication error
P0607	Powertrain Control Module (PCM), performance problem
P060A	Powertrain Control Module (PCM), internal control module monitoring processor performance
P060B	Powertrain Control Module (PCM), internal control module analog-to-digital (A/D) processing performance
P060C	Powertrain Control Module (PCM), internal control module main processor performance
P0610	Powertrain Control Module (PCM), vehicle options error
P061B	Powertrain Control Module (PCM), internal control module, torque calculation error
P061C	Powertrain Control Module (PCM), internal control module engine rpm performance, calculation error
P061D	Powertrain Control Module (PCM), internal control module engine air mass performance, calculation error
P061F	Powertrain Control Module (PCM), internal control module throttle actuator controller, performance error
P0620	Alternator control circuit failure
P0622	Alternator field terminal circuit failure
P0625	Alternator field terminal circuit, low voltage
P0626	Alternator field terminal circuit, high voltage
P062C	Powertrain Control Module (PCM), internal control module vehicle speed performance, calculation error
P0642	Sensor reference voltage circuit, low voltage
P0643	Sensor reference voltage circuit, high voltage
P0645	Air Conditioning Clutch Relay (A/CCR) control circuit malfunction
P0657	Actuator supply voltage circuit open

Code	Probable cause
P0685	Powertrain Control Module (PCM) power relay control circuit, open circuit
P0689	Powertrain Control Module (PCM) power relay sense circuit, low voltage
P0690	Powertrain Control Module (PCM) power relay sense circuit, high voltage
P0703	Brake switch circuit input malfunction
P0704	Clutch Pedal Position (CPP) switch input circuit malfunction
P0705	Transmission Range (TR) sensor, circuit failure
P0707	Transmission Range (TR) sensor circuit, low voltage
P0708	Transmission Range (TR) sensor circuit, open circuit
P0711	Transmission Range (TR) sensor circuit, out of on-board diagnostic range or inoperative sensor
P0712	Transmission Fluid Temperature (TFT) sensor circuit, grounded circuit
P0713	Transmission Fluid Temperature (TFT) sensor circuit, open circuit
P0715	Turbine Speed Sensor (TSS), insufficient input
P0717	Turbine Speed Sensor (TSS), insufficient input
P0718	Turbine Speed Sensor (TSS), noisy signal
P0720	Output Shaft Speed (OSS) sensor circuit open or shorted or insufficient input
P0721	Output Shaft Speed (OSS) sensor circuit, range or performance problem or noisy signal
P0722	Output Shaft Speed (OSS) sensor circuit, insufficient input or no signal
P0723	Output Shaft Speed (OSS) sensor circuit, intermittent failure
P0731	First gear ratio error
P0732	Second gear ratio error
P0733	Third gear ratio error
P0734	Fourth gear ratio error
P0735	Fifth gear ratio error
P0740	Torque Converter Clutch (TCC) solenoid circuit failure during on-board diagnostic
P0741	Torque Converter Clutch (TCC) solenoid circuit, error or stuck in off position
P0743	Torque Converter Clutch (TCC) solenoid circuit, failure during on-board diagnostics
P0745	Pressure Control Solenoid A (PCA), solenoid or circuit fault
P0748	Pressure Control Solenoid A (PCA), solenoid inoperative
P0750	Shift Solenoid A (SSA), solenoid circuit failure
P0753	Shift Solenoid A (SSA), solenoid circuit failure
P0755	Shift Solenoid B (SSB), solenoid circuit failure
P0758	Shift Solenoid B (SSB), solenoid circuit failure
P0760	Shift Solenoid C (SSC), solenoid circuit failure
P0763	Shift Solenoid C (SSC), solenoid circuit failure
P0765	Shift Solenoid D (SSD), solenoid circuit failure

OBD-II trouble codes (continued)

Note: *Not all trouble codes apply to all models.*

Code	Probable cause
P0768	Shift Solenoid D (SSD), solenoid circuit failure
P0775	Pressure Control Solenoid B (PCB), solenoid or circuit fault
P0778	Pressure Control Solenoid B (PCB), solenoid inoperative
P0791	Intermediate Shaft Speed (ISS) sensor signal failure
P0794	Intermediate Shaft Speed (ISS) sensor, intermittent signal
P0795	Pressure Control Solenoid C (PCC), solenoid or circuit fault
P0798	Pressure Control Solenoid C (PCC), solenoid inoperative
P0830	Clutch Pedal Position (CPP) switch A, circuit malfunction
P0833	Clutch Pedal Position (CPP) switch B, circuit malfunction
P0840	Transmission fluid pressure sensor, circuit malfunction
P0960	Pressure Control Solenoid A (PCA), solenoid circuit open
P0962	Pressure Control Solenoid A (PCA) solenoid circuit failure or short to ground
P0963	Pressure Control Solenoid A (PCA) solenoid shorted to battery power or to ground
P0964	Pressure Control Solenoid B (PCB) solenoid circuit open
P0966	Pressure Control Solenoid B (PCB) solenoid circuit failure, shorted to ground
P0967	Pressure Control Solenoid B (PCB) solenoid shorted to battery voltage or to ground
P0968	Pressure Control Solenoid C (PCC) solenoid circuit open
P0970	Pressure Control Solenoid C (PCC) solenoid circuit failure, shorted to ground
P0971	Pressure Control Solenoid C (PCC) solenoid shorted to power or to ground

3 Accelerator Pedal Position Sensor (APPS) - replacement

Refer to illustration 3.3

Note: *The APP sensor is located on the right side of the upper end of the accelerator pedal. If the APP sensor must be replaced, you must replace the accelerator pedal and sensor as a single assembly.*

1   Disconnect the cable from the negative battery terminal (see Chapter 5, Section 1).

2   Remove the knee bolster (see Chapter 11).

3   Disconnect the electrical connector from the APP sensor (**see illustration**).

4   Remove the APP sensor and pedal assembly mounting nuts and detach the assembly from the firewall.

5   Installation is the reverse of removal.

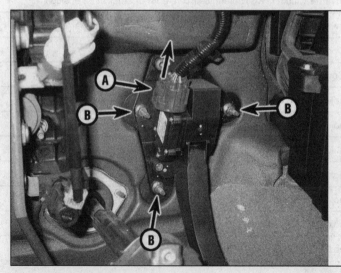

3.3 To disconnect the electrical connector from the Accelerator Pedal Position (APP) sensor assembly, slide the lock up (toward the harness), then depress the release tab (A) and pull off the connector. To remove the APP sensor and pedal assembly, remove the three nuts (B)

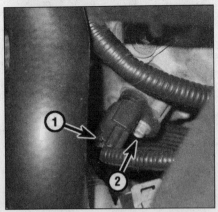

4.7a To remove the CMP sensor from the left cylinder head, depress the release tab (1) and disconnect the electrical connector, then remove the sensor mounting bolt (2)

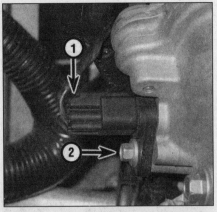

4.7b To remove the CMP sensor from the right cylinder head, depress the release tab (1) and disconnect the electrical connector, then remove the sensor mounting bolt (2)

6.13 To detach the CKP sensor from a 4.6L V8 engine, remove this mounting bolt and pull out the sensor

4 Camshaft Position (CMP) sensor - replacement

4.0L V6 models

Note: *The CMP sensor is located on top of the forward end of the left valve cover.*
1   Disconnect the cable from the negative battery terminal (see Chapter 5, Section 1).
2   Disconnect the electrical connector from the CMP sensor.
3   Remove the CMP sensor mounting bolt and remove the sensor. If the same sensor is to be reinstalled, check the O-ring to make sure it isn't hardened or damaged.
4   Installation is the reverse of removal.

4.6L V8 models

Refer to illustrations 4.7a and 4.7b
Note: *On these models there are two CMP sensors, once for each cylinder head. The CMP sensors are located on the front ends of the cylinder heads.*
5   Disconnect the cable from the negative battery terminal (see Chapter 5, Section 1).
6   Remove the air intake duct (see *Air filter housing - removal and installation* in Chapter 4).
7   Disconnect the electrical connector from the CMP sensor **(see illustrations)**.
8   Remove the CMP sensor mounting bolt and remove the sensor from the cylinder head.
9   Remove and discard the old CMP sensor O-ring.
10   Installation is the reverse of removal. Be sure to use a new O-ring.

5 Clutch Pedal Position (CPP) switch - replacement

Refer to Chapter 8.

6 Crankshaft Position (CKP) sensor - replacement

4.0L V6 models

Note: *The CKP sensor is located on the front of the timing chain cover, at about 10 o'clock in relation to the crankshaft.*
1   Disconnect the cable from the negative battery terminal (see Chapter 5, Section 1).
2   Raise the front of the vehicle and place it securely on jackstands.
3   Disconnect the electrical connector from the CKP sensor.
4   Remove the CKP sensor mounting bolts and remove the sensor from the timing chain cover.
5   Installation is the reverse of removal.

4.6L V8 models

Refer to illustration 6.13
Note: *The CKP sensor is located at the right front corner of the engine block. You must detach the air conditioning compressor to access the CKP sensor.*
6   Disconnect the cable from the negative battery terminal (see Chapter 5, Section 1).
7   Remove the drivebelt (see Chapter 1).
8   Raise the front of the vehicle and place it securely on jackstands.
9   Detach the wiring harness clips from the air conditioning compressor stud bolts.
10   Disconnect the electrical connector from the air conditioning compressor clutch.
11   Unbolt the air conditioning compressor (see Chapter 3) and set it aside to provide access to the CKP sensor. **Warning:** *Don't disconnect the refrigerant lines.*
12   Disconnect the electrical connector from the CKP sensor.
13   Remove the CKP sensor mounting bolt **(see illustration)** and remove the sensor from the block.
14   Installation is the reverse of removal.

7 Cylinder Head Temperature (CHT) sensor - replacement

Refer to illustration 7.3
Note 1: *This procedure applies to V8 models only.*
Note 2: *The CHT sensor is located on the inner wall of the right (passenger's side) cylinder head. To access the CHT sensor, you must remove the intake manifold. The manufacturer says that if you remove the CHT sensor you must replace it with a new one.*
1   Disconnect the cable from the negative battery terminal (see Chapter 5, Section 1).
2   Remove the intake manifold (see Chapter 2B).
3   Disconnect the electrical connector from the CHT sensor **(see illustration)**.

7.3 To remove the CHT sensor from the head, disconnect the electrical connector, then unscrew the sensor

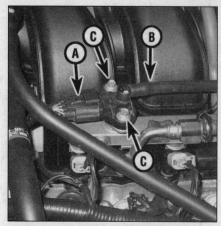

9.3 To remove the FRPT sensor from the fuel rail, relieve the fuel system pressure, depress the release tab (A) and disconnect the electrical connector, then disconnect the vacuum hose (B) from the sensor and remove the sensor mounting bolts (C)

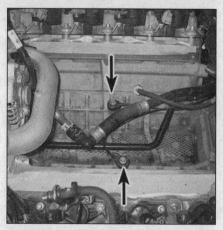

11.9 There are two knock sensors on 4.6L V8 engines. They have separate electrical leads but share a common electrical connector, so they must be removed or replaced as a single assembly

12.2 To remove the MAF sensor, slide out the lock, depress the release tab and disconnect the electrical connector, then remove the two mounting screws

4   Unscrew and remove the CHT sensor. **Note:** *Remember: You cannot reuse the CHT sensor.*
5   Installation is the reverse of removal. Be sure to tighten the CHT sensor to the torque listed in this Chapter's Specifications.

8   Engine Coolant Temperature (ECT) sensor - replacement

Warning: *The engine must be completely cool before beginning this procedure.*
Note: *The ECT sensor, which is used only on 4.0L V6 engines, is located on top of the engine, near the upper radiator hose.*
1   Disconnect the cable from the negative battery terminal (see Chapter 5, Section 1).
2   Drain the cooling system (see Chapter 1).
3   Disconnect the electrical connector from the ECT sensor.
4   Remove the ECT sensor retaining clip and remove the sensor.
5   Installation is the reverse of removal.
6   Refill the cooling system (see Chapter 1).

9   Fuel Rail Pressure and Temperature (FRPT) sensor - replacement

Refer to illustration 9.3
Note: *Both engines are equipped with an FRPT sensor, which is located on the left fuel rail. The photos accompanying this section depict the FRPT sensor on a 4.6L V8 engine, but the FRPT sensor used on the 4.0L V6 is virtually identical.*
1   Relieve the fuel pressure in the fuel system (see Chapter 4).
2   Disconnect the cable from the negative

battery terminal (see Chapter 5, Section 1).
3   Disconnect the electrical connector from the FRPT sensor **(see illustration)**.
4   Disconnect the vacuum hose from the FRPT sensor.
5   Remove the FRPT sensor mounting bolts and remove the sensor from the fuel rail.
6   Remove the old O-ring from the FRPT sensor.
7   Carefully clean off the mating surfaces of the FRPT sensor and the fuel rail.
8   Install a new O-ring on the FRPT sensor.
9   Installation is the reverse of removal. Be sure to tighten the FRPT sensor mounting bolts to the torque listed in this Chapter's Specifications.
10   When you're done start the engine and check for fuel leaks in the vicinity of the FRPT sensor.

10   Intake Air Temperature (IAT) sensor - replacement

The IAT sensor is an integral part of the Mass Air Flow (MAF) sensor on all models covered by this manual. If the IAT sensor must be replaced, you must replace the MAF sensor (see Section 12).

11   Knock sensor - replacement

4.0L V6 models

Note: *The knock sensor is located in the valley between the cylinder heads, underneath the intake manifold, which you must remove to access it.*
1   Disconnect the cable from the negative battery terminal (see Chapter 5, Section 1).
2   Remove the intake manifold (see Chapter 2A).
3   Trace the knock sensor lead to the elec-

trical connector for the sensor and disconnect it.
4   Remove the knock sensor mounting bolt and remove the knock sensor.
5   Installation is the reverse of removal. Be sure to tighten the knock sensor retaining bolt to the torque listed in this Chapter's Specifications.

4.6L V8 models

Refer to illustration 11.9
Note: *The knock sensors are located in the valley between the cylinder heads, underneath the intake manifold, which must be removed to access either sensor. There are two knock sensors on 4.6L V8 engines, but they share the same harness and connector, so they must replaced as a single assembly.*
6   Disconnect the cable from the negative battery terminal (see Chapter 5, Section 1).
7   Remove the intake manifold (see Chapter 2B).
8   Trace the two leads from the knock sensor to their common electrical connector and disconnect it.
9   Remove the knock sensor retaining bolts **(see illustration)** and remove the knock sensors.
10   Installation is the reverse of removal. Be sure to tighten the knock sensor retaining bolts to the torque listed in this Chapter's Specifications.

12   Mass Air Flow (MAF) sensor - replacement

Refer to illustration 12.2
Note: *The MAF sensor is mounted on the air filter housing. The photos accompanying this section depict the MAF sensor on a V8 models, but the MAF sensor used on V6 models is identical.*
1   Disconnect the cable from the negative battery terminal (see Chapter 5, Section 1).
2   Disconnect the electrical connector from the MAF sensor **(see illustration)**.

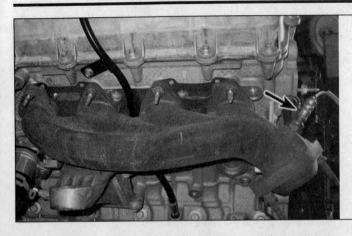

13.4a On 4.6L V8 models, the left upstream oxygen sensor is screwed into the exhaust manifold, right above the manifold flange (engine removed for clarity)

13.4b On 4.6L V8 models, the right upstream oxygen sensor is screwed into the exhaust pipe, right below the exhaust manifold flange. On 4.0L V6 models, both upstream sensors are screwed into the exhaust pipes right below the manifold flanges, at this same location

3   Remove the two MAF sensor mounting screws and remove the sensor from the air filter housing.
4   Installation is the reverse of removal. Be sure to tighten the MAF sensor mounting screws to the torque listed in this Chapter's Specifications.

13 Oxygen sensors - general information and replacement

General information

1   Use special care when servicing an oxygen sensor:

a) *Oxygen sensors have a permanently attached pigtail and electrical connector that can't be removed from the sensor. Damage to or removal of the pigtail or the electrical connector will ruin the sensor.*
b) *Keep grease, dirt and other contaminants away from the electrical connector and the louvered end of the sensor.*
c) *Do not use cleaning solvents of any kind on an oxygen sensor or air/fuel ratio sensor.*
d) *Do not drop or roughly handle an oxygen sensor or air/fuel ratio sensor.*
e) *Be sure to install the silicone boot in the correct position to prevent the boot from melting and to allow the sensor to operate properly.*

Replacement

Note: *Because the oxygen sensors are installed in the exhaust system, which contracts when cool, an oxygen sensor might be difficult to loosen when the engine is cold. Rather than risk damage to the sensor, start and run the engine for a minute or two, then shut it off. Be careful not to burn yourself during the following procedure.*

2   Disconnect the cable from the negative terminal of the battery (see Chapter 5, Section 1).
3   Raise the vehicle and place it securely on jackstands.

Upstream oxygen sensor

Refer to illustrations 13.4a and 13.4b

Note: *There are two upstream oxygen sensors on all models. On 4.0L V6 models, both upstream sensors are screwed into the exhaust pipes right below the manifold flanges. On 4.6L V8 models, the left upstream sensor is screwed into the exhaust manifold, right above the manifold flange, and the right upstream sensor is screwed into the exhaust pipe right below the exhaust manifold flange.*

4   Locate the upstream oxygen sensor that you wish to replace **(see illustrations)**, then trace the lead to the electrical connector and disconnect it.
5   Using a special oxygen sensor socket unscrew the sensor.

6   After removing the old upstream oxygen sensor, clean the threads of the sensor bore in the exhaust manifold.
7   If you're going to install the old sensor, apply anti-seize compound to the threads of the sensor to facilitate future removal. If you're going to install a new oxygen sensor, it's not necessary to apply anti-seize compound to the threads. The threads on new sensors already have anti-seize compound on them.
8   Installation is otherwise the reverse of removal. Be sure to tighten the upstream sensor to the torque listed in this Chapter's Specifications.

Downstream oxygen sensor

Refer to illustration 13.9

Note: *There are two downstream oxygen sensors on all models. They're screwed into the catalytic converters.*

9   Disconnect the downstream oxygen sensor electrical connector **(see illustration)**.
10   Unscrew and remove the downstream oxygen sensor.
11   After removing the old downstream oxygen sensor, clean the threads of the sensor bore in the catalytic converter.
12   If you're going to install the old sensor, apply anti-seize compound to the threads of the sensor to facilitate future removal.
13   If you're going to install a new oxygen sensor, it's not necessary to apply anti-seize compound to the threads. The threads on new sensors already have anti-seize compound on them.
14   Installation is otherwise the reverse of removal. Be sure to tighten the downstream oxygen sensor to the torque listed in this Chapter's Specifications.

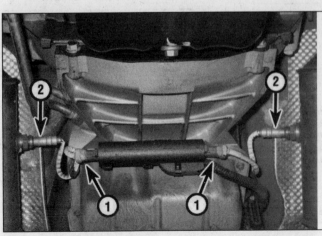

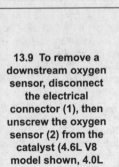

13.9 To remove a downstream oxygen sensor, disconnect the electrical connector (1), then unscrew the oxygen sensor (2) from the catalyst (4.6L V8 model shown, 4.0L V6 models similar)

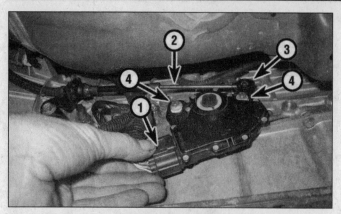

14.5 To remove the Transmission Range (TR) sensor:

1   Disconnect the electrical connector
2   Detach the shift control cable . . .
3   . . . by prying the cable end off the manual control lever
4   Remove the TR sensor mounting bolts

16.2 To disconnect the three electrical connectors from the PCM, flip each hinged connector lock forward, then carefully pull off each connector. To detach the PCM from its mounting bracket, remove the upper mounting bolt (shown) and the lower bolt (below the lower connector, not visible in this photo)

14   Transmission Range (TR) sensor - replacement

Refer to illustration 14.5

Note 1: *The TR sensor is located on the left side of the transmission. To access the TR sensor you'll have to lower the transmission slightly because the sensor is too close to the transmission tunnel to remove when the transmission is in position.*

Note 2: *If you're replacing the TR sensor (not just removing and installing the same sensor on the same transmission), or if you're planning to remove the sensor to replace the transmission with a new or rebuilt unit, then you'll need the special TR sensor alignment tool (307-351), or a suitable substitute, before beginning this procedure.*

1   Put the shift lever in Neutral.
2   Disconnect the cable from the negative battery terminal (see Chapter 5, Section 1).
3   Raise the vehicle and place it securely on jackstands.
4   Support the transmission securely, then remove the transmission crossmember (see Chapter 7B) and lower the transmission far enough to access the TR sensor.
5   Disconnect the electrical connector from the TR sensor **(see illustration)**.
6   Disconnect the shift control cable from the manual control lever **(see illustration 14.5)**.
7   If you're simply removing and installing the same TR sensor, mark the relationship of the sensor to the transmission case.
8   If you're *replacing* the TR sensor, or removing the sensor to replace the transmission with a new or rebuilt unit, marking the sensor's position won't work. You must obtain a special Ford tool (307-351) to adjust the sensor when installing it.
9   Remove the TR sensor mounting bolts **(see illustration 14.5)** and remove the TR sensor. **Note:** *While the TR sensor is removed, do NOT move the manual selector shaft from the Neutral position.*

10   If you or someone else accidentally moved the selector shaft while the TR sensor was removed, you must put the manual selector shaft in the Neutral position before installing the TR sensor. To do so, put a wrench on the two machined flats located at the base of the threaded part of the shaft and rotate the shaft through the gears in a clockwise direction until it stops. Then rotate the shaft counterclockwise three detent positions (three clicks). The selector shaft is now in the Neutral position.
11   Install the TR sensor on the manual selector shaft and loosely install the sensor mounting bolts.
12   If you're installing the old TR sensor, align the marks that you made on the sensor and on the transmission case and hold the sensor in this position while tightening the sensor mounting bolts securely.
13   If you're installing a new TR sensor, install the special sensor alignment tool, then tighten the sensor mounting bolts securely.
14   The remainder of installation is the reverse of removal.

15   Transmission speed sensors - replacement

1   Disconnect the cable from the negative battery terminal (see Chapter 5, Section 1).
2   Raise the vehicle and place it securely on jackstands.

Automatic transmissions

Note: *The automatic transmission speed sensors are located on the left side of the transmission. The front sensor is the Turbine Shaft Speed (TSS) sensor, the middle sensor is the Intermediate Shaft Speed (ISS) sensor and the rear sensor is the Output Shaft Speed (OSS) sensor.*

3   Disconnect the electrical connector from the TSS, ISS or OSS sensor.
4   Remove the TSS, ISS or OSS sensor mounting bolt.

5   Remove the old O-ring from the sensor and discard it. Be sure to install a new O-ring on the sensor, even if you're installing the old sensor.
6   Installation is the reverse of removal. Be sure to tighten the sensor mounting bolt securely.

Manual transmissions

Note: *The Output Shaft Speed (OSS) sensor is located on the right side of the transmission, near the rear end of the transmission extension housing.*

7   Disconnect the electrical connector from the OSS sensor.
8   Remove the OSS sensor mounting bolt and remove the OSS sensor from the transmission extension housing.
9   Remove and discard the old OSS sensor O-ring. Be sure to install a new O-ring on the sensor, even if you're installing the old sensor.
10   Installation is the reverse of removal. Be sure to tighten the OSS sensor mounting bolt securely.

16   Powertrain Control Module (PCM) - removal and installation

Refer to illustration 16.2

Caution: *To avoid electrostatic discharge damage to the PCM, handle the PCM only by its case. Do not touch the electrical terminals during removal and installation. If available, ground yourself to the vehicle with an anti-static ground strap, available at computer supply stores.*

Note: *The Powertrain Control Module (PCM) is located in the right front corner of the compartment, on the left side of the engine compartment fuse and relay box.*

1   Disconnect the cable from the negative battery terminal (see Chapter 5, Section 1).
2   Disconnect the three large electrical connectors from the PCM **(see illustration)**.

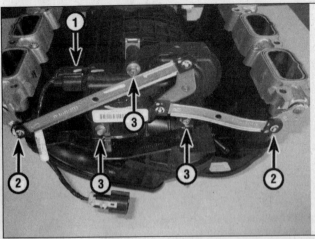

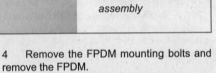

17.2 To detach the CMCV assembly from the intake manifold:

1 *Disconnect the electrical connector and remove the electrical harness*
2 *Remove the E-clips that secure the linkage arms to the actuator arms on the manifold*
3 *Remove the CMCV mounting bolts and remove the CMCV assembly*

3   Remove the upper and lower PCM mounting bolts **(see illustration 16.2)** and carefully remove the PCM from its mounting bracket. **Caution:** *Avoid static electricity damage to the computer by grounding yourself to the vehicle body before touching the PCM and using a special anti-static pad on which to store the PCM once it is removed.*
4   Installation is the reverse of removal.

---

17 Charge Motion Control Valve (CMCV) (4.6L V8 models) - removal and installation

---

Refer to illustration 17.2
Note: *The CMCV is used only on 4.6L V8 engines. The CMCV is located on the backside (facing toward the firewall) of the intake manifold.*
1   Remove the intake manifold (see Chapter 2B).
2   Disconnect the electrical connector from the CMCV **(see illustration)**.
3   Disconnect the outer ends of the CMCV actuator rods from the small actuator arms on the intake manifold.
4   Remove the CMCV mounting bolts and remove the CMCV from the intake manifold. Note the locations of the bolts; the one with the stud must go back into the same location, as it is the mounting point for a large wiring harness.
5   Installation is the reverse of removal.

---

18 Fuel Pump Driver Module (FPDM) - replacement

Refer to illustration 18.3
Note: *The FPDM is located inside the trunk, in the spare tire well.*
1   Disconnect the cable from the negative battery cable (see Chapter 5, Section 1).
2   Remove the spare tire (refer to your owner's manual, if necessary).
3   Disconnect the electrical connector from the FPDM **(see illustration)**.

4   Remove the FPDM mounting bolts and remove the FPDM.
5   Installation is the reverse of removal.

---

19 Variable Camshaft Timing (VCT) oil control solenoid - replacement

---

Refer to illustrations 19.3, 19.4 and 19.5
Note: *The VCT system is used only on 4.6L V8 models. The VCT oil control solenoids (there are two, one per cylinder head) are located on top and near the front of each valve cover. The procedure described below and the accompanying photos (which depict the left VCT oil control solenoid) apply to either solenoid.*
1   Disconnect the cable from the negative battery terminal (see Chapter 5, Section 1).
2   If you're removing the VCT oil control solenoid from the left valve cover, you can give yourself more room to work if you disconnect the electrical connector from the Mass Air Flow (MAF) sensor (see Section 12) and disconnect the Positive Crankcase Ventilation (PCV) hose (see Section 23).
3   Disconnect the electrical connector from the VCT oil control solenoid **(see illustration)**.
4   Remove the grommet that seals the

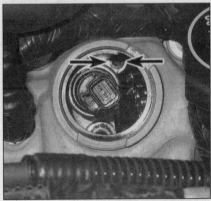

18.3 To remove the Fuel Pump Driver Module (FPDM), depress this release tab (1) and pull off the electrical connector, then remove the FPDM mounting bolts (2) and remove the FPDM from the spare tire well

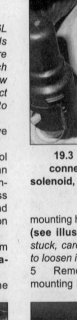

19.3 To disconnect the electrical connector from the VCT oil control solenoid, depress this release tab and pull off the connector

mounting hole for the VCT oil control solenoid **(see illustration)**. **Note:** *If the grommet is stuck, carefully pry around its circumference to loosen it up.*
5   Remove the VCT oil control solenoid mounting bolt **(see illustration)** and remove

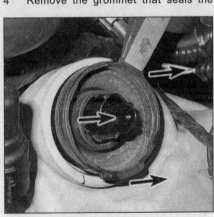

19.4 To remove the VCT oil control solenoid's grommet from the valve cover, carefully pull it out with pliers

19.5 To detach the VCT oil control solenoid from the cylinder head, remove this bolt and pull out the solenoid

the VCT oil control solenoid. **Note:** *You will need a long Torx driver to unscrew this mounting bolt. Also, this bolt is captive; it will stay with the solenoid.*

6   Installation is the reverse of removal. Be sure to tighten the bolt to the torque listed in this Chapter's Specifications.

20   Catalytic converters - description, check and replacement

Note: *Because of the Federally mandated extended warranty which covers emissions-related components such as the catalytic converter, check with a dealer service department before replacing the converter at your own expense.*

General description

1   The catalytic converter is an emission control device installed in the exhaust system that reduces pollutants from the exhaust gas stream. There are two types of converters: The oxidation catalyst reduces the levels of hydrocarbon (HC) and carbon monoxide (CO) by adding oxygen to the exhaust stream to produce water vapor (H_2O) and carbon dioxide (CO_2). The reduction catalyst lowers the levels of oxides of nitrogen (NOx) by removing oxygen from the exhaust gases to produce nitrogen (N) and oxygen. These two types of catalysts are combined into a three-way catalyst that reduces all three pollutants.

2   The amount of oxygen entering the catalyst is critical to its operation because without oxygen it cannot convert harmful pollutants into harmless compounds. The catalyst is most efficient at capturing and storing oxygen when it converts the exhaust gases of an intake charge that's mixed at the ideal (stoichiometric) air/fuel ratio of 14.7:1. If the air/fuel ratio is leaner than stoichiometric for an extended period of time, the catalyst will store even more oxygen. But if the air/fuel ratio is richer than stoichiometric for any length of time, the oxygen content in the catalyst can become totally depleted. If this condition occurs, the catalyst will not convert anything!

3   Because the catalyst's ability to store oxygen is such an important factor in its operation, it can also be considered a factor in the catalyst's eventual inability to do its job. Catalysts do wear out. And one of the reasons that they do so is that they can no longer store oxygen. So the PCM monitors the oxygen content going into and coming out of the catalyst by comparing the voltage signals from the upstream and downstream oxygen sensors. When the catalyst is functioning correctly, there is very little oxygen to monitor at the outlet end of the catalyst because it's capturing, storing and releasing oxygen as needed to convert HC, CO and NOx into more benign substances. But as the catalyst ages, it slowly loses its ability to store oxygen, and the downstream oxygen sensor tells the PCM that the oxygen content in the catalyzed

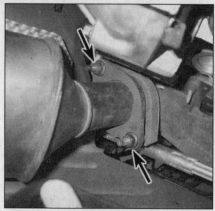

20.10 To disconnect each catalyst inlet pipe from its exhaust manifold, remove these two nuts (left inlet pipe on a 4.6L V8 model shown, other side identical)

exhaust gases is going up. When the amount of oxygen exiting the catalyst reaches a specified threshold, the PCM stores a Diagnostic Trouble Code (DTC) and turns on the Malfunction Indicator Light (MIL).

4   There are two catalytic converters on all models. The inlet pipes, which connect each exhaust manifold to a catalyst, are welded onto the inlet end of each catalyst. The outlet pipes are welded onto the outlet end of each catalyst. On 4.0L V6 models, the outlet pipes are welded into a single pipe right before they connect to the single intermediate pipe. The inlet pipes, catalysts and outlet pipes are a one-piece assembly on 4.0L V6 models. On 4.6L V8 models, each outlet pipe is connected directly to its own intermediate pipe and muffler. But these two outlet pipes are connected by a crossover pipe right before they connect to the intermediate pipes, so the inlet pipes, catalysts and outlet pipes are a single assembly on 4.6L V8 models.

Check

5   The equipment for testing a catalytic converter is expensive. If you suspect that the converter on your vehicle is malfunctioning, take it to a dealer or authorized emissions inspection facility for diagnosis and repair.

6   Whenever the vehicle is raised for servicing underbody components, inspect the converter for leaks, corrosion, dents and other damage. Inspect the welds/flange bolts that attach the front and rear ends of the converter to the exhaust system. If damage is discovered, the converter should be replaced.

7   Although catalytic converters don't break too often, they can become plugged. The easiest way to check for a restricted converter is to use a vacuum gauge to diagnose the effect of a blocked exhaust on intake vacuum.

 a) *Connect a vacuum gauge to an intake manifold vacuum source (see Chapter 2).*

 b) *Warm the engine to operating temperature, place the transaxle in Park (automatic) or Neutral (manual) and apply the parking brake.*

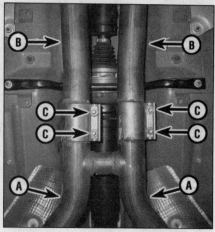

20.11 To disconnect the catalyst assembly (A) from the intermediate exhaust pipes (B), loosen these nuts (C) (4.6L V8 shown; 4.0L V6 models have only one intermediate pipe but use a similar coupler)

 c) *Note and record the vacuum reading at idle.*

 d) *Quickly open the throttle to near full throttle and release it. Note and record the vacuum reading.*

 e) *Perform the test three more times, recording the reading after each test.*

 f) *If the reading after the fourth test is more than one in-Hg lower than the reading recorded at idle, the exhaust system may be restricted (the catalytic converter could be plugged or an exhaust pipe or muffler could be restricted).*

Replacement

Refer to illustrations 20.10 and 20.11

Warning: *Wait until the exhaust system is completely cool before beginning this procedure.*

8   Raise the vehicle and place it securely on jackstands.

9   Disconnect the electrical connectors from the upstream and downstream oxygen sensors and remove the sensors so that neither of them is damaged when removing the catalyst assembly (see Section 13). **Note:** *If you're removing the catalyst assembly from a 4.6L V8 model, its not necessary to disconnect the electrical connector from the left upstream oxygen sensor or to remove the sensor because its installed in the left exhaust manifold, not the inlet pipe for the left catalyst.*

10   Disconnect the upper ends of the catalyst inlet pipes from the exhaust manifold **(see illustration)**.

11   Disconnect the catalyst assembly from the intermediate pipe(s) behind the catalyst assembly **(see illustration)**.

12   Inspect all fasteners for corrosion. Although you can use a thread chaser to clean up the fastener threads, it's usually a good idea to use new fasteners when installing the catalyst assembly.

21.9 To disconnect the electrical connector from the EVAP canister purge valve, depress this release tab and pull off the connector

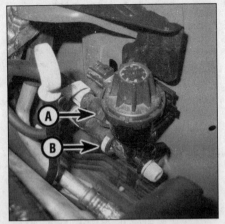

21.10 Disconnect the quick-connect couplings for the EVAP line going to the intake manifold (A) and for the EVAP line coming from the EVAP canister (B)

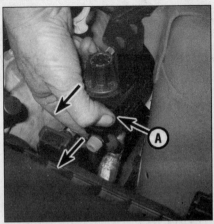

21.11 To disengage the EVAP canister purge valve from its mounting bracket, depress this release tab (A) with your thumb and slide the purge valve forward, out of the bracket

13 Installation is the reverse of removal. Be sure to tighten the exhaust fasteners securely.

21 Evaporative Emission Control (EVAP) system - description and component replacement

General description

1 The Evaporative Emissions Control (EVAP) system absorbs fuel vapors (unburned hydrocarbons) and, during engine operation, releases them into the intake manifold from which they're drawn into the intake ports where they mix with the incoming air-fuel mixture.

2 The EVAP system consists of the dust separator, the EVAP canister, the canister vent solenoid and the canister purge valve. The dust separator, EVAP canister and vent solenoid are located underneath the vehicle, behind the rear axle and below the spare tire well. The EVAP canister purge valve is located in the engine compartment, and is mounted on a bracket bolted to the left strut tower.

3 Other components in the EVAP system include the Fuel Tank Pressure (FTP) sensor, the Fuel Limit Vent Valve (FLVV) assembly, the fuel filler pipe check valve, the fuel filler neck cap and the EVAP system *monitor*, which is part of the PCM's program.

4 The fuel filler neck cap prevents an over-filled tank from spilling out the filler neck and it prevents EVAP vapors from escaping out the filler neck to the atmosphere.

5 The FLVV assembly, which is mounted inside the fuel tank, controls the flow of fuel vapors entering the EVAP system, prevents the tank from overfilling during refueling and prevents liquid fuel from entering the EVAP lines if the vehicle gets out of control or rolls over. The FLVV cannot be serviced separately from the fuel tank.

6 The FTP sensor tells the PCM to energize the EVAP canister purge valve when the

pressure inside the fuel tank is excessive. It also measures the (relative) vacuum conditions inside the fuel tank during an EVAP monitor test.

7 The EVAP system monitor is the test that the PCM runs to check the EVAP system and the fuel tank for leaks. This test is known as the "running loss system leak test." The EVAP system monitor runs this test as soon as the preconditions for purging have been met. There are two stages in the test. During the first stage the system monitor checks the EVAP system for big leaks. The EVAP canister vent solenoid is closed, but the canister purge valve stays open, which increases the (relative) vacuum in the EVAP system so that the FTP sensor can determine whether there is a leak somewhere in the system.

8 During the second stage of the running loss leak test, the canister purge valve closes and the FTP sensor monitors the "decay rate" (how long the system maintains an acceptable vacuum before leaking down) of the EVAP system. A slow decay rate indicates that there are no small leaks; a quick decay rate indicates that there's a small leak somewhere in the system.

Component replacement
EVAP canister purge valve

Refer to illustrations 21.9, 21.10 and 21.11

Note: *The canister purge valve is located in the engine compartment, on a bracket that's bolted to the left strut tower.*

9 Disconnect the electrical connector from the EVAP canister purge valve **(see illustration)**.

10 Disconnect the quick-connect couplings for the EVAP line coming from the EVAP canister and for the EVAP line going to the intake manifold **(see illustration)**. If you're unfamiliar with these types of quick-connect couplings, refer to *Fuel lines and fittings - general information* in Chapter 4.

11 Disengage the EVAP canister purge valve from its mounting bracket **(see illustra-**

tion) and remove it.
12 Installation is the reverse of removal.

EVAP canister assembly

Refer to illustrations 21.15, 21.16 and 21.17

Note 1: *The EVAP canister assembly is located underneath the vehicle, behind the rear axle, below the spare tire well. The canister assembly includes the dust separator, the EVAP canister, the canister vent solenoid, the hoses connecting these components and the canister rock shield, which also serves as the mounting bracket for all of these components.*

Note 2: *This procedure tells you how to remove the entire EVAP canister assembly, which includes not just the canister, but the canister vent solenoid, the filter tube and the dust separator.*

13 Disconnect the cable from the negative battery terminal (see Chapter 5, Section 1).
14 Raise the rear of the vehicle and place it securely on jackstands.
15 Disconnect the electrical connector from the EVAP canister vent solenoid **(see illustration)**.

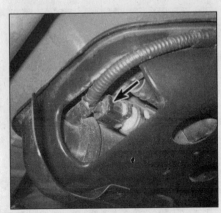

21.15 To disconnect the electrical connector from the EVAP canister vent solenoid, depress this release tab and pull off the connector

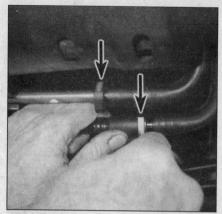

21.16 Disconnect these two quick-connect couplings (if you're unfamiliar with quick-connect couplings, refer to *Fuel lines and fittings - general information* in Chapter 4)

21.17 To detach the EVAP canister assembly from the underside of the vehicle, remove these two bolts and two nuts

16   Disconnect the two EVAP tube quick-connect couplings **(see illustration)**.

17   Remove the two EVAP canister mounting bolts and the two canister mounting nuts **(see illustration)** and remove the EVAP canister assembly.

18   If you're replacing the dust separator, refer to Steps 22 through 24. If you're replacing the canister vent solenoid, refer to Steps 27 through 29. If you're replacing the EVAP canister assembly, remove the dust separator and the canister vent solenoid and install them on the new canister assembly.

19   No further disassembly is possible.

20   Installation is the reverse of removal.

Dust separator

Refer to illustration 21.22

21   Remove the EVAP canister assembly (see Steps 13 through 17).

22   Disconnect the inlet hose from the dust separator **(see illustration)**.

23   Disengage the dust separator from its retaining brackets and lift it off the EVAP canister assembly.

24   Disconnect the outlet hose from the separator.

25   Installation is the reverse of removal.

EVAP canister vent solenoid

Refer to illustration 21.27

26   Remove the EVAP canister assembly (see Steps 13 through 17).

27   Disconnect the inlet hose from the canister vent solenoid **(see illustration)**.

28   Disengage the canister vent solenoid from the EVAP canister assembly.

29   Disconnect the outlet hose from the canister vent solenoid and remove the solenoid.

30   Installation is the reverse of removal.

22   Exhaust Gas Recirculation (EGR) system - description and component replacement

General description

1   When the engine is under load, the temperature inside the combustion chambers heats up dramatically. When the temperature inside the combustion chambers reaches 2500-degrees F., the engine begins to produce oxides of nitrogen (NOx), which is an odorless, colorless and toxic gas. The EGR system reduces NOx by introducing a controlled amount of spent exhaust gases into the intake manifold, which dilutes the air/fuel mixture, lowers combustion chamber temperatures and reduces the creation of NOx.

2   The EGR system, which is used on 4.0L V6 models, is the latest version of Ford's Differential Pressure Feedback EGR (DPFE) system. A DPFE system consists of the Differential Pressure Feedback EGR (DPFE) sensor, the EGR Vacuum Regulator (EVR) solenoid, the EGR valve, the exhaust manifold-to-intake manifold pipe, the Powertrain Control Module (PCM) and the electrical wiring and vacuum hoses connecting these components.

3   The PCM relies on signals from the ECT, CHT, IAT, TP, MAF and CKP sensors to determine when the engine is fully warmed up, stabilized and running at a moderate load and rpm before it energizes the EGR system. The PCM does not turn on the EGR system at idle, or during extended wide-open-throttle conditions, or if it detects a failure of some EGR system component or the failure of some sensor input that it needs to control the EGR system.

4   The exhaust manifold-to-intake manifold pipe provides a path for the spent exhaust gases that are routed back through the intake manifold and into the combustion chambers. When the EGR valve is open, a metering orifice inside the exhaust manifold-to-intake manifold pipe produces a measurable pressure drop as the exhaust gases flow through it. A pair of pipes, one upstream and one downstream in relation to the orifice, send high and low pressure signals, respectively, to the DPFE sensor.

5   The DPFE sensor is a ceramic, capacitive-type pressure transducer that monitors the actual pressure drop (differential pressure) across the metering orifice and provides a proportional feedback signal between 0 and 5 volts to the PCM, which uses this signal

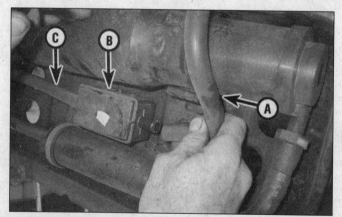

21.22 To remove the dust separator from the EVAP canister assembly, disconnect the inlet hose (A) from the dust separator (B), disengage the separator from its mounting bracket and disconnect the outlet hose (C) from the separator

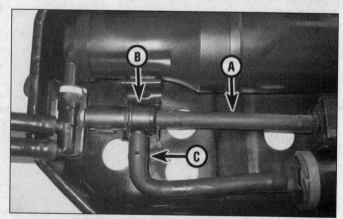

21.27 To remove the canister vent solenoid from the EVAP canister assembly, disconnect the inlet hose (A) from the vent solenoid (B), disengage the solenoid from the its mounting bracket and disconnect the outlet hose (C) from the solenoid

to control the EGR flow rate.

6   The PCM calculates the correct amount of EGR flow for a given engine condition, uses the feedback signal from the DPFE sensor to determine the appropriate pressure drop across the metering orifice inside the exhaust manifold-to-intake manifold pipe that will create the correct flow, then outputs the appropriate signal to the EGR EVR solenoid.

7   The EGR EVR solenoid is controlled by a duty-cycle signal from the PCM that can be anywhere between 0 and 100 percent. The higher the duty cycle the greater the amount of vacuum diverted by the EVR solenoid to the EGR valve's vacuum diaphragm. When the vacuum signal through the EVR solenoid to the EGR valve diaphragm is strong enough to overcome the spring pressure that normally keeps the EGR valve closed, the EGR valve pintle lifts off its seat and allows exhaust gases to flow into the intake manifold.

8   The EGR system used on 4.0L V6 models consists of the EGR system module and the exhaust manifold-to-intake manifold pipe. The EGR system module is similar to a DPFE-type EGR system described above, except that all of the DPFE system components are integrated into a single component.

9   The EGR transducer consists of a PCM-controlled solenoid and a backpressure transducer. When the PCM energizes the solenoid, no vacuum reaches the transducer; when the PCM de-energizes the solenoid, vacuum flows to the transducer. When exhaust backpressure reaches a specified threshold, it closes a bleed valve inside the transducer. When the bleed valve is closed by backpressure and the solenoid is de-energized by the PCM, vacuum flows through the transducer to operate the EGR valve. When exhaust backpressure has not fully closed the bleed valve and the PCM has de-energized the solenoid, a partial vacuum flows to the EGR valve, which reduces the amount of exhaust gases allowed to enter the engine.

Component replacement

Warning: *Make sure that the engine is cool before removing any EGR component.*

Exhaust manifold-to-EGR system module pipe

Warning: *Make sure that the engine is completely cooled off before beginning this procedure.*

Note: *The exhaust manifold-to-EGR system module pipe connects the left exhaust manifold to the mounting base of the EGR system module, which is bolted to the left side of the intake manifold.*

10   Unscrew the threaded fitting that connects the exhaust manifold-to-EGR system module pipe to the left exhaust manifold.

11   Unscrew the threaded fitting that connects the exhaust manifold-to-EGR system module pipe to the EGR system module.

12   Remove the exhaust manifold-to-EGR system module pipe.

13   Blow out the exhaust manifold-to-EGR system module pipe with compressed air.

14   Installation is the reverse of removal. Be sure to coat the threads of the fittings with anti-seize compound, then tighten the fittings securely.

EGR system module

15   Disconnect the cable from the negative battery terminal (see Chapter 5, Section 1).

16   Disconnect the vacuum connector from the EGR system module.

17   Disconnect the electrical connector from the EGR system module.

18   Remove the exhaust manifold-to-EGR system module pipe (see Steps 10 through 12).

19   Remove the two EGR system module mounting bolts and remove the EGR system module.

20   Remove and discard the old EGR system module gasket.

21   Clean off the gasket mating surfaces of the EGR system module and the intake manifold.

22   Installation is the reverse of removal. Be sure to use a new gasket and tighten the EGR system module mounting bolts to the torque listed in this Chapter's Specifications.

23   Positive Crankcase Ventilation (PCV) system

1   The Positive Crankcase Ventilation (PCV) system reduces hydrocarbon emissions by scavenging crankcase vapors. It does this by circulating fresh air from the air cleaner through the crankcase, where it mixes with blow-by gases and is then rerouted through a PCV valve to the intake manifold.

2   On V6 models, a PCV valve restricts the flow when the intake manifold vacuum is high. If abnormal operating conditions (such as piston ring problems) arise, the system is designed to allow excessive amounts of blow-by gases to flow back through the crankcase vent tube into the air cleaner to be consumed by normal combustion.

3   The main components of the PCV system are the fresh air inlet hose, the PCV valve (V6 models only) and the crankcase ventilation hose (PCV hose). The PCV valve used on 4.0L V6 models is *heated*. On these models a heating element inside the PCV valve prevents the valve from freezing in cold weather.

4   Checking and replacement of the PCV valve is covered in Chapter 1 (V6 models).

4.0L V6 models

5   The PCV fresh air inlet hose (the crankcase breather hose) connects the air intake duct to a pipe on the left valve cover. The crankcase ventilation hose (PCV hose) connects the right valve cover to a pipe on the intake manifold. The heated PCV valve is screwed into the valve cover. The heater element is PCM-controlled. When the intake air temperature is less than 32-degrees Fahrenheit, the PCM turns on the heater element in the PCV valve.

4.6L V8 models

Refer to illustrations 23.6a and 23.6b

6   The PCV fresh air inlet hose (crankcase breather hose) connects the air intake duct to a pipe on the right valve cover **(see illustration)**. The crankcase ventilation hose (PCV hose) connects the left valve cover to a pipe on the right side of the intake manifold, right behind the throttle body **(see illustration)**. There is no actual PCV valve on these models.

23.6a On 4.6L V8 models, the PCV fresh air inlet hose (A), or crankcase breather hose, connects the air intake duct to the right valve cover. Quick-connect couplings (B) and (C) allow easy disconnection to remove the air intake duct or the valve cover

23.6b On 4.6L V8 models, the crankcase ventilation hose (A), or PCV hose, connects to a fitting (B) on the left valve cover and a pipe (C) on the left side of the intake manifold

Notes

Chapter 7 Part A
Manual transmission

Contents

Specifications

Torque specifications

	Ft-lbs
Output flange nut*	
Automatic transmission	97
Manual transmission	105
Transmission crossmember-to-chassis bolts	46
Transmission-to-engine bolts	33
Transmission mount-to-crossmember nuts	85
Transmission mount-to-transmission bolt	52

*Use a new nut

1 General information

Vehicles covered by this manual are equipped with either a five-speed manual or a five-speed automatic transmission. The T50D (V6 models) and the TR3650 (V8 models) are fully-synchronized, five-speed manual transmissions with an overdrive fifth gear. Information on the manual transmission is included in this Part of Chapter 7. Information on the automatic transmission can be found in Part B of this Chapter. You'll also find certain procedures common to both transmissions - such as oil seal replacement - in Part A.

Depending on the expense involved in having a transmission overhauled, it might be a better idea to consider replacing it with either a used or rebuilt unit. Your local dealer or transmission shop should be able to supply information concerning cost, availability and exchange policy. Regardless of how you decide to remedy a transmission problem, you can still save a lot of money by removing and installing the unit yourself.

2 Shift lever - removal and installation

1 Position the shift lever into Neutral.
2 Carefully remove the shift lever boot by prying the corner of the boot and moving the assembly (boot and trim bezel) to the rear. Do not pry on the bezel or separate the bezel from the boot.
3 Lift the shift lever boot and the shift knob from the transmission linkage.
4 Raise the vehicle and secure it on jackstands.
5 Remove the driveshaft (see Chapter 8).
6 Remove the two rear gearshift underbody nuts.
7 Remove the shift linkage bolt and the shift arm bolt from the transmission.
8 Remove the shift lever and linkage assembly.
9 Installation is the reverse of removal. Be sure to tighten the nut on the shift lever eccentric stud securely.

3 Oil seal - replacement

Extension housing seal

Refer to illustrations 3.7 and 3.8
Note: *This procedure applies to both manual and automatic transmissions.*
1 Oil leaks frequently occur due to wear of the extension housing oil seal or the transmission speed sensor seal. Replacement of these seals is relatively easy, since the repairs can usually be performed without removing the transmission from the vehicle.
2 If you suspect a leak at the extension housing seal, raise the vehicle and support it securely on jackstands. The extension housing seal is located at the rear end of the transmission, where the driveshaft flange is attached. If the extension housing seal is leaking, transmission lubricant will be dripping from the rear of the transmission.
3 Remove the driveshaft (see Chapter 8).
4 Support the transmission with a jack, remove the crossmember mounting bolts and

3.7 Pry out the extension housing seal with a seal removal tool

3.8 Install the new extension housing seal with a seal driver or a large deep socket

lower the transmission slightly to gain access to the transmission flange.

5   Use a special tool to lock the output flange in place and remove the nut.

6   Remove the flange with a puller. Remove the plastic retainer and O-ring from the flange.

7   Using a screwdriver, pry bar or seal removal tool, carefully pry out the extension housing seal **(see illustration)**. Do not damage the splines on the transmission output shaft.

8   Using a seal driver or a large deep socket, install the new extension housing seal **(see illustration)**. Drive it into the bore squarely and make sure it's completely seated.

9   Install the flange and nut, tightening the nut until the flange is seated. Remove the nut, install a new O-ring and plastic retainer, then reinstall the nut and tighten it to the torque listed in this Chapter's Specifications.

10   Install the crossmember then install the driveshaft (see Chapter 8).

Transmission speed sensor O-ring

11   The transmission speed sensor is located on the transmission extension housing. For information and O-ring replacement for the transmission speed sensor, see Chapter 6.

4   Transmission mount - check and replacement

Check

1   Insert a large screwdriver or prybar into the space between the transmission and the crossmember and try to pry the transmission up. If there is any separation of the rubber, the mount is worn out.

Replacement

2   Remove the transmission crossmember (see Section 5).

3   Remove the nuts attaching the mount to the crossmember **(see illustration 7.15 in Chapter 7B)**. Remove the mount.

4   Installation is the reverse of the removal procedure. Be sure to tighten the nuts/bolts to the torque listed in this Chapter's Specifications.

5   Manual transmission - removal and installation

Removal

Refer to illustration 5.17

1   Disconnect the cable from the negative battery terminal (see Chapter 5, Section 1).

2   Place the transmission in NEUTRAL.

3   Raise the vehicle and support it securely on jackstands.

4   Disconnect the shift linkage from the transmission.

5   Remove the driveshaft (see Chapter 8).

6   Disconnect the clutch hydraulic line (see Chapter 8).

7   Remove the starter motor (see Chapter 5).

8   If you're working on a V8 model, remove the catalytic converter H-pipe (the forward section of the exhaust system).

9   Disconnect the electrical connectors from the two upstream oxygen sensors and the two downstream oxygen sensors (see Chapter 6).

10   Disconnect the back-up light switch connector.

11   Disconnect the transmission speed sensor connector (see Chapter 6).

12   Support the engine with a floor jack. Put a block of wood between the jack head and the engine oil pan to protect the pan.

13   Support the transmission with a transmission jack (available at most auto parts stores and at equipment rental yards) or with a large, heavy duty floor jack. Safety chains should be used to secure the transmission to the jack.

14   Raise the transmission slightly to take the weight off the crossmember.

15   Remove the transmission crossmember bolts.

16   Remove the transmission mount-to-transmission bolt **(see Chapter 7B, illustration 7.15)**.

17   Remove the engine-to-transmission bolts **(see illustration)**. Lowering the jack will make access to the upper transmission bolts easier.

18   Make a final check that all wires have been disconnected from the transmission, then move the transmission and jack toward the rear of the vehicle until the transmission input shaft is clear of the clutch or clutch housing. If the transmission input shaft is difficult to disengage from the clutch hub, use a prybar to separate the transmission from the engine.

5.17 The top transmission-to-engine bolts can be accessed using a long extension attached to the ratchet

Keep the transmission level as you pull it to the rear.

19   Once the input shaft is clear, lower the transmission and remove it from under the vehicle. **Caution:** *Do not depress the clutch pedal while the transmission is out of the vehicle.*

20   Inspect the clutch components. Generally speaking, new clutch components should always be installed whenever the transmission is removed (see Chapter 8).

Installation

21   Install the clutch components, if they were removed (see Chapter 8). Apply a thin film of high-temperature grease to the splines of the transmission input shaft and to the inner surface of the pilot bearing.

22   With the transmission secured to the jack, raise it into position behind the engine and carefully slide it forward, engaging the input shaft with the clutch plate hub. Do not use excessive force to install the transmission - if the input shaft won't slide into place, readjust the angle of the transmission or turn the input shaft so the splines engage properly with the clutch.

23   Once the transmission is flush with the engine, install the transmission-to-engine bolts. Tighten the bolts to the torque listed in this Chapter's Specifications. **Caution:** *Don't use the bolts to force the transmission and engine together. If the transmission doesn't slide up*

to the engine easily, find out what's wrong before proceeding.

24   Install the transmission mount and crossmember. Tighten all nuts and bolts to the torque listed in this Chapter's Specifications.

25   Remove the jacks supporting the transmission and the engine.

26   Install the various components removed previously. To connect the clutch hydraulic line and bleed the clutch hydraulic system, refer to Chapter 8.

27   Make a final check to verify all wires and hoses have been reconnected and the transmission has been filled with lubricant to the proper level (see Chapter 1). Lower the vehicle.

28   Reconnect the shift linkage.

29   Connect the negative battery cable. Fill the transmission with the proper lubricant, then road test the vehicle and check for leaks.

6   Manual transmission overhaul - general information

Overhauling a manual transmission is a difficult job for the do-it-yourselfer. It involves the disassembly and reassembly of many small parts. Numerous clearances must be precisely measured and, if necessary, changed with select fit spacers and snap-rings. As a result, if transmission prob-

lems arise, it can be removed and installed by a competent do-it-yourselfer, but overhaul should be left to a transmission repair shop. Rebuilt transmissions may be available - check with your dealer parts department and auto parts stores. At any rate, the time and money involved in an overhaul is almost sure to exceed the cost of a rebuilt unit.

Nevertheless, it's not impossible for an inexperienced mechanic to rebuild a transmission if the special tools are available and the job is done in a deliberate step-by-step manner so nothing is overlooked.

The tools necessary for an overhaul include internal and external snap-ring pliers, a bearing puller, a slide hammer, a set of pin punches, a dial indicator and possibly a hydraulic press. In addition, a large, sturdy workbench and a vise or transmission stand will be required.

During disassembly of the transmission, make careful notes of how each piece comes off, where it fits in relation to other pieces and what holds it in place.

Before taking the transmission apart for repair, it will help if you have some idea what area of the transmission is malfunctioning. Certain problems can be closely tied to specific areas in the transmission, which can make component examination and replacement easier. Refer to the *Troubleshooting* section at the front of this manual for information regarding possible sources of trouble.

Notes

Chapter 7 Part B
Automatic transmission

Contents

Specifications

General

Transmission fluid type.. See Chapter 1

Torque specifications

Ft-lbs

Torque converter nuts*	
2005	28
2006 and later	30
Transmission crossmember-to-chassis bolts	46
Transmission-to-engine bolts	35
Transmission mount-to-crossmember nuts	32
Transmission mount-to-transmission	52

*Install new torque converter nuts

1 General information

Caution: *If a vehicle with an automatic transmission is disabled, do NOT tow it at speeds greater than 30 mph or distances over 50 miles.*

The automatic transmission used in these models is known as the 5R55S. It is an electronic-shift five-speed transmission with a lock-up torque converter, known as a torque converter clutch, or TCC. The TCC provides a direct connection between the engine and the drive wheels for improved efficiency and fuel economy. The shifting is controlled by the Powertrain Control Module (PCM) using information from various engine and transmission sensors and output actuators (see Chap-

ter 6). Downshifting is also controlled by the PCM in response to certain conditions. Each condition is listed into a specialized category; Coastdown, Torque Demand and Kickdown. Coastdown shifting occurs during a slow vehicle stop. Torque Demand occurs during part throttle acceleration under load. Kickdown occurs when the vehicle requires extra speed during passing or heavy acceleration.

Because of the complexity of the clutches and the electronic and hydraulic control systems, and because of the special tools and expertise needed to overhaul an automatic transmission, diagnosis and repair of this transmission should be handled by a dealer service department or a transmission repair shop. But if the transmission must be rebuilt

or replaced, you can save money by removing and installing it yourself, so instructions for that procedure are included as well.

2 Diagnosis - general

Note: *Automatic transmission malfunctions may be caused by five general conditions: poor engine performance, improper adjustments, hydraulic malfunctions, mechanical malfunctions or malfunctions in the computer or its signal network. Diagnosis of these problems should always begin with a check of the easily repaired items: fluid level and condition (see Chapter 1), and shift cable adjustment. Next, perform a road test to determine if the* *problem has been corrected or if more diagnosis is necessary. If the problem persists after the preliminary tests and corrections are completed, additional diagnosis should be done by a dealer service department or transmission repair shop. Refer to the* Troubleshooting *section at the front of this manual for information on symptoms of transmission problems.*

Preliminary checks

1   Drive the vehicle to warm the transmission to normal operating temperature.
2   Check the fluid level as described in Chapter 1:

a) *If the fluid level is unusually low, add enough fluid to bring the level within the designated area of the dipstick, then check for external leaks (see below).*
b) *If the fluid level is abnormally high, drain off the excess, then check the drained fluid for contamination by coolant. The presence of engine coolant in the automatic transmission fluid indicates that a failure has occurred in the internal radiator walls that separate the coolant from the transmission fluid (see Chapter 3).*
c) *If the fluid is foaming, drain it and refill the transmission, then check for coolant in the fluid or a high fluid level.*

3   Check for any stored trouble codes (see Chapter 6). **Note:** *If the engine is malfunctioning, do not proceed with the preliminary checks until it has been repaired and runs normally.*
4   Inspect the shift cable (see Section 4). Make sure it's properly adjusted and operates smoothly.

Fluid leak diagnosis

5   Most fluid leaks are easy to locate visually. Repair usually consists of replacing a seal or gasket. If a leak is difficult to find, the following procedure may help.
6   Identify the fluid. Make sure it's transmission fluid and not engine oil or brake fluid (automatic transmission fluid is a deep red color).
7   Try to pinpoint the source of the leak. Drive the vehicle several miles, then park it over a large sheet of cardboard. After a minute or two, you should be able to locate the leak by determining the source of the fluid dripping onto the cardboard.
8   Make a careful visual inspection of the suspected component and the area immediately around it. Pay particular attention to gasket mating surfaces. A mirror is often helpful for finding leaks in areas that are hard to see.
9   If the leak still cannot be found, clean the suspected area thoroughly with a degreaser or solvent, then dry it.
10   Drive the vehicle for several miles at normal operating temperature and varying speeds. After driving the vehicle, visually inspect the suspected component again.
11   Once the leak has been located, the cause must be determined before it can be properly repaired. If a gasket is replaced but the sealing flange is bent, the new gasket will

not stop the leak. The bent flange must be straightened.
12   Before attempting to repair a leak, check to make sure the following conditions are corrected or they may cause another leak. **Note:** *Some of the following conditions cannot be fixed without highly specialized tools and expertise. Such problems must be referred to a transmission repair shop or a dealer service department.*

Gasket leaks

13   Check the pan periodically. Make sure the bolts are tight, no bolts are missing, the gasket is in good condition and the pan is flat (dents in the pan may indicate damage to the valve body inside).
14   If the pan gasket is leaking, the fluid level or the fluid pressure may be too high, the vent may be plugged, the pan bolts may be too tight, the pan sealing flange may be warped, the sealing surface of the transmission housing may be damaged, the gasket may be damaged or the transmission casting may be cracked or porous. If sealant instead of gasket material has been used to form a seal between the pan and the transmission housing, it may be the wrong sealant.

Seal leaks

15   If a transmission seal is leaking, the fluid level or pressure may be too high, the vent may be plugged, the seal bore may be damaged, the seal itself may be damaged or improperly installed, the surface of the shaft protruding through the seal may be damaged or a loose bearing may be causing excessive shaft movement.
16   Make sure the dipstick tube seal is in good condition and the tube is properly seated. Periodically check the area around the speedometer gear or sensor for leakage. If transmission fluid is evident, check the O-ring for damage.

Case leaks

17   If the case itself appears to be leaking, the casting is porous and will have to be repaired or replaced.
18   Make sure the oil cooler hose fittings are tight and in good condition.

Fluid comes out vent pipe or fill tube

19   If this condition occurs, the transmission is overfilled, there is coolant in the fluid, the case is porous, the dipstick is incorrect, the vent is plugged or the drain-back holes are plugged.

---

3   Shift lever - removal and installation

Removal

Refer to illustration 3.7

1   Set the parking brake, then place the shift lever in the Neutral position.

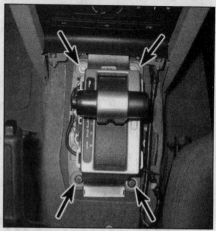

3.7 Location of the shift lever assembly mounting bolts and nuts

2   Remove the center console shifter bezel and the center console (see Chapter 11). Remove the driver and passenger center console kick panels.
3   Raise the vehicle and secure it on jackstands.
4   Separate the shift cable end from the shift lever using a screwdriver or trim panel tool (see Section 4).
5   Disconnect the shift cable from the shift assembly mounting bracket (see Section 4).
6   Disconnect the shift lock solenoid electrical connector.
7   Remove the two nuts and two bolts and remove the shift lever assembly from the floor **(see illustration)**.

Installation

8   Installation is the reverse of the removal procedure. Adjust the shift cable if necessary (see Section 4).

---

4   Shift cable - check, replacement and adjustment

Warning: *The models covered by this manual are equipped with a Supplemental Restraint System (SRS), more commonly known as airbags. Always disable the airbag system before working in the vicinity of any airbag system component to avoid the possibility of accidental deployment of the airbag(s), which could cause personal injury (see Chapter 12). Do not use a memory saving device to preserve the PCM or radio memory when working on or near airbag system components.*

Check

1   Firmly apply the parking brake, depress the brake pedal and try to momentarily operate the starter in each shift lever position. The starter should only operate when the manual shift lever is in the PARK or NEUTRAL positions. If the starter operates in any position other than PARK or NEUTRAL, adjust the shift cable (see below). If, after adjustment,

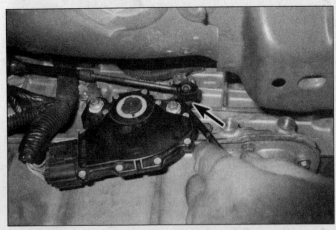

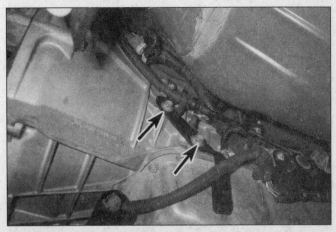

4.6 Pry the cable end off the transmission shift lever using a screwdriver or trim panel tool

4.7 Location of the shift cable bracket mounting bolts

the starter still operates in positions other than PARK or NEUTRAL, the transmission range (TR) sensor is defective (see Chapter 6).

Replacement

Refer to illustrations 4.6, 4.7 and 4.8

2   Place the shift lever in the NEUTRAL position.

3   Raise the vehicle and support it on jackstands.

4   Remove the driveshaft (see Chapter 8).

5   Secure the transmission with a transmission jack, remove the transmission crossmember bolts **(see illustration 7.15)** and partially lower the transmission to gain access to the shift cable components.

6   Pry the cable end off the transmission shift lever **(see illustration)**.

7   Remove the two shift cable bracket bolts and separate the shift cable and bracket from the transmission **(see illustration)**.

8   Disconnect the cable end from the shift lever. Release the mechanism by pushing the lock upward and disconnect the lock (cable end) from the shift lever **(see illustration)**.

9   Remove the clip and separate the shift cable from the bracket **(see illustration 4.8)**.

10   Installation is the reverse of removal. When you're done, adjust the cable (see below).

11   When installing, attach the shift cable to the bracket using a new clip and move the cable end to adjust the position to the shift lever. The shift lever should remain stationary.

12   Slide the lock tab DOWN to lock the shift cable end into position.

Adjustment

Note: *The adjusting mechanism is built into the lock mechanism located on the cable end.*

13   Working in the passenger's compartment, move the shifter into the 1st position and back into the DRIVE position. Place a three-pound weight (approximate) on the lever to hold it in position, or have an assistant hold it there.

14   Raise the vehicle and support it securely on jackstands.

15   Working under the vehicle, disconnect the cable end from the shift lever **(see illustration 4.8)**. Release the mechanism by pushing the lock upward and disconnect the lock (cable end) from the shift lever.

16   Rotate the shift lever on the transmission clockwise to its full travel and then back three detents to the Drive position.

17   Connect the shift cable to the shift lever and rotate the lock (cable end) down to lock the adjuster.

18   Remove the jackstands and lower the vehicle.

19   Remove the weight from the shifter, if used.

20   Verify that the engine will start only in PARK and NEUTRAL, and that the back-up lights come on when the shifter is placed in REVERSE. If necessary, readjust the cable until these conditions are met. It may also be necessary to adjust the transmission range sensor afterwards (see Chapter 6).

5   Brake Transmission Shift Interlock (BTSI) system - description, check and solenoid replacement

Warning: *The models covered by this manual are equipped with a Supplemental Restraint System (SRS), more commonly known as airbags. Always disable the airbag system before working in the vicinity of any airbag system component to avoid the possibility of accidental deployment of the airbag(s), which could cause personal injury (see Chapter 12). Do not use a memory saving device to preserve the PCM or radio memory when working on or near airbag system components.*

Description

1   The Brake Transmission Shift Interlock (BTSI) system is a solenoid-operated device located under the shift lever in the center con-

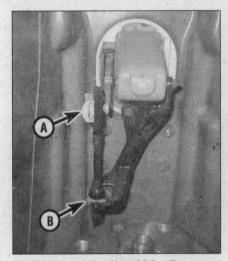

4.8 Working under the vehicle, disconnect the cable end from the shift lever (A) then remove the clip (B) and separate the shift cable from the bracket

sole. The solenoid locks the shift lever in the PARK position when the ignition key is in the LOCK or ACCESSORY position. When the ignition key is in the RUN position, a magnetic holding device is energized. When the system is functioning correctly, the only way to unlock the shift lever and move it out of PARK is to depress the brake pedal. The BTSI system also prevents the ignition key from being turned to the LOCK or ACCESSORY position unless the shift lever is fully locked into the PARK position.

Check

2   Verify that the ignition key can be removed *only* in the PARK position.

3   When the shift lever is in the PARK position, you *should* be able to rotate the ignition key from OFF to LOCK. But when the shift lever is in any gear position other than PARK (including NEUTRAL), you should not be able to rotate the ignition key to the LOCK position.

5.11 Pry the plastic cover up using a small screwdriver to access the BTSI override button

6.2 Location of the engine splash shield fasteners

4   You should *not* be able to move the shift lever out of the PARK position when the ignition key is turned to the OFF position.

5   You should *not* be able to move the shift lever out of the PARK position when the ignition key is turned to the RUN or START position until you depress the brake pedal.

6   You should *not* be able to move the shift lever out of the PARK position when the ignition key is turned to the ACC or LOCK position.

7   Once in gear, with the ignition key in the RUN position, you *should* be able to move the shift lever between gears, or put it into NEUTRAL or PARK, without depressing the brake pedal.

8   If the BTSI system doesn't operate as described, try adjusting it as follows.

Solenoid replacement

9   The BTSI system on these models is not repairable. In the event of failure, replace the entire shift lever assembly.

Shift lock override feature

Refer to illustration 5.11

Note: *If the shift lever is non-operational and the shift override feature must be activated*

to move the shift lever, be sure to check the brake light fuse and the brake light system for a possible short that could have deactivated the BTSI system.

10   In the event the Brake Transmission Shift Interlock (BTSI) system fails and the shift lever cannot be moved out of gear, the system is equipped with an override feature. The BTSI system can be bypassed and the shift lever can be used in manual operation.

11   Apply the parking brake and remove the access cover using a small screwdriver **(see illustration)**. Locate the button for the BTSI solenoid, push the button down, depress the brake pedal and shift the console select lever into NEUTRAL.

---

6   Transmission fluid cooler - removal and installation

Refer to illustrations 6.2 and 6.4

1   Raise the front of the vehicle and support it securely on jackstands.

2   Remove the engine splash shield **(see illustration)**.

3   Put a drain pan underneath the trans-

mission fluid cooler line fittings to catch any spilled transmission fluid.

4   Disconnect the transmission fluid cooler hoses from the metal lines **(see illustration)**. Plug the lines and hoses to prevent fluid spills.

5   Remove the transmission fluid cooler mounting bolts and remove the transmission fluid cooler. Be careful not to damage the tubes or fins of the cooler or the condenser cooling fins.

6   If you removed the cooler in order to flush it after a transmission failure, have it flushed at a transmission shop.

7   Installation is the reverse of removal.

8   When you're done, check the transmission fluid level and add some if necessary (see Chapter 1).

---

7   Automatic transmission - removal and installation

Caution: *The transmission and torque converter must be removed as a single assembly. If you try to leave the torque converter attached to the driveplate, the converter driveplate, pump bushing and oil seal will be damaged. The driveplate is not designed to support the load, so none of the weight of the transmission should be allowed to rest on the plate during removal.*

Removal

Refer to illustrations 7.5, 7.9, 7.10, 7.12 and 7.15

1   Place the shift lever in the NEUTRAL position. Apply the parking brake and block the rear wheels.

2   Disconnect the cable from the negative terminal of the battery (see Chapter 5, Section 1).

3   Raise the vehicle and support it securely on jackstands.

4   Drain the transmission fluid (see Chapter 1), then reinstall the fluid pan.

6.4 To separate the transmission fluid cooler flexible lines from the metal lines, first compress the spring clamps (A) and move them over the metal lines, then loosen the screw-type hose clamps (B) and slide the hoses off

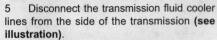

7.5 Disconnect the transmission fluid cooler lines from the transmission using a flare-nut wrench on the tube nut. Hold the fitting with an open-end wrench to prevent it from turning

7.9 Location of the rubber inspection hole cover for access to the torque converter nuts on V8 models

5   Disconnect the transmission fluid cooler lines from the side of the transmission (see illustration).
6   Remove the exhaust system and the forward heat shields (see Chapter 4).
7   Remove the driveshaft (see Chapter 8).
8   Support the transmission with a transmission jack (available at most equipment rental facilities) and secure the transmission to the jack with safety chains. Detach the shift cable from the transmission (see Section 4).
9   On V8 models, remove the rubber plug at the left rear of the engine block (see illustration) and mark the relationship of the torque converter to the driveplate so they can be installed in the same position. Note: The torque converter nuts on V6 models are removed through the starter access hole.
10   Remove the inspection cover at the bottom of the bellhousing (see illustration).
11   Remove the starter motor (see Chapter 5). If you're working on a V6 model, you can now mark the relationship of the torque converter to the driveplate through the hole left by the starter.
12   Remove the driveplate-to-torque con-

verter nuts (see illustration). Discard the nuts and replace them with new ones during installation.
13   Disconnect the electrical connectors from the two upstream oxygen sensors and the two downstream oxygen sensors (see Chapter 6), the transmission speed sensor, the output shaft speed sensor, the digital transmis-

sion range (DTR) sensor (see Chapter 6) and the solenoid body assembly.
14   Support the engine with a floor jack. Place a wood block between the jack head and the engine oil pan.
15   Raise the transmission slightly to take the weight off the crossmember, then remove the crossmember (see illustration).

7.10 Location of the inspection cover mounting bolts

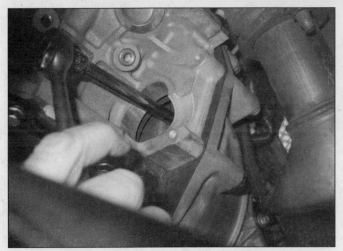

7.12 Remove the torque converter nuts through the hole left by the starter (V6 models) or the rubber plug (V8 models)

7.15 Location of the transmission crossmember mounting bolts (A) and the transmission mount-to-transmission bolt (B)

16   On V6 models, position the fuel line and bracket away from the transmission bellhousing.

17   Remove the transmission-to-engine bolts. Be sure to note the length and the location of each bolt for correct reassembly.

18   Make a final check that all wires have been disconnected from the transmission, then move the transmission and jack toward the rear of the vehicle until the torque converter is separated from the driveplate. Clamp a pair of locking pliers to the bellhousing so the torque converter won't fall out during removal.

Installation

19   Prior to installation, make sure the torque converter is securely engaged in the pump. If you've removed the converter, apply a small amount of transmission fluid on the torque converter rear hub, where the transmission front seal rides. Install the torque converter onto the input shaft of the transmission while rotating the converter back and forth. It should engage with the transmission front pump and input shaft in stages. Make sure it's completely engaged and rotates freely.

20   With the transmission secured to the jack, raise it into position. Be sure to keep it level so the torque converter doesn't fall out and disengage itself from the pump gear.

21   Turn the torque converter to line up the studs with the holes in the driveplate. The marks on the torque converter and driveplate must line up.

22   Move the transmission forward carefully until the dowel pins on the engine are engaged with the holes on the transmission. Make sure the transmission mates with the engine with no gap. If there's a gap, make sure there are no wires or other objects pinched between the engine and transmission.

23   Install the transmission-to-engine bolts and tighten them to the torque listed in this Chapter's Specifications. As you're tightening the bolts, make sure that the engine and transmission mate completely at all points. If not, find out why. Never try to force the engine and transmission together with the bolts or you'll break the transmission case!

24   Attach the shift cable to the manual lever (see Section 4).

25   Install the transmission mount and crossmember and tighten the bolts/nuts to the torque listed in this Chapter's Specifications.

26   Install the torque converter nuts and tighten them to the torque listed in this Chapter's Specifications. **Caution:** *Be sure to install new torque converter nuts.*

27   The remainder of installation is the reverse of removal.

28   Fill the transmission with the specified fluid (see Chapter 1), run the engine and check for fluid leaks.

9   Automatic transmission overhaul - general information

In the event of a fault occurring, it will be necessary to establish whether the fault is electrical, mechanical or hydraulic in nature, before repair work can be contemplated. Diagnosis requires detailed knowledge of the transmission's operation and construction, as well as access to specialized test equipment, and so is deemed to be beyond the scope of this manual. It is therefore essential that problems with the automatic transmission are referred to a dealer service department or other qualified repair facility for assessment.

Note that a faulty transmission should not be removed before the vehicle has been assessed by a knowledgeable technician equipped with the proper tools, as troubleshooting must be performed with the transmission installed in the vehicle.

Chapter 8
Clutch and driveline

Contents

Specifications

General
Differential ring gear size
- V6 models ... 7.5 inches
- V8 models ... 8.8 inches

Differential pinion bearing preload ... 16 to 29 in-lbs

Torque specifications
Ft-lbs (unless otherwise indicated)

Clutch master cylinder mounting bolts	89 in-lbs
Clutch pressure plate-to-flywheel bolts*	
Stage one	33
Stage two	Tighten an additional 60 degrees
Clutch release cylinder mounting bolts	15
Differential pinion shaft lock bolt	22
Driveshaft-to-transmission flange	
V6 engine	76
V8 engine	41
Driveshaft-to-rear axle flange	
V6 engine	76
V8 engine	76
Center support bearing mounting bolts	35
Wheel lug nuts	See Chapter 1

Use new pressure plate bolts

1 General information

The information in this Chapter deals with the components from the rear of the engine to the front wheels, except for the transmission, which is dealt with in Chapter 7. For the purposes of this Chapter, these components are grouped into three categories - clutch, driveshaft and axle(s).

Since nearly all the procedures covered in this Chapter involve working under the vehicle, make sure it's securely supported on sturdy jackstands or on a hoist where the vehicle can be easily raised and lowered.

2 Clutch - description and check

1   All vehicles with a manual transmission have a single dry plate, diaphragm spring-type clutch. The clutch disc has a splined hub which allows it to slide along the splines of the transmission input shaft. The clutch and pressure plate are held in contact by spring pressure exerted by the diaphragm in the pressure plate.

2   The clutch release system is operated by hydraulic pressure. The hydraulic release system consists of the clutch pedal, a master cylinder and a shared common reservoir with the brake master cylinder, a release (or slave) cylinder and the hydraulic line connecting the two components.

3   When the clutch pedal is depressed, a pushrod pushes against brake fluid inside the master cylinder, applying hydraulic pressure to the release cylinder, which pushes the release bearing against the diaphragm fingers of the clutch pressure plate.

4   Terminology can be a problem when discussing the clutch components because common names are in some cases different from those used by the manufacturer. For example, the driven plate is also called the clutch plate

or disc, the clutch release bearing is sometimes called a throwout bearing, the release cylinder is sometimes called the slave cylinder.

5   Unless you're replacing components with obvious damage, do these preliminary checks to diagnose clutch problems:

  a) *The first check should be of the fluid level in the master cylinder. If the fluid level is low, add fluid as necessary and inspect the hydraulic system for leaks. If the master cylinder reservoir is dry, bleed the system as described in Section 8 and recheck the clutch operation.*

  b) *To check "clutch spin-down time," run the engine at normal idle speed with the transmission in Neutral (clutch pedal up - engaged). Disengage the clutch (pedal down), wait several seconds and shift the transmission into Reverse. No grinding noise should be heard. A grinding noise would most likely indicate a bad pressure plate or clutch disc.*

  c) *To check for complete clutch release, run the engine (with the parking brake applied to prevent vehicle movement) and hold the clutch pedal approximately 1/2-inch from the floor. Shift the transmission between 1st gear and Reverse several times. If the shift is rough, component failure is indicated.*

  d) *Visually inspect the pivot bushing at the top of the clutch pedal to make sure there's no binding or excessive play.*

3   Clutch master cylinder - removal and installation

Removal

1   Clamp a pair of locking pliers onto the clutch fluid feed hose, a couple of inches downstream of the brake fluid reservoir (the clutch master cylinder is supplied with fluid from the brake fluid reservoir). The pliers should be just tight enough to prevent fluid flow when the hose is disconnected. Disconnect the reservoir hose from the clutch master cylinder.

2   Working under the dashboard, disconnect the clutch master cylinder pushrod from the pedal by removing the clip and the washer from the clutch pedal pin.

3   Working under the dash, remove the mounting bolts and detach the cylinder from the clutch pedal bracket.

4   Pull the clutch master cylinder forward and release the line clip at the hydraulic line fitting. Separate the hydraulic line from the cylinder. Also detach the reservoir feed hose from the master cylinder. Have rags handy, as some fluid will be lost as the line is removed. Cap or plug the ends of the lines to prevent fluid leakage and the entry of contaminants. **Caution:** *Don't allow brake fluid to come into contact with the paint, as it will damage the finish.*

Installation

5   Connect the hydraulic line fitting to the clutch master cylinder and install the retaining clip at the fitting (junction).

6   Attach the fluid feed hose from the reservoir to the clutch master cylinder and tighten the hose clamp. Remove the locking pliers.

7   Place the master cylinder in position on the clutch pedal bracket and install the mounting bolts finger tight. Install the pushrod, the washer and the clip onto the clutch pedal pin.

8   Tighten the mounting bolts to the torque listed in this Chapter's Specifications.

9   Fill the reservoir with brake fluid conforming to DOT 3 specifications and bleed the clutch system as outlined in Section 5.

4   Clutch release cylinder - removal and installation

Warning: *Dust produced by clutch wear is hazardous to your health. DO NOT blow it out with compressed air and DO NOT inhale it. DO NOT use gasoline or petroleum-based solvents to remove the dust. Brake system cleaner should be used to flush the dust into a drain pan. After the clutch components are wiped clean with a rag, dispose of the contaminated rags and cleaner in a covered, marked container.*

Removal

1   Remove the clip and disconnect the hydraulic line at the transmission. Have a small can and rags handy, as some fluid will be spilled as the line is removed. Plug the line to prevent excessive fluid loss and contamination.

2   Remove the transmission (see Chapter 7, Part A).

3   Remove the release cylinder mounting bolts.

4   Remove the release cylinder.

Installation

5   Install the release cylinder into the transmission. Tighten the bolts to the torque listed in this Chapter's Specifications.

6   Install the transmission (see Chapter 7, Part A).

7   Connect the hydraulic line fitting to the transmission and install the clip.

8   Check the fluid level in the brake fluid reservoir, adding brake fluid conforming to DOT 3 specifications until the level is correct.

9   Bleed the system as described in Section 5, then recheck the brake fluid level.

5   Clutch hydraulic system - bleeding

1   Bleed the hydraulic system whenever any part of the system has been removed or the fluid level has fallen so low that air has been drawn into the master cylinder. The bleeding procedure is very similar to bleeding a brake system.

2   Fill the brake master cylinder reservoir with new brake fluid conforming to DOT 3 specifications. **Caution:** *Do not re-use any of the fluid coming from the system during the bleeding operation or use fluid which has been inside an open container for an extended period of time.*

3   Remove the cap from the bleeder valve and attach a length of clear hose to the valve. Place the other end of the hose into a container partially filled with clean brake fluid.

4   Have an assistant depress the clutch pedal and hold it. Open the bleeder valve on the hydraulic line, allowing fluid and any air to escape. Close the bleeder valve when the flow of fluid (and bubbles) ceases. Once closed, have your assistant release the pedal.

5   Continue this process until all air is evacuated from the system, indicated by a solid stream of fluid being ejected from the bleeder valve each time with no air bubbles. Keep a close watch on the fluid level inside the brake master cylinder reservoir - if the level drops too far, air will get into the system and you'll have to start all over again. **Note:** *Wash the area with water to remove any spilled brake fluid.*

6   Check the brake fluid level again, and add some, if necessary, to bring it to the appropriate level. Check carefully for proper operation before placing the vehicle into normal service.

6   Clutch release bearing - removal, inspection and installation

The clutch release bearing and bearing hub are integral components of the clutch release cylinder. Replace the release cylinder as a single assembly (see Section 4).

7   Clutch components - removal, inspection and installation

Warning: *Dust produced by clutch wear is hazardous to your health. DO NOT blow it out with compressed air and DO NOT inhale it. DO NOT use gasoline or petroleum-based solvents to remove the dust. Brake system cleaner should be used to flush the dust into a drain pan. After the clutch components are wiped clean with a rag, dispose of the contaminated rags and cleaner in a covered, marked container.*

Removal

Refer to illustration 7.5

1   Access to the clutch components is normally accomplished by removing the transmission, leaving the engine in the vehicle. If the engine is being removed for major overhaul, check the clutch for wear and replace worn components as necessary. However, the relatively low cost of the clutch components

7.5 Mark the relationship of the pressure plate to the flywheel (if you're planning to re-use the old pressure plate)

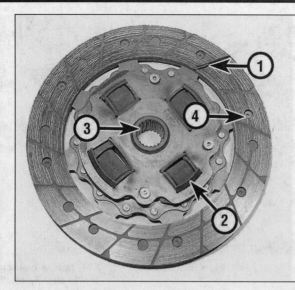

7.9 The clutch disc

1 **Lining** - this will wear down in use
2 **Springs or dampers** - check for cracking and deformation
3 **Splined hub** - the splines must not be worn and should slide smoothly on the transmission input shaft splines
4 **Rivets** - these secure the lining and will damage the flywheel or pressure plate if allowed to contact the surfaces

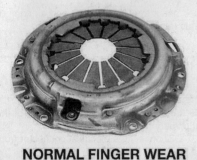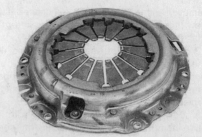

NORMAL FINGER WEAR **EXCESSIVE FINGER WEAR** **BROKEN OR BENT FINGERS**

EXCESSIVE WEAR

7.11a Replace the pressure plate if excessive wear or damage are noted

compared to the time and trouble spent gaining access to them warrants their replacement anytime the engine or transmission is removed, unless they are new or in near-perfect condition. The following procedures are based on the assumption the engine will stay in place.

2 Remove the transmission from the vehicle (see Chapter 7, Part A). Support the engine while the transmission is out. Preferably, an engine support fixture or a hoist should be used to support it from above.

3 The release bearing can remain attached to the transmission housing for the time being.

4 To support the clutch disc during removal, install a clutch alignment tool through the clutch disc hub.

5 Carefully inspect the flywheel and pressure plate for indexing marks. The marks are usually an X, an O or a black mark. If they cannot be found, scribe or paint marks yourself so the pressure plate and the flywheel will be in the same alignment during installation **(see illustration)**.

6 Turning each bolt a little at a time, loosen the pressure plate-to-flywheel bolts. Work in a criss-cross pattern until all spring pressure is relieved. Then hold the pressure plate

securely and completely remove the bolts, followed by the pressure plate and clutch disc. Discard the bolts; new ones should be used during installation.

Inspection

Refer to illustrations 7.9, 7.11a, 7.11b and 7.13

7 Ordinarily, when a problem occurs in the clutch, it can be attributed to wear of the clutch driven plate assembly (clutch disc). However, all components should be inspected at this time.

8 Inspect the flywheel for cracks, heat checking, grooves and other obvious defects. If the imperfections are slight, a machine shop can machine the surface flat and smooth, which is highly recommended regardless of the surface appearance. Refer to Chapter 2 for the flywheel removal and installation procedure.

9 Inspect the lining on the clutch disc. There should be at least 1/16-inch of lining above the rivet heads. Check for loose rivets, distortion, cracks, broken springs and other obvious damage **(see illustration)**. As mentioned above, ordinarily the clutch disc is routinely replaced, so if in doubt about the condition, replace it with a new one.

7.11b Inspect the pressure plate surface for excessive score marks, cracks and signs of overheating

10 The release bearing and release cylinder should also be replaced along with the clutch disc (see Section 4).

11 Check the machined surfaces and the diaphragm spring fingers of the pressure plate **(see illustrations)**. If the surface is grooved or otherwise damaged, replace the pressure plate. Also check for obvious damage, distortion, cracking, etc. Light glazing can be

7.13 A small slide hammer is handy for removing a pilot bearing

7.14 Tap the bearing into place with a bearing driver or a socket that is slightly smaller than the outside diameter of the bearing

removed with emery cloth or sandpaper. If a new pressure plate is required, new and re-manufactured units are available.

12   Check the pilot bearing in the end of the crankshaft for excessive wear, scoring, dryness, roughness and any other obvious damage. If any of these conditions are noted, replace the bearing.

13   Removal can be accomplished with a slide hammer and puller attachment **(see illustration),** which are available at most auto parts stores or tool rental yards.

Installation

Refer to illustrations 7.14 and 7.16

14   To install a new pilot bearing, lightly lubri-cate the outside surface with grease, then drive it into the recess with a bearing driver or a socket **(see illustration). Note:** *The seal end of the bearing must be facing toward the transmission.*

15   Before installation, clean the flywheel and pressure plate machined surfaces with brake cleaner, lacquer thinner or acetone. It's important that no oil or grease is on these sur-

faces or the lining of the clutch disc. Handle the parts only with clean hands.

16   Position the clutch disc and pressure plate against the flywheel with the clutch held in place with an alignment tool **(see illustra-tion).** Make sure the disc is installed properly (most replacement clutch discs will be marked "flywheel side" or something similar - if not marked, install the clutch disc with the damper springs toward the transmission).

17   Install the new pressure plate-to-fly-wheel bolts only finger tight, working around the pressure plate.

18   Center the clutch disc by ensuring the alignment tool extends through the splined hub and into the pilot bearing in the crank-shaft. Wiggle the tool up, down or side-to-side as needed to center the disc. Tighten the pressure plate-to-flywheel bolts a little at a time, working in a criss-cross pattern to pre-vent distorting the cover. After all of the bolts are snug, tighten them to the torque listed in this Chapter's Specifications. Remove the alignment tool.

19   Install the clutch release bearing and release cylinder (see Section 4).

20   Install the transmission and all compo-nents removed previously.

8   Clutch pedal position (CPP) switch - replacement

1   Locate the CPP switch at the top of the clutch pedal.

2   Disconnect the electrical connector from the CPP switch.

3   Remove the CPP switch by releasing the tangs from its bracket.

4   Installation is the reverse of removal.

9   Driveshaft - inspection

1   Raise the rear of the vehicle and support it securely on jackstands.

2   Crawl under the vehicle and visually inspect the driveshaft. Look for any dents or cracks in the tubing. If any are found, the driveshaft must be replaced.

3   Check for any oil leakage at the front and rear of the driveshaft. Leakage where the driveshaft enters the transmission indicates a defective transmission rear seal. Leakage where the driveshaft enters the differential indicates a defective pinion seal.

4   While under the vehicle, have an assis-tant turn the rear wheel so the driveshaft will rotate. As it does, make sure the universal joints are operating properly without binding, noise or looseness.

5   The universal joints can also be checked with the driveshaft motionless, by gripping your hands on either side of the joint and attempting to twist the joint. Any movement at all in the joint is a sign of considerable wear. Lifting up on the shaft will also indicate move-ment in the universal joints.

6   Finally, check the driveshaft mounting bolts at the ends to make sure they are tight.

7.16 Center the clutch disc in the pressure plate with a clutch alignment tool

10.2 After marking the relationship of the driveshaft to the differential companion flange, remove the bolts and separate the driveshaft from the differential companion flange

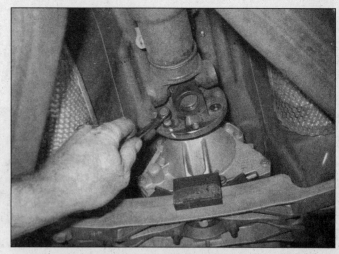

10.4 After marking the relationship of the driveshaft to the differential companion flange, remove the bolts and separate the driveshaft flange from the transmission flange

10 Driveshaft - removal and installation

Refer to illustrations 10.2 and 10.4

1   Raise the rear of the vehicle an support it securely on jackstands. Block the front wheels.

2   Mark the relationship of the driveshaft to the differential companion flange, then remove the bolts and separate the driveshaft from the differential companion flange **(see illustration)**.

3   If the vehicle is equipped with a two-piece driveshaft, remove the center support bearing mounting bolts.

4   Mark the relationship of the driveshaft to the transmission companion flange, then remove the bolts securing the driveshaft flange to the transmission flange. Lower the driveshaft **(see illustration)**.

5   Installation is the reverse of removal. If the shaft cannot be lined up due to the components of the differential or transmission having been rotated, put the vehicle in Neutral or rotate one wheel to allow the original alignment to be achieved. Tighten the fasteners to the torque listed in this Chapter's Specifications.

11 Universal joints - replacement

Refer to illustrations 11.2, 11.4 and 11.9

Note: *A press or large vise will be required for this procedure. It may be advisable to take the driveshaft to a local dealer service department, service station or machine shop where the universal joints can be replaced for you, normally at a reasonable charge.*

1   Remove the driveshaft as outlined in Section 10.

2   On U-joints with external snap-rings, use a small pair of pliers to remove the snap-rings from the spider **(see illustration)**.

3   Supporting the driveshaft, place it in position on a workbench equipped with a vise.

4   Place a piece of pipe or a large socket, having an inside diameter slightly larger than the outside diameter of the bearing caps, over one of the bearing caps. Position a socket with an outside diameter slightly smaller than that of the opposite bearing cap against the cap **(see illustration)** and use the vise or press to force the bearing cap out (inside the pipe or large socket). Use the vise or large pliers to work the bearing cap the rest of the way out.

5   Transfer the sockets to the other side and press the opposite bearing cap out in the same manner.

6   Pack the new universal joint bearings with grease. Ordinarily, specific instructions for lubrication will be included with the universal joint servicing kit and should be followed carefully.

11.2 A pair of needle-nose pliers can be used to remove the universal joint snap-rings

11.4 To press the universal joint out of the driveshaft yoke, set it up in a vise with the small socket pushing the joint and bearing cap into the large socket

7   Position the spider in the yoke and partially install one bearing cap in the yoke.

8   Start the spider into the bearing cap and then partially install the other cap. Align the spider and press the bearing caps into position, being careful not to damage the dust seals.

9   Install the snap-rings. If difficulty is encountered in seating the snap-rings, strike the driveshaft yoke sharply with a hammer. This will spring the yoke ears slightly and allow the snap-rings to seat in the groove **(see illustration)**.

10   Install the grease fitting and fill the joint with grease. Be careful not to overfill the joint, as this could blow out the grease seals.

11   Install the driveshaft (see Section 10).

11.9 If the snap-ring will not seat in the groove, strike the yoke with a brass hammer - this will relieve the tension that has set up in the yoke and slightly spring the yoke ears (this should also be done if the joint feels tight when assembled)

12   Differential pinion oil seal - replacement

Refer to illustrations 12.4, 12.5, 12.6a, 12.6b, 12.7 and 12.8

1   Raise the rear of the vehicle and place it securely on jackstands.

2   Remove the rear wheels, brake calipers and brake discs (see Chapter 9).

3   Mark the driveshaft and companion flange for ease of realignment during reassembly, then remove the driveshaft (see Section 10).

4   Mark the relationship between the pinion and companion flange **(see illustration)**.

5   Using an inch-pound torque wrench, measure and record the torque required to turn the pinion nut through several revolutions **(see illustration)**. This value, known as pinion bearing preload, will be used when the pinion flange is reinstalled.

6   Using a suitable tool, hold the compan-

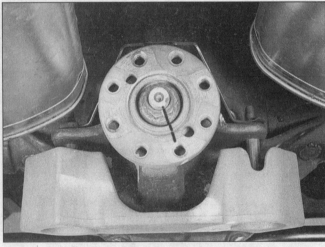

12.4 Mark the relationship between the pinion and flange as shown

12.5 Using an inch-pound torque wrench, measure the pinion bearing preload (turning torque) and jot down this measurement

12.6a To keep the flange from turning while you're loosening and removing the pinion nut, hold the flange with a punch jammed through a hole in the flange and wedged under the reinforcing rib on the side of the differential housing

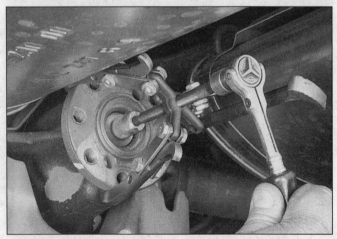

12.6b If necessary, use a small puller to separate the pinion flange from the pinion shaft

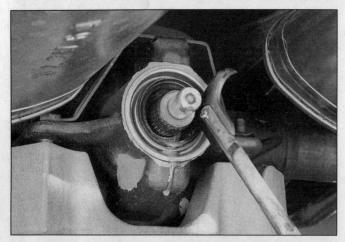

12.7 Pry out the old seal with a seal removal tool

12.8 Using a seal driver or a large socket, tap the new pinion seal into place with the seal square to the bore

ion flange and remove the pinion nut **(see illustration)**. Using a suitable puller **(see illustration)**, remove the companion flange.

7   Pry out the old seal with a seal removal tool **(see illustration)**.

8   Clean the oil seal mounting surface, then tap the new seal into place, taking care to insert it squarely as shown **(see illustration)**.

9   Inspect the splines on the pinion shaft for burrs and nicks. Remove any rough areas with a crocus cloth. Wipe the splines clean.

10   Install the companion flange, aligning it with the marks made during removal. Gently tap the flange on with a soft-faced hammer until you can start the pinion nut on the pinion shaft.

11   Using a suitable tool, hold the companion flange while tightening the pinion nut, stopping frequently to take rotational torque measurements, using the inch-pound torque wrench, until the measurement recorded in Step five is reached. **Caution:** *If the measurement recorded in Step five was less than the pinion bearing preload torque listed in this Chapter's Specifications, continue tightening until the specified torque is reached. If it was*

more than specified, continue tightening until the recorded measurement is reached. Under no circumstances should the pinion nut be backed off to reduce pinion bearing preload. Increase the nut torque in small increments and check the preload after each increase.

12   Reinstall the driveshaft, brake discs, brake calipers and wheels.

13   Check the differential oil level and fill as necessary.

14   Lower the vehicle and take a test drive to check for leaks.

13   Axleshaft - removal and installation

Refer to illustrations 13.4a, 13.4b, 13.4c and 13.5

1   Loosen the wheel lug nuts, raise the rear of the vehicle, support it securely on jackstands and remove the wheel.

2   Remove the brake caliper and disc (see Chapter 9).

3   Remove the cover from the differential

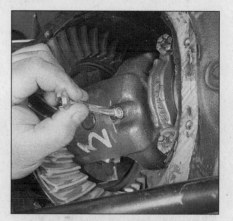

13.4a Position a large screwdriver between the rear axle case and a ring gear bolt to keep the differential case from turning, then loosen the pinion shaft lock bolt . . .

carrier and allow the lubricant to drain into a container (see Chapter 1).

4   Remove the lock bolt from the differential pinion shaft. Slide the notched end of the pin-

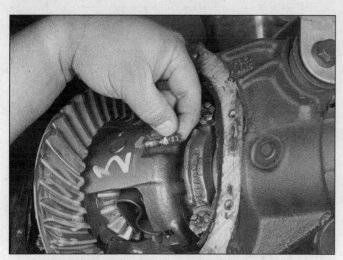

13.4b . . . remove the lock bolt . . .

13.4c . . . and slide out the pinion shaft

13.5 Push in on the axle flange and remove the C-lock from the inner end of the axleshaft (a magnet is handy for this job)

14.2a Use a seal removal tool to remove the old seal from the axle housing

ion shaft out of the differential case as far as it will go **(see illustrations)**.

5   Push in the outer (flanged) end of the axleshaft and remove the C-lock from the inner end of the shaft **(see illustration)**.

6   Withdraw the axleshaft, taking care not to damage the oil seal in the end of the axle housing as the splined end of the axleshaft passes through it.

7   Installation is the reverse of removal. Tighten the differential pinion shaft lock bolt to the torque listed in this Chapter's Specifications.

8   Install the differential cover, then refill the axle with the correct quantity and grade of lubricant (see Chapter 1).

14   Axleshaft oil seal - replacement

Refer to illustrations 14.2a, 14.2b and 14.3

1   Remove the axleshaft (see Section 13).
2   Pry the old oil seal out of the end of the axle housing, using a seal removal tool or the

inner end of the axleshaft itself as a lever **(see illustrations)**.

3   Using a seal driver or a large socket, tap the seal into position so that the lips are facing in and the metal face is visible from the end of the axle housing **(see illustration)**. When correctly installed, the face of the oil seal should be flush with the end of the axle housing. Lubricate the lips of the seal with gear oil.

4   Install the axleshaft (see Section 13).

15   Axleshaft bearing - replacement

Refer to illustrations 15.2a, 15.2b and 15.4

1   Remove the axleshaft (see Section 13) and the oil seal (see Section 14).
2   You'll need a bearing puller **(see illustrations)**, or you'll need to fabricate a similar tool.
3   Attach a slide hammer and pull the bearing out of the axle housing.
4   Clean out the bearing recess and drive

14.2b If a seal removal tool isn't available, a prybar or even the end of the axle can be used to pry the seal out of the housing

in the new bearing with a bearing driver or a large socket **(see illustration)**. Lubricate the new bearing with gear lubricant. Make sure

14.3 Use a seal driver or a large socket to tap the new axleshaft oil seal into place

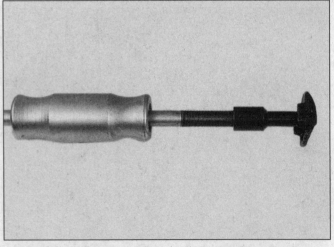

15.2a A typical axleshaft bearing removal tool (available at stores which carry automotive tools)

15.2b To remove the axle bearing, insert a bearing removal tool attached to a slide hammer through the center, pull the tool up against the back side and use the slide hammer to drive the bearing from the axle housing

15.4 A correctly-sized bearing driver or large socket must be used to drive the bearing into the housing

that the bearing is seated into the full depth of its recess.

5   Discard the old oil seal and install a new one (see Section 14), then install the axleshaft (see Section 13).

---

16   Rear axle assembly - removal and installation

Removal

1   Disconnect the cable from the negative battery terminal (See Chapter 5, Section 1).
2   Loosen, but do not remove the rear wheel lug nuts. Block the front wheels and raise the rear of the vehicle. Support it securely on jackstands placed under the frame. Remove the rear wheels.
3   Remove the brake calipers and discs (see Chapter 9).
4   Remove the differential cover and allow the lubricant to drain (see Chapter 1).
5   If the vehicle is equipped with ABS, remove the rear wheel speed sensors (see

Chapter 9).
6   Disconnect all clips for the ABS sensor wires, hydraulic brake lines and parking brake cables from the axle housing.
7   If you're working on a convertible model, remove the front, upper and rear support brace mounting bolts.
8   Unbolt the driveshaft from the differential companion flange (see Section 10) and secure it out of the way.
9   Disconnect the stabilizer bar from the axle assembly (see Chapter 10).
10   Remove the trackbar (see Chapter 10).
11   Place a floor jack under the differential and raise the rear axle slightly.
12   Remove the shock absorber lower mounting bolts (see Chapter 10).
13   Remove the upper suspension arm-to-rear axle housing nuts and bolts (see Chapter 10).
14   Lower the axle housing until the coil springs are fully extended, then remove them.
15   Disconnect the trailing arms from the rear axle assembly (see Chapter 10).
16   Remove the lower suspension arm-to-

axle housing nuts and bolts (see Chapter 10).
17   Lower the axle housing and guide it out from underneath the vehicle.

Installation

18   Installation is the reverse of removal, noting the following points:

* *Raise the axle housing to simulate normal ride height and tighten the upper and lower suspension arm-to-axle housing nuts to the torque listed in the Chapter 10 Specifications.*
* *Connect the lower ends of the shock absorbers to the rear axle housing and tighten the bolts/nuts to the torque listed in the Chapter 10 Specifications.*
* *Install the brake discs and calipers (see Chapter 9).*
* *Install the differential cover and fill the differential with the recommended oil (see Chapter 1).*
* *Install the wheels and lug nuts. Lower the vehicle and tighten the lug nuts to the torque listed in the Chapter 1 Specifications.*

Notes

Chapter 9
Brakes

Contents

Specifications

Brake fluid type	DOT 3 brake fluid

Disc brakes

Minimum brake pad thickness	See Chapter 1
Disc minimum thickness	Cast into disc
Disc runout limit	0.003-inch

Torque specifications

	Ft-lbs
Caliper mounting bolts	
Front	25
Rear	24
Caliper mounting bracket bolts	
Front	85
Rear	76
Brake hose-to-caliper banjo bolt	52
Master cylinder mounting nuts	18
Power brake booster mounting nuts	18
Wheel speed sensor mounting bolt	132 in-lbs
Wheel lug nuts	See Chapter 1

1   General information

Hydraulic system

The hydraulic system consists of two separate circuits. The master cylinder has separate reservoirs for the two circuits and, in the event of a leak or failure in one hydraulic circuit, the other circuit will remain operative.

Power brake booster

The power brake booster, utilizing engine manifold vacuum and atmospheric pressure to provide assistance to the hydraulically operated brakes, is mounted on the firewall in the engine compartment.

Parking brake

The parking brake lever mechanically operates the rear brakes through cables which actuate the brake calipers. A series of cables lead from the brake calipers to the parking brake lever located in the center console. The parking brake cable tension is adjustable and is performed at the parking brake lever.

Service

After completing any operation involving disassembly of any part of the brake system, always test drive the vehicle to check for proper braking performance before resuming normal driving. When testing the brakes, perform the tests on a clean, dry, flat surface. Conditions other than these can lead to inaccurate test results.

Test the brakes at various speeds with both light and heavy pedal pressure. The vehicle should stop evenly without pulling to one side or the other. Tires, vehicle load and wheel alignment are factors which also affect braking performance.

Precautions

There are some general cautions and warnings involving the brake system on this vehicle:

a) Use only brake fluid conforming to DOT 3 specifications.
b) The brake pads and linings may contain fibers which are hazardous to your health if inhaled. Whenever you work on brake system components, clean all parts with brake system cleaner. Do not allow the fine dust to become airborne. Also, wear an approved filtering mask.
c) Safety should be paramount whenever any servicing of the brake components is performed. Do not use parts or fasteners which are not in perfect condition, and be sure that all clearances and torque specifications are adhered to. If you are at all unsure about a certain procedure, seek professional advice. Upon completion of any brake system work, test the brakes carefully in a controlled area before putting the vehicle into normal service. If a problem is suspected in the brake system, don't drive the vehicle until it's fixed.

2   Anti-lock Brake System (ABS) - general information

General information

Refer to illustration 2.2

1   The anti-lock brake system is designed to maintain vehicle steerability, directional stability and optimum deceleration under severe braking conditions on most road surfaces. It does so by monitoring the rotational speed of each wheel and controlling the brake line pressure to each wheel during braking. This prevents the wheels from locking up.

2   The ABS system has three main components - the wheel speed sensors, the anti-lock brake control module and the hydraulic control unit **(see illustration)**. Wheel speed sensors - one at each wheel - send a variable voltage signal to the control unit, which monitors these signals, compares them to its program and determines whether a wheel is about to lock up. When a wheel is about to lock up, the control unit signals the hydraulic unit to reduce hydraulic pressure (or not increase it further) at that wheel's brake caliper. Pressure modulation is handled by electrically-operated solenoid valves.

3   If a problem develops within the system, an ABS warning light will glow on the dashboard. Sometimes, a visual inspection of the ABS system can help you locate the problem. Carefully inspect the ABS wiring harness. Pay particularly close attention to the harness and connections near each wheel. Look for signs of chafing and other damage caused by incorrectly routed wires. If a wheel sensor harness is damaged, the sensor must be replaced. **Warning:** *Do NOT try to repair an ABS wiring harness. The ABS system is sensitive to even the smallest changes in resistance. Repairing the harness could alter resistance values and cause the system to malfunction. If the ABS wiring harness is damaged in any way, it must be replaced.* **Caution:** *Make sure the ignition is turned off before unplugging or reattaching any electrical connections.*

Diagnosis and repair

4   If a dashboard warning light comes on and stays on while the vehicle is in operation, the ABS system requires attention. Although special electronic ABS diagnostic testing tools are necessary to properly diagnose the system, you can perform a few preliminary checks before taking the vehicle to a dealer service department.

a) Check the brake fluid level in the reservoir.
b) Verify that the computer electrical connectors are securely connected.
c) Check the electrical connectors at the hydraulic control unit.
d) Check the fuses.
e) Follow the wiring harness to each wheel and verify that all connections are secure and that the wiring is undamaged.

5   If the above preliminary checks do not

2.2 The ABS hydraulic control unit is located in the right front corner of the engine compartment

rectify the problem, the vehicle should be diagnosed by a dealer service department or other qualified repair shop. Due to the complex nature of this system, all actual repair work must be done by a qualified automotive technician.

Wheel speed sensor - removal and installation

Refer to illustrations 2.9a and 2.9b

6   Loosen the wheel lug nuts, raise the vehicle and support it securely on jackstands. Remove the wheel.

7   Make sure the ignition key is turned to the Off position.

8   Trace the wiring back from the sensor, detaching all brackets and clips while noting its correct routing, then disconnect the electrical connector.

9   Remove the mounting bolt and carefully pull the sensor out from the knuckle or plate **(see illustrations)**.

10   Installation is the reverse of the removal procedure. Tighten the mounting bolt securely.

11   Install the wheel and lug nuts, tightening them securely. Lower the vehicle and tighten the lug nuts to the torque listed in the Chapter 1 Specifications.

3   Brake pads - replacement

Warning: *Disc brake pads must be replaced on both wheels at the same time - never replace the pads on only one wheel. Also, the dust created by the brake system is harmful to your health. Never blow it out with compressed air and don't inhale any of it. An approved filtering mask should be worn when working on the brakes. Do not, under any circumstances, use petroleum-based solvents to clean brake parts. Use brake system cleaner only!*

1   Remove the cover from the brake fluid reservoir and siphon out about 1/2 of the brake fluid. Replace the cover.

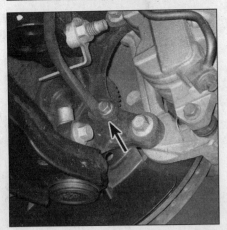

2.9a The front ABS wheel speed sensor

2.9b The rear ABS wheel speed sensor

3.5 Before removing the caliper, be sure to depress the piston into the bottom of its bore in the caliper with a large C-clamp to make room for the new pads

Front disc brake pads

Refer to illustrations 3.5 and 3.6a through 3.6m

Caution: *Install new caliper spring clips onto the caliper mounting bracket when installing new brake pads.*

2   Loosen the wheel lug nuts, raise the front of the vehicle and support it securely on jackstands. Apply the parking brake.

3   Remove the wheels. Work on one brake assembly at a time, using the assembled brake for reference if necessary.

4   Inspect the brake disc carefully as outlined in Section 5. If machining is necessary, follow the information in that Section, at which time the pads can be removed as well.

5   Push the piston back into its bore to provide room for the new brake pads. A C-clamp can be used to accomplish this **(see illustration)**. As the piston is depressed to the bottom of the caliper bore, the fluid in the master cylinder will rise. Make sure that it doesn't overflow. If necessary, siphon off some of the fluid.

6   Follow the accompanying photos **(illustrations 3.6a through 3.6m)**, for the actual

pad replacement procedure. Be sure to stay in order and read the caption under each illustration.

7   When reinstalling the caliper, be sure to tighten the mounting bolts to the torque listed in this Chapter's Specifications. After the job has been completed, firmly depress the brake

pedal a few times to bring the pads into contact with the disc. Check the level of the brake fluid, adding some if necessary. Check the operation of the brakes carefully before placing the vehicle into normal service.

3.6a Wash down the brake caliper assembly and the disc with brake cleaner to remove brake dust; DO NOT blow-off brake dust with compressed air

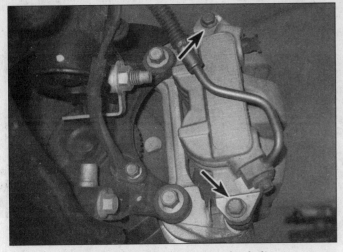

3.6b Remove the caliper mounting bolts

3.6c Remove the caliper and hang it aside with a piece of coat-hanger wire; DO NOT let the caliper hang from its brake hose!

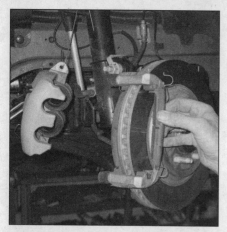

3.6d Remove the outer brake pad

3.6e Remove the inner brake pad

3.6f Remove all of the spring clips from
the caliper mounting bracket

3.6g Remove the guide pins and clean them thoroughly.
Inspect the boots and pins for wear or damage
and replace them if necessary

3.6h Lubricate the guide pins with high-temperature
grease and reinstall them

Rear disc brake pads

Refer to illustrations 3.12a through 3.12o

Caution: *Install new caliper spring clips onto the caliper mounting bracket when installing new brake pads.*

8   Loosen the wheel lug nuts, raise the rear of the vehicle and support it securely on jackstands. Block the wheels at the opposite end.

9   Remove the wheels. Work on one brake assembly at a time, using the assembled brake for reference if necessary.

10   Clean the brake caliper assembly and disc with brake cleaner to remove the brake dust; DO NOT blow-off the brake dust with compressed air **(see illustration 3.6a)**.

11   Inspect the brake disc carefully as out-

3.6i Install new spring clips onto the
caliper mounting bracket

3.6j Install the inner brake pad

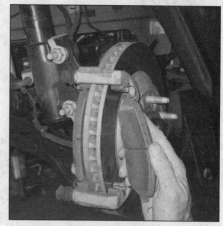

3.6k Install the outer brake pad

3.6l While holding the brake pads against the disc, place the caliper into position

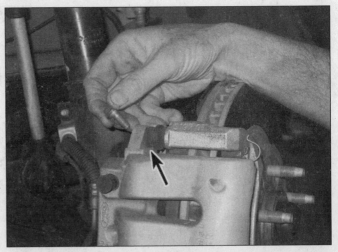

3.6m Make sure that the guide pins are properly matched to the caliper, install the caliper mounting bolts and tighten them to the torque listed in this Chapter's Specifications

3.12a Remove the caliper mounting bolts

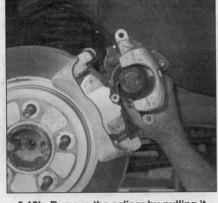

3.12b Remove the caliper by pulling it directly away from the mounting bracket

3.12c Suspend the caliper by hanging it aside with a piece of coat-hanger wire; DO NOT let the caliper hang from its brake hose!

lined in Section 5. If machining is necessary, follow the information in that Section, at which time the pads can be removed as well.

12 Follow the accompanying photos **(illustrations 3.12a through 3.12o),** for the actual pad replacement procedure. Be sure to stay in order and read the caption under each illustration.

13 When reinstalling the caliper, be sure to tighten the mounting bolts to the torque listed in this Chapter's Specifications. After the job has been completed, firmly depress the brake pedal a few times to bring the pads into con-

tact with the disc. Check the level of the brake fluid, adding some if necessary. Check the operation of the brakes carefully before placing the vehicle into normal service.

3.12d Remove the outer brake pad

3.12e Remove the inner brake pad

3.12f Remove both spring clips from the caliper mounting bracket . . .

3.12g . . . and replace them with new ones

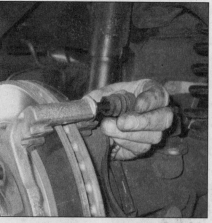

3.12h Remove the guide pins and clean them thoroughly. Inspect the boots and pins for wear or damage and replace them if necessary

3.12i Lubricate the guide pins with high-temperature grease and reinstall them

3.12j Compare the new brake pads with the original ones and make sure that they have an outer shim installed on them

3.12k Install the new inner brake pad

3.12l Install the new outer brake pad

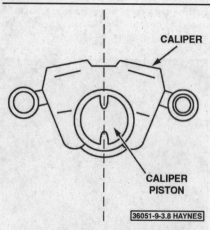

CALIPER

CALIPER PISTON

36051-9-3.8 HAYNES

3.12m Rotate the piston clockwise with a brake piston tool or needle-nose pliers - screw the piston in all the way and align the notches in the piston with the openings in the caliper, as shown

3.12n Position the brake caliper onto the mounting bracket while aligning the pin on the inner brake pad with the notch in the caliper piston

3.12o Make sure that the guide pins are properly matched to the caliper; install the caliper mounting bolts and tighten them to the torque listed in this Chapter's Specifications

4 Brake caliper - removal and installation

Warning: *Dust created by the brake system is harmful to your health. Never blow it out with compressed air and don't inhale any of it. An approved filtering mask should be worn when working on the brakes. Do not, under any circumstances, use petroleum-based solvents to clean brake parts. Use brake system cleaner only.*

Note: *If replacement is indicated (usually because of fluid leakage), it is recommended that the calipers be replaced, not overhauled. New and factory rebuilt units are available on an exchange basis, which makes this job quite easy. Always replace the calipers in pairs - never replace just one of them.*

Removal

Refer to illustrations 4.2 and 4.4

1   Loosen the wheel lug nuts, raise the vehicle (front or rear) and place it securely on jackstands. Remove the wheels.

2   Remove the inlet fitting bolt and disconnect the brake hose from the caliper **(see illustration)**. Discard the sealing washers from each side of the hose fitting. Plug the brake hose to keep contaminants out of the brake system and to prevent losing any more brake fluid than is necessary. A piece of vacuum hose can be placed inside the fitting to accomplish this. **Note:** *If the caliper is being removed for access to another component, don't disconnect the hose.*

3   Refer to Section 3 for the caliper removal procedure. If the caliper is being removed just to access another component, use a piece of wire to securely hang it out of the way **(see illustration 3.6c)**. **Caution:** *Do not let the caliper hang by the brake hose.*

4   If you are replacing a rear caliper, disconnect the parking brake cable from the caliper by removing the retaining clip **(see illustration)**.

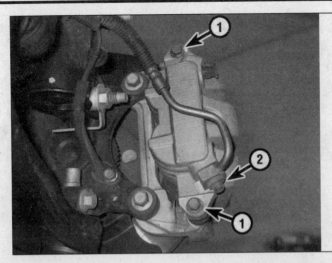

4.2 Caliper mounting details:

1   *Caliper mounting bolts*
2   *Banjo bolt and fitting*

Installation

5   Install the caliper by reversing the removal procedure.

6   Bleed the brake circuit according to the procedure in Section 8. Make sure there are no leaks from the hose connections. Test the brakes carefully before returning the vehicle to normal service.

5 Brake disc - inspection, removal and installation

Warning: *Dust created by the brake system is harmful to your health. Never blow it out with compressed air and don't inhale any of it. An approved filtering mask should be worn when working on the brakes. Do not, under any circumstances, use petroleum-based solvents to clean brake parts. Use brake system cleaner only.*

Inspection

Refer to illustrations 5.4, 5.5a, 5.5b, 5.6a and 5.6b

Note: *The manufacturer recommends that* any disc refinishing (machining) be performed with a specialized brake lathe that can cut the disc while it is still installed on the vehicle. They caution against using any other type of machining process to refinish the brake disc.

1   Loosen the wheel lug nuts, raise the vehicle and support it securely on jackstands. Remove the wheel.

2   Remove the brake caliper as outlined in Section 4. It's not necessary to disconnect the brake hose for this procedure. After removing the caliper bolts, suspend the caliper out of the way with a piece of wire. Don't let the caliper hang by the hose and don't stretch or twist the hose.

3   Reinstall the lug nuts (inverted) to hold the disc against the hub. It may be necessary to install washers between the disc and the lug nuts to take up space.

4   Visually check the disc surface for score marks, cracks and other damage. Light scratches and shallow grooves are normal after use and may not always be detrimental to brake operation. Deep score marks or cracks may require disc refinishing by an automotive machine shop or disc replacement **(see illustration)**. Be sure to check both

4.4 Remove the retaining clip for the parking brake cable and then remove the cable from the linkage on the caliper

5.4 The brake pads on this vehicle were obviously neglected, as they wore down completely and cut deep grooves into the disc - wear this severe means the disc must be replaced

5.5a To check disc runout, mount a dial indicator as shown and rotate the disc

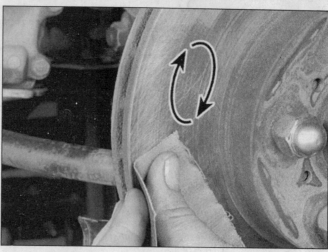

5.5b Using a swirling motion, remove the glaze from the disc with sandpaper or emery cloth

sides of the disc. If pulsating has been noticed during application of the brakes, suspect disc runout. **Note:** *The most common symptoms of damaged or worn brake discs are pulsation in the brake pedal when the brakes are applied or loud grinding noises caused from severely worn brake pads. If these symptoms are extreme, it is very likely that the disc(s) will need to be replaced.*

5   To check disc runout, place a dial indicator at a point about 1/2-inch from the outer edge of the disc **(see illustration)**. Set the indicator to zero and turn the disc. An indicator reading that exceeds 0.003-inch could cause pulsation upon brake application and will require disc refinishing by an automotive machine shop or disc replacement. **Note:** *If disc refinishing or replacement is not necessary, you can deglaze the brake pad surface on the disc with emery cloth or sandpaper (use a swirling motion to ensure a non-directional finish)* **(see illustration)**.

6   The disc must not be machined to a thickness less than the specified minimum refinish thickness. The minimum wear (or discard) thickness is cast into either the front or backside of the disc **(see illustration)**. The disc thickness can be checked with a micrometer **(see illustration)**.

Removal and installation

Refer to illustrations 5.7 and 5.8

7   Remove the caliper mounting bracket **(see illustration)**.

8   Mark the disc in relation to the hub so that it can be installed in it's original position on the hub and then remove the disc. If it's stuck, make sure you have removed any lug nuts installed during inspection. If the disc has never been removed, there may be retaining washers on the wheel stud securing it to the hub flange. Simply cut them off and discard them **(see illustration)**.

5.6a The minimum thickness limit is cast into the brake disc like the one shown in the photo. Refer to the specific dimension cast into your disc

5.6b Use a micrometer to measure disc thickness at several points, about 1/2-inch from the edge

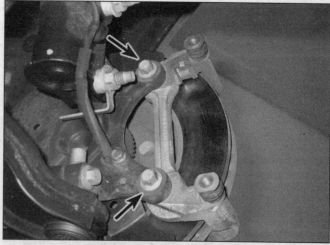

5.7 Front caliper mounting bracket bolts (mounting bracket bolts for the rear caliper are similar)

5.8 Use cutting pliers to remove any retaining washers from the wheel studs

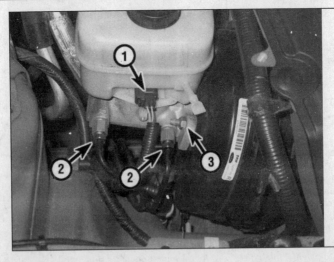

6.3 Master cylinder mounting details:

1 Brake fluid level sensor electrical connector
2 Brake line fitting tube nuts
3 Master cylinder mounting nut (one nut visible in photo)

9   Clean the hub flange and the inside of the brake disc thoroughly, removing any rust or corrosion, then Install the disc onto the hub assembly.

10   Install the caliper onto the caliper mounting bracket (see Section 3). Install the caliper mounting bolts and tighten them to the torque listed in this Chapter's Specifications.

11   Install the wheel, then lower the vehicle to the ground. Tighten the wheel lug nuts to the torque listed in the Chapter 1 Specifications. Depress the brake pedal a few times to bring the brake pads into contact with the disc. Bleeding of the system will not be necessary unless the brake hose was disconnected from the caliper. Check the operation of the brakes carefully before placing the vehicle into normal service.

6   Master cylinder - removal and installation

Warning: *If air finds its way into the hydraulic control unit on ABS models, the system must be bled with the use of a scan tool.*
Caution: *Brake fluid will damage paint. Cover all painted surfaces and avoid spilling fluid during this procedure.*

Removal

Refer to illustration 6.3

1   Remove the mounting fasteners for the air conditioning accumulator/drier (see Chapter 3).

2   Using a large syringe or equivalent, siphon the brake fluid from the master cylinder reservoir and dispose of it properly.

3   Unplug the electrical connector for the fluid level sensor **(see illustration)**.

4   On models equipped with a manual transmission, remove the hose from the fluid reservoir that leads to the clutch master cylinder.

5   Place rags under the brake line fittings and prepare caps or plastic bags to cover the ends of the lines once they are disconnected. Unscrew the tube nuts at the master cylinder

(see illustration 6.3). To prevent rounding off the flats on these nuts, a flare-nut wrench, which wraps around the fitting, should be used. Pull the brake lines away from the master cylinder slightly and plug the ends to prevent contamination.

6   Remove the two master cylinder mounting nuts **(see illustration 6.3)** and detach the master cylinder from the power brake booster. Move the air conditioning accumulator/drier as necessary to clear the master cylinder from the air conditioning lines leading to the accumulator.

7   Remove the reservoir cap, then discard any fluid remaining in the reservoir.

8   If a new master cylinder is being installed and it is not equipped with a reservoir, transfer the reservoir from the old master cylinder to the new one using new seals. Remove the reservoir retaining pin and carefully pry the reservoir away from the old master cylinder. Use clean brake fluid to ease installation of the new seals and reservoir.

Installation

Refer to illustrations 6.10 and 6.19

9   Bench bleed the new master cylinder before installing it. Mount the master cylinder in a vise, with the jaws of the vise clamping on

the mounting flange.

10   Attach a pair of master cylinder bleeder tubes to the outlet ports of the master cylinder **(see illustration)**. On models equipped with a manual transmission, plug the fitting on the reservoir for the clutch master cylinder hose.

11   Fill the reservoir with brake fluid of the recommended type (see Chapter 1).

12   Slowly push the pistons into the master cylinder (a large Phillips screwdriver can be used for this) - air will be expelled from the pressure chambers and into the reservoir. Because the tubes are submerged in fluid, air can't be drawn back into the master cylinder when you release the pistons.

13   Repeat the procedure until no more air bubbles are present.

14   Remove the bleed tubes, one at a time, and install plugs in the open ports to prevent fluid leakage and air from entering. Install the reservoir cap.

15   Install the master cylinder over the studs on the power brake booster and tighten the mounting nuts finger tight at this time.

16   On models equipped with a manual transmission, attach the clutch master cylinder hose to the fluid reservoir.

17   Thread the brake line fittings into the master cylinder. Since the master cylinder is still a bit loose, it can be moved slightly in

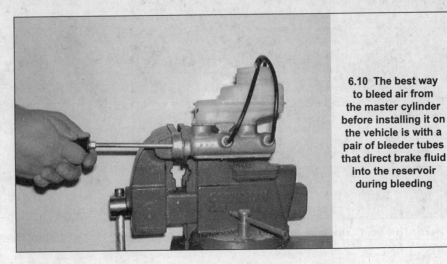

6.10 The best way to bleed air from the master cylinder before installing it on the vehicle is with a pair of bleeder tubes that direct brake fluid into the reservoir during bleeding

order for the fittings to thread in easily. Be careful not to cross-thread or strip the fittings as they are installed.

18 Fully tighten the mounting nuts to the torque listed in this Chapter's Specifications and then tighten the brake line fittings securely.

19 Fill the master cylinder reservoir with fluid, then bleed the master cylinder and the brake system as described in Section 8. To bleed the cylinder on the vehicle, have an assistant depress the brake pedal and hold the pedal to the floor. Loosen the fitting just enough to allow air and fluid to escape then tighten it lightly **(see illustration)**. Repeat this procedure on both fittings until the fluid is clear of air bubbles and then tighten the fittings securely. **Caution:** *Have plenty of rags on hand to catch the fluid - brake fluid will ruin painted surfaces. After the bleeding procedure is completed, rinse the area under the master cylinder with clean water.*

20 The remainder of installation is the reverse of removal. Test the operation of the brake system carefully before placing the vehicle into normal service. **Warning:** *If you are in doubt about the effectiveness of the brake system, DO NOT OPERATE THE VEHICLE. On models equipped with ABS, it is possible for air to become trapped in the anti-lock brake system hydraulic control unit. If the brake pedal continues to feel spongy after repeated bleedings or the BRAKE or ANTI-LOCK light stays on, have the vehicle towed to a dealer service department or other qualified shop to be bled with the aid of a scan tool.*

---

7 Brake hoses and lines - inspection and replacement

Warning: *If air has found its way into the hydraulic control unit on ABS models, the system must be bled with the use of a scan tool. If the brake pedal feels "spongy" even after bleeding the brakes, or the ABS light on* the instrument panel does not go off, or if you have any doubts whatsoever about the effectiveness of the brake system, have the vehicle towed to a dealer service department or other repair shop equipped with the necessary tools for bleeding the system.

1 About every six months, the rubber hoses which connect the steel brake lines to the front and rear calipers should be inspected. These are critical yet vulnerable parts of the brake system, so your inspection should be thorough. Raise the vehicle and support it securely on jackstands. Look for cracks, chafing of the outer cover, leaks, blisters and any other damage (a light and mirror are helpful). If a hose exhibits any of the above conditions, replace it with a new one.

Flexible hose

Refer to illustrations 7.2a and 7.2b

Note: *The brake hoses have integrated brackets which mount to the chassis.*

2 Use a flare nut wrench to disconnect the metal brake line from the hose fitting, be careful not to bend the frame bracket or brake line **(see illustrations)**. **Caution:** *On ABS-equipped models, plug the metal brake line immediately to prevent air from getting into the hydraulic control unit (HCU).*

3 Separate the ABS sensor harness from the brake hose. On front brakes, remove the hose bracket from the front strut and the chassis. On rear brake lines, remove the ABS harness connector that is mounted to the bracket for the brake hose, then unbolt the bracket from the chassis.

4 Remove the banjo bolt from the caliper **(see illustration 4.2)** and discard the sealing washers.

5 Connect the replacement hose to the caliper using new sealing washers. Tighten the banjo bolt to the torque listed in this Chapter's Specifications.

6 Mount the hose bracket to the chassis (and, on front brakes, to the strut).

7 Connect the metal brake line tube nut(s) to the hose fitting by hand, then, using a flare

6.19 Have an assistant depress the brake pedal and hold it down, then loosen the fitting nut, allowing the air and fluid to escape; repeat this procedure on both fittings until the fluid is clear of air bubbles

nut wrench, tighten the fitting(s) securely.

8 Attach the ABS brake harness to the brake hose. For rear brakes, reattach the ABS harness connector to the rear hose bracket.

9 When the brake hose installation is complete, there should be no kinks in the hose. Make sure the hose doesn't contact any part of the suspension. Check this by turning the wheels to the extreme left and right positions. If the hose makes contact, remove it and correct the installation as necessary.

Metal brake line

10 When replacing brake lines be sure to use the correct parts. Don't use copper tubing for any brake system components. Purchase steel brake lines from a dealer or auto parts store.

11 Prefabricated brake line, with the tube ends already flared and fittings installed, is available at auto parts stores and dealers. These lines are also sometimes bent to the proper shapes.

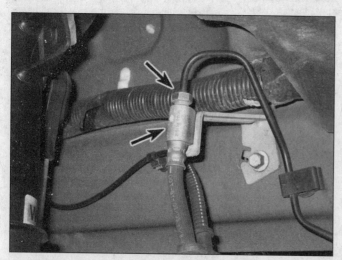

7.2a The front brake hose fitting and metal line tube nut

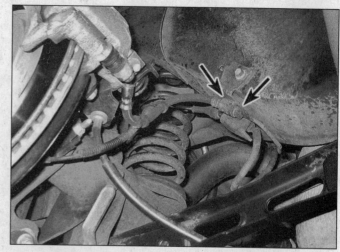

7.2b The rear brake hose fitting and metal line tube nut

8.8 When bleeding the brakes, a hose is connected to the bleed screw at the caliper and the other end is submerged in brake fluid - air will be seen as bubbles in the tube and container (all air must be expelled before moving to the next wheel)

9.9 Power brake booster mounting details:

1   *Plastic retaining pin cover (retaining pin underneath)*
2   *Pushrod pin*
3   *Booster mounting nut locations (only one nut visible in photo)*

12   When installing the new line make sure it's supported securely in the brackets and has plenty of clearance between moving or hot components.

13   After installation, check the master cylinder fluid level and add fluid as necessary. Bleed the brake system as outlined in the next Section and test the brakes carefully before driving the vehicle in traffic.

8   Brake hydraulic system - bleeding

Refer to illustration 8.8

Warning: *Wear eye protection when bleeding the brake system. If the fluid comes in contact with your eyes, immediately rinse them with water and seek medical attention.*

Note: *Bleeding the brake system is necessary to remove any air that's trapped in the system when it's opened during removal and installation of a hose, line, caliper, wheel cylinder or master cylinder.*

1   It will probably be necessary to bleed the system at all four brakes if air has entered the system due to low fluid level, or if the brake lines have been disconnected at the master cylinder.

2   If a brake line was disconnected only at a wheel, then only that caliper must be bled.

3   If a brake line is disconnected at a fitting located between the master cylinder and any of the brakes, that part of the system served by the disconnected line must be bled.

4   Remove any residual vacuum from the brake power booster by applying the brake several times with the engine off.

5   Remove the master cylinder reservoir cap and fill the reservoir with brake fluid. Reinstall the cap. **Note:** *Check the fluid level often during the bleeding operation and add fluid as necessary to prevent the fluid level from falling low enough to allow air bubbles into the master cylinder.*

6   Have an assistant on hand, as well as a supply of new brake fluid, an empty clear plastic container, a length of plastic, rubber or vinyl tubing to fit over the bleeder valve and a wrench to open and close the bleeder valve.

7   Beginning at the right rear wheel, loosen the bleeder screw slightly, then tighten it to a point where it's snug but can still be loosened quickly and easily.

8   Place one end of the tubing over the bleeder screw fitting and submerge the other end in brake fluid in the container **(see illustration)**.

9   Have the assistant slowly depress the brake pedal and hold it in the depressed position.

10   While the pedal is held depressed, open the bleeder screw just enough to allow a flow of fluid to leave the valve. Watch for air bubbles to exit the submerged end of the tube. When the fluid flow slows after a couple of seconds, tighten the screw and have your assistant release the pedal.

11   Repeat Steps 9 and 10 until no more air is seen leaving the tube, then tighten the bleeder screw and proceed to the left rear wheel, the right front wheel and the left front wheel, in that order, and perform the same procedure. Be sure to check the fluid level in the master cylinder reservoir frequently.

12   Never use old brake fluid. It contains moisture which can boil, rendering the brake system inoperative.

13   Refill the master cylinder with fluid at the end of the operation.

14   Check the operation of the brakes. The pedal should feel solid when depressed, with no sponginess. If necessary, repeat the entire process. **Warning:** *If you are in doubt about the effectiveness of the brake system, DO NOT OPERATE THE VEHICLE. On models equipped with ABS, it is possible for air to become trapped in the anti-lock brake system hydraulic control unit. If the brake pedal continues to feel spongy after repeated bleedings or the BRAKE or ANTI-LOCK light stays on,* have the vehicle towed to a dealer service department or other qualified shop to be bled with the aid of a scan tool.

9   Power brake booster - check, removal and installation

Operating check

1   Depress the brake pedal several times with the engine off and make sure that there is no change in the pedal reserve distance.

2   Depress the pedal and start the engine. If the pedal goes down slightly, operation is normal.

Airtightness check

3   Start the engine and turn it off after one or two minutes. Depress the brake pedal several times slowly. If the pedal goes down farther the first time but gradually rises after the second or third depression, the booster is airtight.

4   Depress the brake pedal while the engine is running, then stop the engine with the pedal depressed. If there is no change in the pedal reserve travel after holding the pedal for 30 seconds, the booster is airtight.

5   The power brake booster is not serviceable and must be replaced if found to be faulty.

Removal and installation

Refer to illustration 9.9

6   Remove the mounting fasteners for the air conditioning accumulator/drier (see Chapter 3).

7   Remove the master cylinder (see Section 6).

8   Disconnect the vacuum hose from the fitting on the power brake booster.

9   Inside the vehicle, remove the pushrod retaining pin cover and the retaining pin and then disconnect the pushrod from the upper end of the brake pedal **(see illustration)**.

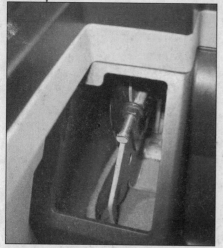

10.2 With the cover removed and the parking brake lever four clicks up, the parking brake cable adjuster nut can be accessed

11.1 The brake light switch location

10   Remove the nuts attaching the booster to the firewall.

11   Carefully lift the booster unit away from the firewall and out of the engine compartment.

12   To install the booster, place it into position and tighten the retaining nuts to the torque listed in the Chapter's Specifications. Connect the pushrod to the brake pedal and install the retaining clip and cover.

13   Install the master cylinder (see Section 6). Reconnect the vacuum hose to the booster.

14   Carefully test the operation of the brakes before placing the vehicle in normal service.

10   Parking brake cable - adjustment

Refer to illustration 10.2

1   Firmly apply the parking brake lever four times.

2   Open the center console storage compartment and remove the parking brake cable adjuster cover **(see illustration)**.

3   Raise the lever four clicks to gain access to the adjuster nut.

4   Turn the adjuster nut clockwise to tighten the cable or counterclockwise to loosen it. The correct adjustment is achieved when the adjuster nut is loosely touching the lever when the lever is in its lowest (fully released) position.

5   Apply the parking brake lever four times and recheck the adjustment.

6   Replace the small cover in the center console storage compartment.

11   Brake light switch - replacement

Refer to illustration 11.1

1   Rotate the brake light switch clockwise to release it from its bracket **(see illustration)**.

2   Disconnect the brake light switch electrical connector.

3   Connect the replacement switch with the electrical connector.

4   Position the switch into the mounting bracket and turn it counterclockwise to secure it.

5   Check the brake lights for proper operation.

Chapter 10
Suspension and steering systems

Contents

Specifications

Torque specifications

	Ft-lbs
Front suspension	
Balljoint-to-steering knuckle pinch bolt nut	76
Hub and bearing retaining nut	221
Control arm	
Pivot bolts/nuts	129
Rear mounting bolts/nuts	129
Splash shield	15
Stabilizer bar	
Stabilizer bar bracket nuts	52
Stabilizer bar link nuts	85
Strut	
Strut-to-steering knuckle bolts	148
Strut-to-vehicle nuts	26
Damper rod nut	46
Rear suspension	
Lateral stiffener fasteners	
Right side	85
Left side	46
Lower control arm	
Lower arm-to-axle bolt/nut	129
Lower arm-to-frame bolt/nut	129
Panhard rod fasteners	129
Rear support braces	
Mounting bolt on brace extension near differential housing	26
All other mounting bolts	46

Torque specifications (continued)

Ft-lbs (unless otherwise indicated)

Shock absorber
    Shock absorber-to-axle bolt/nut 85
    Shock absorber-to-chassis nut 30
Stabilizer bar
    Link-to-chassis bolts 85
    Bracket nuts 52
Upper control arm
    Upper arm-to-frame bolts
        Interior (1) 129
        Exterior (2) 85
    Upper arm-to-axle bolt/nut 129

Steering

Airbag module retaining bolts .... 62 in-lbs
Power steering line clamp bolt 17
Steering gear to-crossmember bolts 85
Steering shaft-to-steering gear U-joint pinch bolt 18
Steering wheel retaining bolt 41
Tie-rod end ballstud nut 59

1.1 Front suspension and steering components

1	Strut assembly	3	Balljoint	5	Steering gear
2	Tie-rod end	4	Control arm	6	Stabilizer bar

1 General information

Front suspension

Refer to illustration 1.1

The front suspension **(see illustration)** on these vehicles is a MacPherson strut design. The upper end of each strut is attached to the vehicle's body strut support. The lower end of the strut is connected to the steering knuckle. The steering knuckle is attached to the outer end of a control arm by a balljoint. The control arm is mounted to a front suspension crossmember. A stabilizer bar, mounted to the front crossmember and linked to the strut, reduces body roll while cornering.

Rear suspension

Refer to illustration 1.2

The rear suspension **(see illustration)** utilizes coil springs, shock absorbers, a Panhard rod, a lateral stiffener and one upper and two lower suspension arms. Body roll is controlled by a stabilizer bar linked to the chassis and attached to the rear axle.

Steering

All models are equipped with power assisted rack-and-pinion steering systems. The steering gear is bolted to the crossmember and is connected to the steering knuckles by a pair of tie-rods.

Precautions

Frequently, when working on the suspension or steering system components, you may come across fasteners which seem impossible to loosen. These fasteners on the underside of the vehicle are continually subjected to water, road grime, mud, etc., and can become rusted or "frozen," making them extremely difficult to remove. In order to unscrew these stubborn fasteners without damaging them (or other components), be sure to use lots of penetrating oil and allow it to soak in for awhile. Using a wire brush to clean exposed threads will also ease removal of the nut or bolt and prevent damage to the threads. Sometimes a sharp blow with a hammer and punch is effective in breaking the bond between a nut and bolt threads, but care must be taken to prevent the punch from slipping off the fastener and ruining the threads. Heating the stuck fastener and surrounding area with a torch sometimes helps too, but isn't recommended because of the obvious dangers associated with fire. Long breaker bars and extensions, or "cheater," pipes will increase leverage, but never use an extension pipe on a ratchet - the ratcheting mechanism could be damaged. Sometimes, turning the nut or bolt in the tightening (clockwise) direction first will help to break it loose. Fasteners that require drastic measures to remove should always be replaced with new ones. Most manufacturers now recommend replacing all suspension fasteners that are removed with new ones.

Since most of the procedures that are dealt with in this Chapter involve jacking up the vehicle and working underneath it, a good pair of jackstands will be needed. A hydraulic floor jack is the preferred type of jack to lift the vehicle, and it can also be used to support certain components during various operations. **Warning:** *Never, under any circumstances, rely on a jack to support the vehicle while working on it. Also, whenever any of the suspension or steering fasteners are loosened or removed they must be inspected and, if necessary, replaced with new ones of the same part number or of original equipment quality and design. Torque specifications must be followed for proper reassembly and component retention. Never attempt to heat or straighten suspension or steering components. Instead, replace bent or damaged parts with new ones.*

1.2 Rear suspension components

1 Shock absorber	4 Stabilizer bar	6 Lateral stiffener
2 Lower control arm	5 Rear axle	7 Panhard rod
3 Coil spring		

2.6 To detach the lower end of the strut from the steering knuckle, remove these bolts; be sure to first make alignment marks on the strut, knuckle and fasteners to ensure that the alignment is restored when the strut is installed

2 Strut assembly (front) - removal and installation

Removal

Refer to illustrations 2.6 and 2.7

1   Loosen the front wheel lug nuts, raise the vehicle and support it securely on jackstands. Remove the wheels.

2   Place a floor jack under the lower control arm and raise it slightly. The lower control arm must be supported throughout the entire procedure.

3   Remove the ABS harness from the strut and remove the ABS sensor, if equipped (see Chapter 9).

4   Remove the brake line bracket from the strut.

5   Remove the stabilizer bar link from the strut (see Section 4).

6   Mark the positions of the bolt heads to the strut, then remove the strut-to-steering knuckle fasteners by loosening the bolts **(see illustration)**.

7   Mark the upper strut mount in relation to the body, lower the control arm slightly and

then remove the upper mount-to-body fasteners **(see illustration)**.

8   Separate the strut assembly from the steering knuckle and remove it from the vehicle. Don't allow the steering knuckle to fall outward, as the brake hose could be damaged.

Inspection

9   Check the strut body for leaking fluid, dents, cracks and other obvious damage that would warrant repair or replacement.

10   Check the coil spring for chips or cracks in the spring coating (this can cause premature spring failure due to corrosion). Inspect the spring seat for cuts, hardness and general deterioration.

11   If any undesirable conditions exist, proceed to the strut disassembly procedure (see Section 3).

Installation

12   Guide the strut assembly up into the fenderwell and insert the strut mount studs through the holes in the strut tower. Once seated, install the nuts so the strut won't fall back through and then tighten the nuts to the torque listed in this Chapter's Specifications. **Note:** *An assistant would be helpful with this*

procedure because the strut is heavy and awkward and must be held while the mounting nuts are installed.

13   Slide the steering knuckle into the strut flange and insert the two bolts. Install the nuts, then tighten the bolts to the torque listed in this Chapter's Specifications.

14   Reattach the brake hose and the ABS harness (if equipped) to the strut bracket.

15   Remove the jack from under the lower control arm, install the wheels, lower the vehicle and tighten the lug nuts to the torque listed in the Chapter 1 Specifications.

16   Have the front wheel alignment checked and, if necessary, adjusted.

3   Strut/coil spring assembly - replacement

1   If the struts or coil springs exhibit the telltale signs of wear (leaking fluid, loss of damping capability, chipped, sagging or cracked coil springs) explore all options before beginning any work. The struts or coil springs themselves are not serviceable and must be replaced if a problem develops. However, complete strut assemblies (struts and springs) may be available on an exchange basis, which eliminates much time and work. Whichever route you choose to take, check on the cost and availability of parts before disassembling your vehicle. **Warning:** *Disassembling a strut assembly is potentially dangerous and utmost attention must be directed to the job, or serious injury may result. Use only a high-quality spring compressor and carefully follow the manufacturer's instructions furnished with the tool. After removing the coil spring from the strut assembly, set it aside in a safe, isolated area.*

Disassembly

Refer to illustrations 3.3 and 3.6

2   Remove the strut assembly (see Section 2). Mount the strut assembly in a vise. Line the vise jaws with wood or rags to pre-

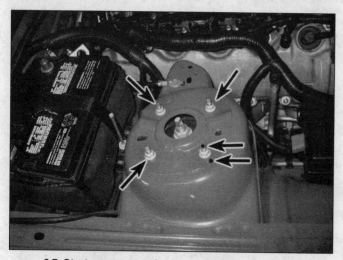

2.7 Strut upper mounting nuts and alignment marks

3.3 Install the spring compressor in accordance with the tool manufacturer's instructions and compress the spring until all pressure is relieved from the upper bearing assembly

3.6 Remove the compressed spring assembly - keep the ends of the spring pointed away from your body

3.9 When installing the spring, make sure the end fits into the recessed portion of the lower spring seat

vent damage to the unit and don't tighten the vise excessively.

3   Following the tool manufacturer's instructions, install the spring compressor (which can be obtained at most auto parts stores or equipment yards on a daily rental basis) on the spring and compress it sufficiently to relieve all pressure from the upper bearing assembly **(see illustration)**. This can be verified by wiggling the spring.

4   Use two wrenches to remove the damper rod nut; hold the damper rod with one wrench to keep it from turning while removing the damper rod nut with the other.

5   Remove the upper bearing assembly. Inspect the bearing in the assembly for smooth operation. If it doesn't turn smoothly, replace it. Check the rubber portion of the assembly for cracking and general deterioration. If there is any separation of the rubber, replace it.

6   Carefully lift the compressed spring from the assembly **(see illustration)** and set it in a safe place. **Warning:** *Never place your head*

near the end of the spring!

7   Slide the dust cover and rubber bumper off the damper rod.

Reassembly

Refer to illustration 3.9

8   Install the rubber bumper and dust cover.

9   Carefully place the coil spring onto the strut with the end of the spring positioned correctly onto the lower spring seat **(see illustration)**.

10   Install the upper bearing assembly, making sure that the notch on the bearing assembly lines up with the clevis (bracket) on the bottom of the strut. **Note:** *There is a notch and an arrow etched on the top of the upper bearing assembly that will point away from the vehicle when the strut assembly is properly assembled and installed on the vehicle.*

11   Install the damper rod nut and tighten it to the torque listed in this Chapter's Specifications. Remove the spring compressor tool.

12   Install the strut/spring assembly (see Section 2).

---

4   Stabilizer bar, bushings and links (front) - removal and installation

---

Removal

Refer to illustrations 4.2 and 4.3

Note: *The manufacturer recommends using new fasteners for the installation of the stabilizer bar and links.*

1   Raise the front of the vehicle and support it securely on jackstands. Apply the parking brake.

2   Remove the stabilizer bar-to-link nuts and discard them **(see illustration)**. If you are removing the link entirely, remove the strut-to-link nuts also.

3   Remove the stabilizer bar bracket nuts and detach the bar from the vehicle **(see illustration)**.

4.2 The stabilizer bar link nuts are fastened to ballstuds on the link; hold the ballstud with a wrench while loosening the link nuts to remove them

4.3 To detach the stabilizer bar from the frame, remove the bracket nuts

5.2 Remove the pinch bolt and separate the balljoint from the steering knuckle with a pry bar

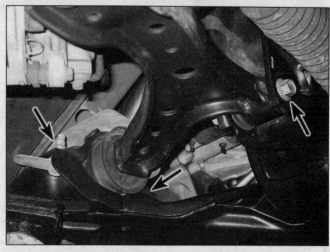

5.3 The front control arm pivot bolt and rear mounting bracket fasteners (one fastener hidden - vicinity shown)

4   Pull the brackets off the stabilizer bar and inspect the bushings for cracks, hardening and other signs of deterioration. If the bushings are damaged, remove and discard them.

Installation

5   Lubricate the new stabilizer bar bushings with a rubber lubricant (such as a silicone spray) and slide the bushings onto the bar.
6   Push the brackets over the bushings and raise the bar up to the frame. Install new bracket nuts finger tight.
7   Install the stabilizer bar links using new nuts.
8   Tighten the bracket nuts and the stabilizer bar link nuts to the torque listed in this Chapter's Specifications.

5   Control arm (front) - removal and installation

Refer to illustrations 5.2 and 5.3
1   Loosen the wheel lug nuts, raise the vehicle and support it securely on jackstands. Remove the wheel.
2   Remove the steering knuckle pinch bolt and carefully separate the steering knuckle from the balljoint and control arm **(see illustration)**.
3   Remove the steering gear mounting fasteners and move the gear enough to remove the forward pivot bolt (see Section 17) **(see illustration)**. **Caution:** *Be careful not to damage the steering gear boot while removing the pivot bolt.*
4   Remove the rear mounting fasteners, then remove the control arm and rear mounting bracket **(see illustration 5.3)**.
5   Installation is the reverse of the removal procedure. Be sure to tighten all of the fasteners to the torque listed in this Chapter's Specifications. Tighten the lug nuts to the torque listed in the Chapter 1 Specifications. **Note:**

The rear mounting fasteners can be reversed to ease installation; the flag bolts can be inserted from the top of the rear mounting bracket and the nuts can go on the bottom.
6   Have the front wheel alignment checked and, if necessary, adjusted.

6   Balljoint - check and replacement

Check

Refer to illustration 6.2
1   Raise the front end of the vehicle and place it securely on jackstands.
2   Grasp the top and bottom of the tire, then try to simultaneously push in at the top and pull out at the bottom of the tire **(see illustration)**, then the opposite (push in on the bottom while pulling out on the top). Replace the balljoint if there is any movement in the balljoint area.
3   Push up and pull down on the control arm. If the balljoint has more than 0.012-inch of play, it's worn out.

Replacement

4   The balljoints on these models are not serviceable. If the balljoint is faulty, the control arm and balljoint are replaced as a single assembly.

7   Hub and bearing assembly (front) - removal and installation

Refer to illustrations 7.3 and 7.4
Warning: *The manufacturer recommends using a new retaining nut for the installation of the hub and bearing assembly.*
Note: *The front wheel bearings are part of the hub and are not serviced separately.*
1   Loosen the wheel lug nuts, apply the parking brake and raise the vehicle. Support it securely on jackstands placed under the frame. Remove the wheel.
2   Remove the brake caliper and hang it out of the way with a length of wire, then remove the caliper mounting bracket and the brake disc (see Chapter 9).

6.2 To check a balljoint, raise the front of the vehicle and place it securely on jackstands, support the lower control arm with a floor jack, then try to rock the wheel in and out; if there's any play in the balljoint, replace it

7.3 To remove the hub and bearing retaining nut, first remove the grease cap with a hammer and chisel

7.4 To detach the hub and bearing assembly from the steering knuckle, remove the retaining nut and pull the hub from the spindle (this nut is very tight, so make sure the vehicle is supported securely)

3   Remove the grease cap from the hub and bearing assembly (see illustration).
4   Remove the hub and bearing retaining nut and discard it (see illustration).
5   Remove the hub and bearing assembly. If the hub does not come off easily, use an appropriate puller.
6   Installation is the reverse of removal. Be sure to clean off the spindle and lubricate it with wheel bearing grease, install a new hub and bearing retaining nut and then tighten it to the torque listed in this Chapter's Specifications. Tighten the lug nuts to the torque listed in the Chapter 1 Specifications.

8   Steering knuckle - removal and installation

1   Loosen the wheel lug nuts, apply the parking brake and raise the vehicle. Support it securely on jackstands placed under the frame. Remove the wheel.

2   Remove the ABS wheel speed sensor and the brake disc (see Chapter 9).
3   Remove the hub and bearing assembly (see Section 7).
4   Loosen the strut-to-steering knuckle fasteners, but don't remove them yet (see Section 2). **Note:** *Be sure to mark the lower strut in relation to the steering knuckle before loosening the fasteners.*
5   Separate the tie-rod end from the steering knuckle (see Section 15).
6   Separate the lower control arm from the steering knuckle by removing the steering knuckle pinch bolt for the ballstud (see Section 5).
7   Remove the strut-to-steering knuckle fasteners and then remove the steering knuckle.
8   Remove the splash shield from the steering knuckle.
9   Installation is the reverse of removal. Be sure to tighten all fasteners to the torque listed in this Chapter's Specifications.
10   Install the wheel and lug nuts. Lower the vehicle to the ground and tighten the nuts to

the torque listed in the Chapter 1 Specifications.
11   Have the front wheel alignment checked and, if necessary, adjusted.

9   Stabilizer bar bushings and links (rear) - removal and installation

Refer to illustration 9.2
Note 1: *The manufacturer recommends using new fasteners for the installation of the stabilizer bar and links.*
Note 2: *Replacement bushings for coupes are different than those for convertibles. The stabilizer bar is different as well. Be sure that the replacement part is the right one for the vehicle you're working on.*
1   Raise the rear of the vehicle and support it securely on jackstands. Place the jack and jackstands under the frame, not under the axle housing.
2   Remove the link-to-chassis fasteners and discard them (see illustration).
3   Remove the fasteners that attach the stabilizer bar brackets to the brackets on the rear axle and remove the bar.
4   Pull the brackets off the stabilizer bar and inspect the bushings for cracks, hardening and other signs of deterioration. If the bushings are damaged, remove and discard them.
5   Inspect the link bushings and replace them if necessary.
6   Lubricate the replacement bushings with a rubber lubricant (such as a silicone spray) and slide them onto the bar.
7   Installation is the reverse of removal. Tighten all fasteners to the torque listed in this Chapter's Specifications.
8   Remove the jackstands and lower the vehicle.

9.2 The rear stabilizer bar link-to-chassis fastener (A) and the stabilizer bar bracket fasteners (B)

10   Shock absorber (rear) - removal and installation

Refer to illustrations 10.1, 10.4 and 10.6

Caution: *Remove and install shock absorbers one side at a time.*

Note: *The manufacturer recommends using new fasteners for shock absorber installation.*

1   With the vehicle on the ground and at normal ride height, mark the shock absorbers in relation to the dust sleeve **(see illustration)**.

2   Raise the rear of the vehicle and support it securely on jackstands placed under the frame, not under the differential housing or axle.

3   Place a floor jack under the axle tube on the side of the vehicle that you're working on to support it.

4   Locate the upper retaining nut for the shock absorber inside the trunk **(see illustration)**. To remove the upper mounting nut, hold the shock's piston rod with a wrench to prevent it from moving while loosening the nut. Remove the nut, washer and upper insulator. Discard the nut.

5   For models equipped with a rear stabilizer bar, remove the stabilizer bar links from the chassis (see Section 9) and move the stabilizer bar to gain access to the lower shock mounting fasteners.

6   Remove the lower shock mounting fasteners **(see illustration)** and the shock absorber. Move the axle up or down as necessary to free the shock from the vehicle. Discard the mounting fasteners.

7   If you are replacing shock absorbers with new ones, transfer the index mark made in step one to the new shocks.

8   Place the shock in position. Raise or lower the axle as needed to install the upper and lower mounting fasteners and leave them finger tight.

9   With the mounting fasteners in place,

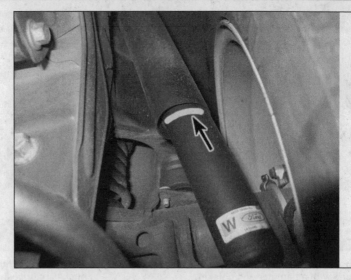

10.1 Mark the shock in relation to the dust sleeve while the vehicle is at normal ride height

carefully raise the axle so that the mark on the shock lines up with the dust sleeve and then tighten the mounting fasteners to the torque listed in this Chapter's Specifications.

Note: *Raising the axle so that the shock compresses to the reference mark simulates the position of the vehicle's suspension when it's sitting at normal ride height.*

10   Remove the jackstands and lower the vehicle.

11   Coil springs (rear) - removal and installation

Removal

Note 1: *Rear coil springs should always be replaced in pairs.*

Note 2: *The manufacturer recommends using new fasteners for suspension component installation.*

1   With the vehicle on the ground and normal ride height, mark the shock absorb-

ers in relation to the dust sleeve (see Section 10).

2   Loosen the wheel lug nuts, raise the rear of the vehicle and support it securely on jackstands placed under the frame. Remove the wheels.

3   On convertible models, remove the rear support braces (see Section 13).

4   Detach the rear stabilizer bar links from the chassis, if equipped (see Section 9).

5   Place two floor jacks under the axle tubes (one on each side of the differential housing) to support it. They will be used to lower the axle in the following steps.

6   Remove the shock absorber lower mounting fasteners (see Section 10).

7   Remove the rear brake hose brackets (see Chapter 9).

8   Slowly lower the jacks under the axle until all tension is removed from the coil springs. Note the position of the spring ends on their seats and then remove them and the insulators. Inspect the insulators for wear and damage and replace them if necessary.

10.4 The shock absorber upper mounting fastener can be found in the trunk compartment. Hold the shock's piston shaft with a wrench while loosening the mounting nut

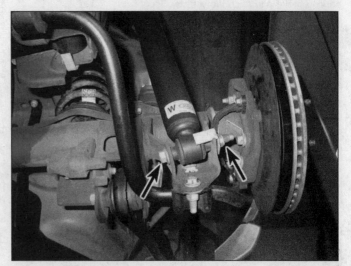

10.6 The shock absorber lower mounting fasteners

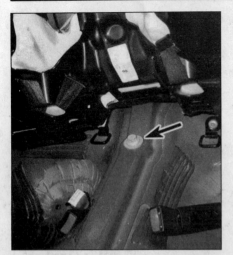

12.4 Location of the upper control arm front bolt under the rear seat

12.6 Upper control arm mounting details:

1 *Rear mounting bolts (one hidden)*
2 *Front mounting bolt (removed from interior)*
3 *Control arm-to-rear axle pivot bolt (must be removed by loosening the nut on other side)*
4 *Upper control arm front pivot bolt (the manufacturer recommends NOT to remove this fastener).*

Installation

9   Reinstall the insulators. Use tape to hold the top insulator in place, if necessary.
10   Place the spring in its original position between the frame and axle tube seats.
11   Raise the axle enough to install the shock absorber's lower mounting fasteners but don't tighten them yet (see Section 10).
12   With the mounting fasteners in place, carefully raise the axle so that the marks on the shocks line up with the dust sleeves and then tighten the shock's lower mounting fasteners to the torque listed in this Chapter's Specifications. **Note:** *Raising the axle so that the shocks compress to the reference marks, simulates the position of the vehicle's suspension when it's sitting at normal ride height.*
13   Reinstall the brake hose bracket (see Chapter 9).
14   Reattach the stabilizer bar links to the chassis (see Section 9).
15   On convertible models, install the rear support braces (see Section 13). **Caution:** *The brace fasteners must be sequentially torqued* (see illustration 13.3).
16   Remove the jackstands and lower the vehicle.

12 Suspension arms, lateral stiffener and Panhard rod (rear) - removal and installation

Note: *The manufacturer recommends using new fasteners for suspension component installation.*
1   With the vehicle on the ground and at normal ride height, mark the shock absorbers in relation to the dust sleeve (see Section 10).
2   Loosen the wheel lug nuts, raise the rear of the vehicle and support it securely on jackstands placed under the frame. Remove the wheels.
3   On convertible models, remove the rear support braces (see Section 13).

Upper control arm

Refer to illustrations 12.4 and 12.6

4   Remove the rear seat (see Chapter 11) and then remove the upper control arm front bolt (see illustration). Discard the bolt.
5   Support the fuel tank and remove the two rear fuel tank strap fasteners (see Chapter 4). Lower the fuel tank enough for access to the upper control arm mounting fasteners (see Chapter 4).
6   Remove the upper control arm-to-rear axle pivot bolt and nut (see illustration). **Caution:** *Do not remove the other pivot bolt that attaches the control arm to the mount through the arm bushing. The angle of the arm and mount are set during production and affect the ride height of the vehicle. If the bushing between the control arm and mount is damaged, replace the bushing, arm and mount as a single unit.*
7   Unscrew the fasteners for the control arm assembly and remove the arm (see illustration 12.6).
8   Inspect the bushings in the control arm assembly and at the differential housing. If either bushing is damaged or worn, the control arm assembly or the bushing at the dif-

ferential housing will have to be replaced.
9   Position the control arm assembly and install new mounting fasteners to the chassis. Tighten the fasteners to the torque listed in this Chapter's Specifications.
10   Carefully raise the axle so that the marks on the shock absorbers line up with the dust sleeves, then install the control arm-to-rear axle pivot bolt. Tighten the fasteners to the torque listed in this Chapter's Specifications. **Note:** *Raising the axle so that the shocks compress to the reference marks simulates the position of the vehicle's suspension when it's sitting at normal ride height.*

Lower control arm

Refer to illustration 12.13

11   Detach the parking brake cable from the caliper (see Chapter 9). Remove the parking brake cable bracket fastener and remove the bracket (located in front of where the lower control arm attaches to the chassis).
12   Place two floor jacks under the axle tubes (one on each side of the differential housing) to support it.
13   Remove the lower control arm mounting fasteners, then remove the lower arm (see

12.13 The lower control arm mounting fasteners

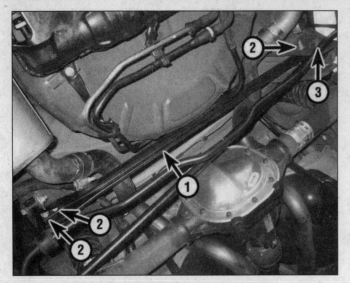

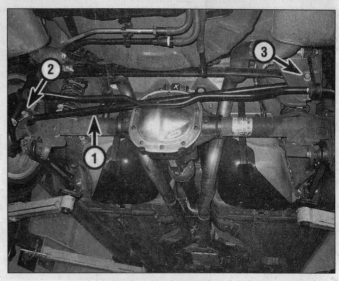

12.18 Lateral stiffener mounting details:

1   Lateral stiffener
2   Lateral stiffener mounting fasteners
3   Panhard rod-to-body mounting fasteners

12.24 Panhard rod mounting details:

1   Panhard rod
2   Panhard rod-to-axle mounting fasteners
3   Panhard rod-to-body mounting fasteners

illustration). Discard the fasteners.

14   Inspect the bushings at both ends of the arm. If either bushing is damaged or worn, the control arm will require replacement.

15   To install the arm, position it in the frame and axle mounting brackets and install new fasteners. Do not tighten the fasteners at this time.

16   Carefully raise the axle so that the marks on the shock absorbers line up with the dust sleeves, then tighten the lower control arm mounting fasteners to the torque listed in this Chapter's Specifications. **Note:** *Raising the axle so that the shocks compress to the reference marks, simulates the position of the vehicle's suspension when its sitting at normal ride height.*

Lateral stiffener

Refer to illustration 12.18

Note: *New fasteners for the Panhard rod as well as the stiffener will be necessary for this procedure.*

17   Place two floor jacks under the axle tubes (one on each side of the differential housing) to support it.

18   Remove the Panhard rod-to-body mounting fasteners, then remove the lateral stiffener mounting fasteners **(see illustration)**. Discard the fasteners.

19   Remove the lateral stiffener.

20   To install the stiffener, position it to the frame and install new fasteners. Tighten the fasteners to the torque listed in this Chapter's Specifications.

21   Install new Panhard rod-to-body fasteners but do not tighten them yet.

22   Carefully raise the axle so that the marks

on the shock absorbers line up with the dust sleeves and then tighten the Panhard rod-to-body mounting fasteners to the torque listed in this Chapter's Specifications. **Note:** *Raising the axle so that the shocks compress to the reference marks simulates the position of the vehicle's suspension when it's sitting at normal ride height.*

Panhard rod

Refer to illustration 12.24

23   Place two floor jacks under the axle tubes (one on each side of the differential housing) to support it.

24   Remove the Panhard rod-to-axle and Panhard rod-to-body mounting fasteners and then remove the rod **(see illustration)**. Discard the fasteners. **Note:** *Remove the plastic cover over the Panhard rod-to-axle mounting fasteners by inserting screwdrivers into the access slots to release the retaining clips, if equipped.*

25   To install the Panhard rod, position it to the frame and axle and install new mounting fasteners but do not tighten them yet.

26   Carefully raise the axle so that the marks on the shock absorbers line up with the dust sleeves, then tighten the Panhard rod-to-axle and Panhard rod-to-body mounting fasteners to the torque listed in this Chapter's Specifications. **Note:** *Raising the axle so that the shocks compress to the reference marks simulates the position of the vehicle's suspension when it's sitting at normal ride height.*

27   Replace the plastic cover over the Panhard rod-to-axle mounting fasteners, if equipped.

All components

28   On convertible models, reinstall the rear support braces (see Section 13).

29   Install the wheel and lug nuts. Remove the jackstands and floor jacks, lower the vehicle and tighten the lug nuts to the torque listed in the Chapter 1 Specifications.

13   Rear support braces (convertibles only) - removal and installation

Refer to illustrations 13.3 and 13.5

Note 1: *The two rear support braces must be removed and installed together, not one side at a time.*

Note 2: *The manufacturer recommends using new fasteners for the installation of the rear support braces.*

1   The rear support braces add stiffness to the frame, extra protection in the event of collision and stability and handling to convertible models. They are recognized by two long metal braces that form a V under the rear portion of the vehicle.

2   Raise the vehicle and support it on jackstands. Do not place the jackstands under the rear axle.

3   Follow the sequence in the accompanying illustration to remove the mounting fasteners **(see illustration)**.

4   To install, position the support braces in place and install the new fasteners by hand. **Note:** *Install all of the fasteners before tightening any of them.*

5   Follow the sequence in the accompany-

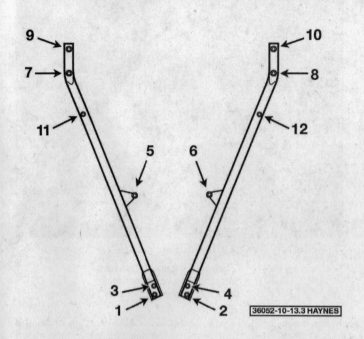

13.3 Follow this sequence to remove the brace fasteners

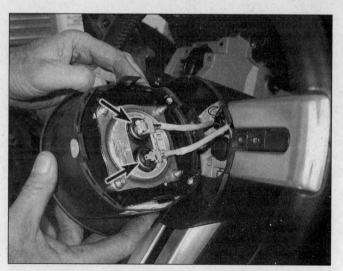

13.5 Follow this sequence to tighten the brace fasteners

ing illustration and tighten the fasteners to the torque listed in this Chapter's Specifications **(see illustration)**.

6   Remove the jackstands and lower the vehicle.

14 Steering wheel - removal and installation

Refer to illustrations 14.3, 14.4, 14.5 and 14.8
Warning: *The models covered by this manual are equipped with Supplemental Restraint systems (SRS), more commonly known as airbags. Always disable the airbag system before working in the vicinity of any airbag system component to avoid the possibility of accidental deployment of the airbag, which could cause personal injury (see Chapter 12).*

1   Park the vehicle with the front wheels pointed straight ahead and the steering wheel centered. Disconnect the cable from the negative terminal of the battery (see Chapter 5).

2   Refer to Chapter 12 and disable the airbag system.

3   Pry out the two small covers, one on each side of the steering wheel, and unscrew the two fasteners that retain the airbag module **(see illustration)**.

4   Lift the airbag module off the steering wheel and disconnect the airbag electrical connectors **(see illustration)**. **Warning:** *When handling the airbag module, hold it with the trim side facing away from you. Set the airbag module down in a safe location with the trim side facing up.*

5   Remove the steering wheel retaining bolt and discard it. Mark the position of the steer-

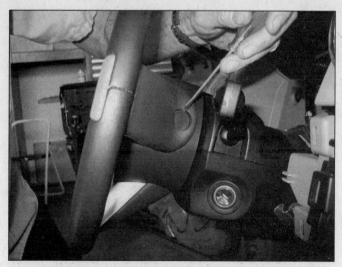

14.3 Pry off the covers on the sides of the steering wheel and remove the fasteners that retain the airbag module

14.4 Disconnect the electrical connectors from the airbag module (squeeze the locking tabs on the sides of each connector, then pull the connectors out)

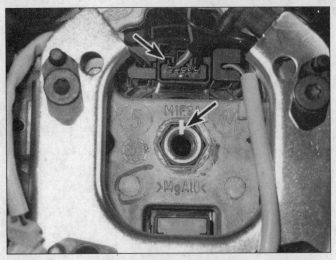

14.5 Remove the steering wheel bolt and mark the relationship of the steering wheel to the steering shaft. Also, disconnect the electrical connector in the center of the clockspring terminal

14.8 Clockspring details:

1   *Clockspring mounting fasteners*
2   *Clockspring electrical connector*
3   *Clockspring alignment marks*

ing wheel to the shaft **(see illustration).**

6   Unplug the electrical connector in the center of the clockspring terminal **(see illustration 14.5).**

7   Lift the steering wheel off the shaft while guiding the airbag module electrical connectors through the steering wheel hub. Tape the clockspring in the centered position to keep it from moving while the steering wheel is off. **Warning:** *While the steering wheel is removed, DO NOT turn the steering shaft. If the steering shaft or clockspring are accidentally turned, the clockspring must be centered before installing the steering wheel. Failing to center the clockspring will damage it and render the driver's airbag inoperative.*

8   Make sure the clockspring is centered, as follows: Verify that the front wheels are pointing straight ahead. Turn the clockspring rotor counterclockwise by hand until it becomes

hard to turn (don't apply too much force, though, because the cable could break). Then turn the clockspring clockwise about 3 turns and align the marks **(see illustration).**

9   If it's necessary to remove the clockspring from the steering column, apply two pieces of tape across the hub of the clockspring to the housing to prevent it from rotating. Remove the steering column covers (see Chapter 11). Disconnect the electrical connector from the bottom of the clockspring, unscrew the mounting fasteners and then remove it **(see illustration 14.8).** Reverse the removal procedure to install the clockspring and be sure it's centered.

10   To install the steering wheel, align the mark on the hub with the mark on the shaft and slide the wheel onto the shaft. Install a new bolt and tighten it to the torque listed in this Chapter's Specifications.

11   Installation is otherwise the reverse of removal. Be sure the electrical connectors are securely connected to the airbag module and tighten the airbag module mounting fasteners to the torque listed in this Chapter's Specifications.

15   Tie-rod ends - removal and installation

Refer to illustrations 15.2a, 15.2b and 15.3

Warning: *The manufacturer recommends installing a new tie-rod end ballstud nut during installation.*

1   Loosen the wheel lug nuts, raise the vehicle and place it securely on jackstands. Remove the wheel.

2   Loosen the tie-rod end locknut and mark the position of the tie-rod end on the threaded

15.2a Loosen the tie-rod end locknut

15.2b Mark the position of the tie-rod end on the threaded portion of the tie-rod

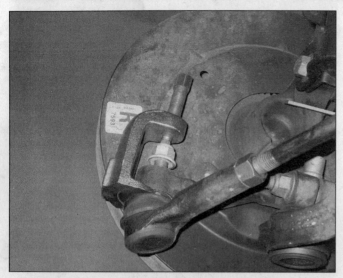 

15.3 With the ballstud nut loosened, separate the tie-rod end from the steering knuckle using a suitable tool or puller

16.3 The outer clamp on the steering gear boot can be removed with pliers - the inner clamp must be cut off

portion of the tie-rod **(see illustrations)**.

3   Loosen the nut from the tie-rod end balljoint stud, then separate the tie-rod end from the steering knuckle **(see illustration)**. Remove the nut and detach the tie-rod end from the steering knuckle arm. **Note:** *Hold the ballstud on the tie-rod end with the appropriate tool if it turns while removing the nut.*

4   Unscrew the old tie-rod end and install the new one. Make sure the new tie-rod end is aligned with the mark you made on the threads of the tie-rod.

5   Installation is the reverse of removal. Be sure to tighten the new tie-rod end ballstud nut to the torque listed in this Chapter's Specifications. Tighten the locknut securely.

6   Have the front wheel alignment checked and, if necessary, adjusted.

16 Steering gear boots - replacement

Refer to illustration 16.3

1   Loosen the lug nuts, raise the vehicle and support it securely on jackstands. Remove the wheel.

2   Remove the tie-rod end (see Section 15).

3   Remove the steering gear boot clamps **(see illustration)** and slide off the boot. **Note:** *Some boots may have small vents or tube fittings molded into them. Make sure to note the position of these vents so that the replacement boots can be installed in the same position.*

4   Before installing the new boot, wrap the threads and serrations on the end of the inner tie rod with a layer of tape so the small end of the new boot isn't damaged.

5   Slide the new boot into position on the steering gear until it seats in the groove in the steering gear and install new clamps.

6   Remove the tape from the inner tie rod and install the tie-rod end (see Section 15).

7   Install the wheel and lug nuts. Lower the vehicle and tighten the lug nuts to the torque listed in this Chapter's Specifications.

8   Have the front wheel alignment checked and, if necessary, adjusted.

17 Steering gear - removal and installation

Refer to illustrations 17.4, 17.5 and 17.6

Warning 1: *The models covered by this manual are equipped with Supplemental Restraint systems (SRS), more commonly known as airbags. Always disable the airbag system before working in the vicinity of any airbag system component to avoid the possibility of*

accidental deployment of the airbag, which could cause personal injury (see Chapter 12). **Warning 2:** *Make sure the steering column shaft is not turned while the steering gear is removed or you could damage the airbag system clockspring. To prevent the shaft from turning, turn the ignition key to the lock position before beginning work, and run the seat belt through the steering wheel and clip it into its latch.*

1   Park the vehicle with the front wheels pointing straight ahead.

2   Loosen the wheel lug nuts. Raise the front of the vehicle and support it securely on jackstands. Apply the parking brake. Remove the wheels.

3   Disconnect the tie-rod ends from the steering knuckles (see Section 15).

4   Remove the power steering fluid line brackets **(see illustration)**.

5   Mark the relationship of the lower steering shaft U-joint to the steering gear input shaft, remove the U-joint pinch bolt, then slide

17.4 Steering gear mounting details:

1   *Power steering line bracket locations*

2   *Steering gear mounting fasteners*

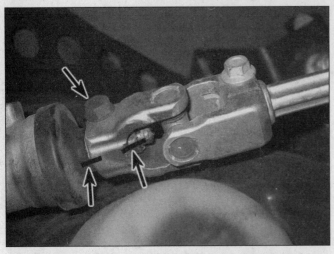

17.5 Mark the relationship of the steering shaft U-joint before removing the pinch bolt

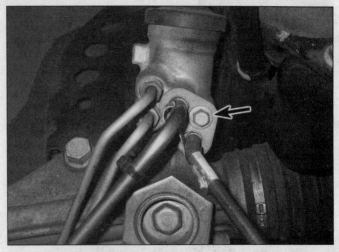

17.6 The pressure and return line clamp fastener

the U-joint up and off the input shaft **(see illustration)**.

6   Disconnect the pressure and return lines from the steering gear and discard any O-ring seals **(see illustration)**.

7   Remove the steering gear mounting fasteners **(see illustration 17.4)**.

8   Remove the steering gear assembly.

9   Installation is the reverse of removal. Be sure to align the matchmarks made in Step 5 and tighten all suspension and steering gear fasteners to the torque listed in this Chapter's Specifications. Also, use new O-rings when assembling the pressure and return lines to the steering gear. Tighten the wheel lug nuts to the torque listed in the Chapter 1 Specifications.

10   Bleed the power steering system when you're done (see Section 20).

18   Steering column - removal and installation

Warning 1: *The models covered by this manual are equipped with Supplemental Restraint systems (SRS), more commonly known as*

18.4 The panel reinforcement mounting fasteners

airbags. Always disable the airbag system before working in the vicinity of any airbag system component to avoid the possibility of accidental deployment of the airbag, which could cause personal injury (see Chapter 12).
Warning 2: *When the steering wheel or column is removed, it is important that the front wheels not be moved to point in a different direction. Moving the wheels will change the*

original position of the steering shafts coming from the steering gear. The steering shafts must be returned to their original positions in relation to the clockspring when the steering column or wheel is reinstalled. This is crucial for the proper operation of the Supplemental Restraint systems (SRS). If the steering shafts have been turned for some reason, refer to Section 14 for the procedure on re-centering the clockspring.

Removal

Refer to illustrations 18.4, 18.6 and 18.7

1   Park the vehicle with the wheels pointing straight ahead. Disconnect the cable from the negative battery terminal (see Chapter 5, Section 1).

2   Remove the steering wheel (see Section 11).

3   Remove the knee bolster from under the steering column (see Chapter 11).

4   Remove the panel reinforcement from under the steering column **(see illustration)**.

5   Remove the steering column covers (see Chapter 11).

6   Disconnect the electrical connectors for the steering column harness **(see illustration)**.

18.6 The steering column harness and electrical connectors (some connectors are hidden)

18.7 Steering column mounting details:

1 *Alignment mark*
2 *Intermediate shaft coupler pinch bolt*
3 *Steering column mounting fasteners*

19.3 Remove the pulley from the power steering pump with a pulley removal tool

7   Mark the steering column shaft in relation to the intermediate shaft coupler, then remove the coupler pinch bolt and discard it **(see illustration)**.

8   Remove the steering column mounting nuts **(see illustration 18.7)**. Lower the column and pull it to the rear, making sure nothing is still connected. Separate the intermediate shaft from the steering shaft and remove the column.

Installation

9   Guide the steering column into position, connect the intermediate shaft, then install the mounting fasteners, but don't tighten them yet.

10   Install the coupler pinch bolt, tightening it to the torque listed in this Chapter's Specifications. **Warning:** *The manufacturer recommends using a new bolt during reassembly.*

11   Tighten the column mounting fasteners to the torque listed in this Chapter's Specifications.

12   The remainder of installation is the reverse of removal.

19 Power steering pump - removal and installation

Refer to illustrations 19.3, 19.4 and 19.7

Note: *On V8 engine models, it's necessary to remove and install the power steering pump pulley. This requires special tools that can be found at most auto parts stores.*

1   Remove the drivebelt (see Chapter 1).

2   Raise the front of the vehicle and place it securely on jackstands.

3   On V8 engine models, remove the power steering pump pulley **(see illustration)**.

4   Position a drain pan under the power steering pump. Disconnect the pressure and return hoses from the pump **(see illustra-**

tion). Plug the hoses to prevent contaminants from entering. Discard the seal from the pressure line - a new one should be used during installation.

5   Remove the small shield that covers the pulley that is mounted to the side of the pump, if equipped.

6   Remove the pump mounting fasteners and remove the pump from the engine compartment, taking care not to spill fluid on the painted surfaces.

7   Installation is the reverse of removal. Use a new seal on the pressure line fitting. Tighten the pressure line fitting and the pump mounting bolts to the torque listed in this Chapter's Specifications. On V8 engine models, use a special tool to reinstall the pulley onto the pump shaft **(see illustration)**.

8   Fill the power steering reservoir with the recommended fluid (see Chapter 1) and bleed the system following the procedure described in the next Section.

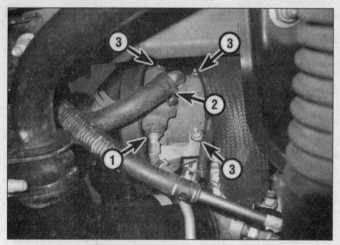

19.4 Power steering pump details (V8 model shown, V6 model similar):

1   *Pressure line fitting*
2   *Return hose*
3   *Mounting fasteners*

19.7 Press the pulley onto the shaft using a pulley installation tool - don't attempt to drive it on with a hammer or push it on with a traditional press!

20   Power steering system - bleeding

1   Following any operation in which the power steering fluid lines have been disconnected, the power steering system must be bled to remove all air and obtain proper steering performance.

2   With the front wheels in the straight ahead position, check the power steering fluid level and, if low, add fluid until it reaches the Cold mark on the reservoir.

3   Start the engine and allow it to run at fast idle. Recheck the fluid level and add more if necessary to reach the Cold mark on the reservoir.

4   Bleed the system by turning the wheels from side-to-side, without hitting the stops. This will work the air out of the system. Keep the reservoir full of fluid as this is done.

5   When the air is worked out of the system, return the wheels to the straight ahead position and leave the vehicle running for several more minutes before shutting it off.

6   Road test the vehicle to be sure the steering system is functioning normally and noise free.

7   Recheck the fluid level to be sure it is up to the Hot mark on the reservoir while the engine is at normal operating temperature. Add fluid if necessary (see Chapter 1).

21   Wheels and tires - general information

Refer to illustration 21.1

All models covered by this manual are equipped with metric-sized radial tires **(see illustration)**. Use of other size or type of tires may affect the ride and handling of the vehicle. Don't mix different types of tires, such as radials and bias belted, on the same vehicle as handling may be seriously affected. It's recommended that tires be replaced in pairs on the same axle, but if only one tire is being replaced, be sure it's the same size, structure and tread design as the other.

Because tire pressure has a substantial effect on handling and wear, the pressure on all tires should be checked at least once a month or before any extended trips (see Chapter 1).

Wheels must be replaced if they are bent, dented, leak air, have elongated bolt holes, are heavily rusted, out of vertical symmetry or if the lug nuts won't stay tight. Wheel repairs that use welding or peening are not recommended.

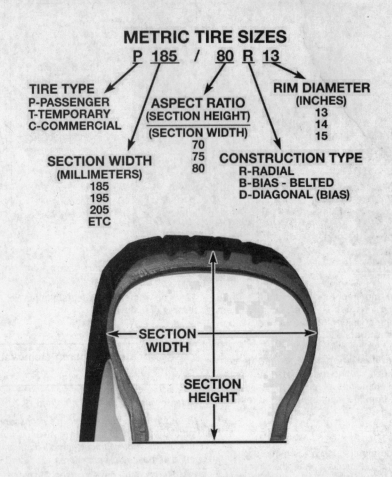

21.1 Metric tire size code

Tire and wheel balance is important to the overall handling, braking and performance of the vehicle. Unbalanced wheels can adversely affect handling and ride characteristics as well as tire life. Whenever a tire is installed on a wheel, the tire and wheel should be balanced by a shop with the proper equipment.

22   Front end alignment - general information

Refer to illustration 22.1

A front end alignment refers to the adjustments made to the front wheels so they are in proper angular relationship to the suspension

and the ground. Front wheels that are out of proper alignment not only affect steering control, but also increase tire wear **(see illustration)**.

Getting the proper front wheel alignment is a very exacting process, one in which complicated and expensive machines are necessary to perform the job properly. Because of this, you should have a technician with the proper equipment perform these tasks. We will, however, use this space to give you a basic idea of what is involved with front end alignment so you can better understand the process and deal intelligently with the shop that does the work.

Toe-in is the turning in of the front wheels. The purpose of a toe specification is to ensure parallel rolling of the front wheels.

In a vehicle with zero toe-in, the distance between the front edges of the wheels will be the same as the distance between the rear edges of the wheels. The actual amount of toe-in is normally only a fraction of an inch. Toe-in adjustment is controlled by the tie-rod end position on the tie-rod. Incorrect toe-in will cause the tires to wear improperly by making them scrub against the road surface.

Camber is the tilting of the front wheels from the vertical when viewed from the front of the vehicle. When the wheels tilt out at the top, the camber is said to be positive (+). When the wheels tilt in at the top the camber is negative (-). The amount of tilt is measured in degrees from the vertical and this measurement is called the camber angle. This angle affects the amount of tire tread which contacts the road and compensates for changes in the suspension geometry when the vehicle is cornering or traveling over an undulating surface.

Caster is the tilting of the top of the front steering axis from the vertical. A tilt toward the rear is positive caster and a tilt toward the front is negative caster.

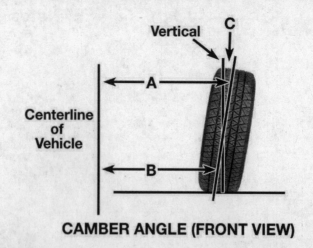

CAMBER ANGLE (FRONT VIEW)

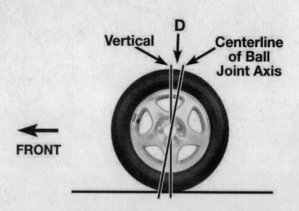

CASTER ANGLE (SIDE VIEW)

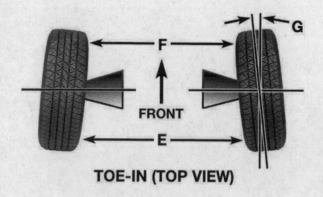

TOE-IN (TOP VIEW)

22.1 Front end alignment details

A minus B = C (degrees camber)
D = degrees caster
E minus F = toe-in (measured in inches)
G = toe-in (expressed in degrees)

Notes

Chapter 11
Body

Contents

1   General information

These models feature a "unibody" construction, using a floor pan with front and rear frame side rails which support the body components, front and rear suspension systems and other mechanical components. Certain components are particularly vulnerable to accident damage and can be unbolted and repaired or replaced. Among these parts are the body moldings, bumpers, hood and trunk lids and all glass.

Only general body maintenance practices and body panel repair procedures within the scope of the do-it-yourselfer are included in this Chapter.

2   Body - maintenance

1   The condition of your vehicle's body is very important, because the resale value depends a great deal on it. It's much more dif-ficult to repair a neglected or damaged body than it is to repair mechanical components. The hidden areas of the body, such as the wheel wells, the frame and the engine compartment, are equally important, although they don't require as frequent attention as the rest of the body.

2   Once a year, or every 12,000 miles, it's a good idea to have the underside of the body steam cleaned. All traces of dirt and oil will be removed and the area can then be inspected carefully for rust, damaged brake lines, frayed electrical wires, damaged cables and other problems. The front suspension components should be greased after completion of this job.

3   At the same time, clean the engine and the engine compartment with a steam cleaner or water soluble degreaser.

4   The wheel wells should be given close attention, since undercoating can peel away and stones and dirt thrown up by the tires can cause the paint to chip and flake, allowing rust to set in. If rust is found, clean down to the bare metal and apply an anti-rust paint.

5   The body should be washed about once a week. Wet the vehicle thoroughly to soften the dirt, then wash it down with a soft sponge and plenty of clean soapy water. If the surplus dirt is not washed off very carefully, it can wear down the paint.

6   Spots of tar or asphalt thrown up from the road should be removed with a cloth soaked in solvent.

7   Once every six months, wax the body and chrome trim. If a chrome cleaner is used to remove rust from any of the vehicle's plated parts, remember that the cleaner also removes part of the chrome, so use it sparingly.

3   Vinyl trim - maintenance

Don't clean vinyl trim with detergents, caustic soap or petroleum-based cleaners. Plain soap and water works just fine, with a soft brush to clean dirt that may be ingrained. Wash the vinyl as frequently as the rest of the vehicle.

After cleaning, application of a high quality rubber and vinyl protectant will help prevent oxidation and cracks. The protectant can also be applied to weatherstripping, vacuum lines and rubber hoses (which often fail as a result of chemical degradation) and to the tires.

4 Upholstery and carpets - maintenance

1　Every three months remove the carpets or mats and clean the interior of the vehicle (more frequently if necessary). Vacuum the upholstery and carpets to remove loose dirt and dust.

2　Leather upholstery requires special care. Stains should be removed with warm water and a very mild soap solution. Use a clean, damp cloth to remove the soap, then wipe again with a dry cloth. Never use alcohol, gasoline, nail polish remover or thinner to clean leather upholstery.

3　After cleaning, regularly treat leather upholstery with a leather wax. Never use car wax on leather upholstery.

4　In areas where the interior of the vehicle is subject to bright sunlight, cover leather seats with a sheet if the vehicle is to be left out for any length of time.

5 Body repair - minor damage

See photo sequence

Repair of minor scratches

1　If the scratch is superficial and does not penetrate to the metal of the body, repair is very simple. Lightly rub the scratched area with a fine rubbing compound to remove loose paint and built-up wax. Rinse the area with clean water.

2　Apply touch-up paint to the scratch, using a small brush. Continue to apply thin layers of paint until the surface of the paint in the scratch is level with the surrounding paint. Allow the new paint at least two weeks to harden, then blend it into the surrounding paint by rubbing with a very fine rubbing compound. Finally, apply a coat of wax to the scratch area.

3　If the scratch has penetrated the paint and exposed the metal of the body, causing the metal to rust, a different repair technique is required. Remove all loose rust from the bottom of the scratch with a pocket knife, then apply rust inhibiting paint to prevent the formation of rust in the future. Using a rubber or nylon applicator, coat the scratched area with glaze-type filler. If required, the filler can be mixed with thinner to provide a very thin paste, which is ideal for filling narrow scratches. Before the glaze filler in the scratch hardens, wrap a piece of smooth cotton cloth

around the tip of a finger. Dip the cloth in thinner and then quickly wipe it along the surface of the scratch. This will ensure that the surface of the filler is slightly hollow. The scratch can now be painted over as described earlier in this Section.

Repair of dents

4　When repairing dents, the first job is to pull the dent out until the affected area is as close as possible to its original shape. There is no point in trying to restore the original shape completely as the metal in the damaged area will have stretched on impact and cannot be restored to its original contours. It is better to bring the level of the dent up to a point which is about 1/8-inch below the level of the surrounding metal. In cases where the dent is very shallow, it is not worth trying to pull it out at all.

5　If the back side of the dent is accessible, it can be hammered out gently from behind using a soft-face hammer. While doing this, hold a block of wood firmly against the opposite side of the metal to absorb the hammer blows and prevent the metal from being stretched.

6　If the dent is in a section of the body which has double layers, or some other factor makes it inaccessible from behind, a different technique is required. Drill several small holes through the metal inside the damaged area, particularly in the deeper sections. Screw long, self-tapping screws into the holes just enough for them to get a good grip in the metal. Now the dent can be pulled out by pulling on the protruding heads of the screws with locking pliers.

7　The next stage of repair is the removal of paint from the damaged area and from an inch or so of the surrounding metal. This is done with a wire brush or sanding disk in a drill motor, although it can be done just as effectively by hand with sandpaper. To complete the preparation for filling, score the surface of the bare metal with a screwdriver or the tang of a file, or drill small holes in the affected area. This will provide a good grip for the filler material. To complete the repair, see the subsection on filling and painting later in this Section.

Repair of rust holes or gashes

8　Remove all paint from the affected area and from an inch or so of the surrounding metal using a sanding disk or wire brush mounted in a drill motor. If these are not available, a few sheets of sandpaper will do the job just as effectively.

9　With the paint removed, you will be able to determine the severity of the corrosion and decide whether to replace the whole panel, if possible, or repair the affected area. New body panels are not as expensive as most people think and it is often quicker to install a new panel than to repair large areas of rust.

10　Remove all trim pieces from the affected area except those which will act as a guide to the original shape of the damaged body, such

as headlight shells, etc. Using metal snips or a hacksaw blade, remove all loose metal and any other metal that is badly affected by rust. Hammer the edges of the hole in to create a slight depression for the filler material.

11　Wire brush the affected area to remove the powdery rust from the surface of the metal. If the back of the rusted area is accessible, treat it with rust inhibiting paint.

12　Before filling is done, block the hole in some way. This can be done with sheet metal riveted or screwed into place, or by stuffing the hole with wire mesh.

13　Once the hole is blocked off, the affected area can be filled and painted. See the following subsection on filling and painting.

Filling and painting

14　Many types of body fillers are available, but generally speaking, body repair kits which contain filler paste and a tube of resin hardener are best for this type of repair work. A wide, flexible plastic or nylon applicator will be necessary for imparting a smooth and contoured finish to the surface of the filler material. Mix up a small amount of filler on a clean piece of wood or cardboard (use the hardener sparingly). Follow the manufacturer's instructions on the package, otherwise the filler will set incorrectly.

15　Using the applicator, apply the filler paste to the prepared area. Draw the applicator across the surface of the filler to achieve the desired contour and to level the filler surface. As soon as a contour that approximates the original one is achieved, stop working the paste. If you continue, the paste will begin to stick to the applicator. Continue to add thin layers of paste at 20-minute intervals until the level of the filler is just above the surrounding metal.

16　Once the filler has hardened, the excess can be removed with a body file. From then on, progressively finer grades of sandpaper should be used, starting with a 180-grit paper and finishing with 600-grit wet-or-dry paper. Always wrap the sandpaper around a flat rubber or wooden block, otherwise the surface of the filler will not be completely flat. During the sanding of the filler surface, the wet-or-dry paper should be periodically rinsed in water. This will ensure that a very smooth finish is produced in the final stage.

17　At this point, the repair area should be surrounded by a ring of bare metal, which in turn should be encircled by the finely feathered edge of good paint. Rinse the repair area with clean water until all of the dust produced by the sanding operation is gone.

18　Spray the entire area with a light coat of primer. This will reveal any imperfections in the surface of the filler. Repair the imperfections with fresh filler paste or glaze filler and once more smooth the surface with sandpaper. Repeat this spray-and-repair procedure until you are satisfied that the surface of the filler and the feathered edge of the paint are perfect. Rinse the area with clean water and allow it to dry completely.

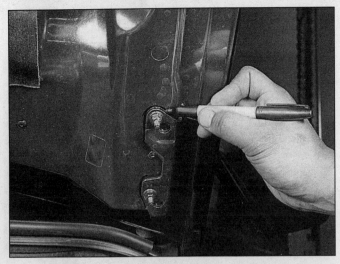

9.2 Before removing the hood, draw a mark around the hinge plate

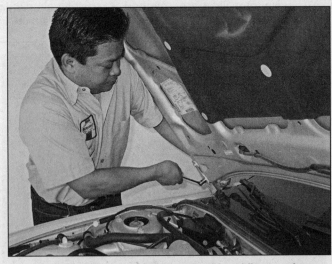

9.4 Support the hood with your shoulder while removing the hood bolts

19   The repair area is now ready for painting. Spray painting must be carried out in a warm, dry, windless and dust free atmosphere. These conditions can be created if you have access to a large indoor work area, but if you are forced to work in the open, you will have to pick the day very carefully. If you are working indoors, dousing the floor in the work area with water will help settle the dust which would otherwise be in the air. If the repair area is confined to one body panel, mask off the surrounding panels. This will help minimize the effects of a slight mismatch in paint color. Trim pieces such as chrome strips, door handles, etc., will also need to be masked off or removed. Use masking tape and several thickness of newspaper for the masking operations.

20   Before spraying, shake the paint can thoroughly, then spray a test area until the spray painting technique is mastered. Cover the repair area with a thick coat of primer. The thickness should be built up using several thin layers of primer rather than one thick one. Using 600-grit wet-or-dry sandpaper, rub down the surface of the primer until it is very smooth. While doing this, the work area should be thoroughly rinsed with water and the wet-or-dry sandpaper periodically rinsed as well. Allow the primer to dry before spraying additional coats.

21   Spray on the top coat, again building up the thickness by using several thin layers of paint. Begin spraying in the center of the repair area and then, using a circular motion, work out until the whole repair area and about two inches of the surrounding original paint is covered. Remove all masking material 10 to 15 minutes after spraying on the final coat of paint. Allow the new paint at least two weeks to harden, then use a very fine rubbing compound to blend the edges of the new paint into the existing paint. Finally, apply a coat of wax.

6   Body repair - major damage

1   Major damage must be repaired by an auto body shop specifically equipped to perform these repairs. Most shops have the specialized equipment required to do the job properly.

2   If the damage is extensive, the body must be checked for proper alignment or the vehicle's handling characteristics may be adversely affected and other components may wear at an accelerated rate.

3   Due to the fact that all of the major body components (hood, fenders, etc.) are separate and replaceable units, any seriously damaged components should be replaced rather than repaired. Sometimes the components can be found in a wrecking yard that specializes in used vehicle components, often at considerable savings over the cost of new parts.

7   Hinges and locks - maintenance

Once every 3000 miles, or every three months, the hinges and latch assemblies on the doors, hood and trunk should be given a few drops of light oil or lock lubricant. The door latch strikers should also be lubricated with a thin coat of grease to reduce wear and ensure free movement. Lubricate the door and trunk locks with spray-on graphite lubricant.

8   Windshield and fixed glass - replacement

Replacement of the windshield and fixed glass requires the use of special fast-setting adhesive/caulk materials and some specialized tools. It is recommended that these operations be left to a dealer or a shop specializing in glass work.

9   Hood - removal, installation and adjustment

Note: *The hood is heavy and somewhat awkward to remove and install - at least two people should perform this procedure.*

Removal and installation

Refer to illustrations 9.2 and 9.4

1   Use blankets or pads to cover the cowl area of the body and fenders. This will protect the body and paint as the hood is lifted off.

2   Make marks or scribe a line around the hood hinge to ensure proper alignment during installation (**see illustration**).

3   Detach the windshield washer hose. Also disconnect any cables or wires that will interfere with removal.

4   Have an assistant support the hood. Remove the hinge-to-hood bolts (**see illustration**).

5   Lift off the hood.

6   Installation is the reverse of removal.

Adjustment

Refer to illustrations 9.10 and 9.11

7   Fore-and-aft and side-to-side adjustment of the hood is done by moving the hinge plate slot after loosening the bolts or nuts.

8   Scribe a line around the entire hinge plate so you can determine the amount of movement (**see illustration 9.2**).

9   Loosen the bolts or nuts and move the hood into correct alignment. Move it only a little at a time. Tighten the hinge bolts and carefully lower the hood to check the position.

10   If necessary after installation, the entire hood latch assembly can be adjusted up-and-down as well as from side-to-side on the radiator support so the hood closes securely and flush with the fenders. To make the adjustment, scribe a line or mark around the

These photos illustrate a method of repairing simple dents. They are intended to supplement *Body repair - minor damage* in this Chapter and should not be used as the sole instructions for body repair on these vehicles.

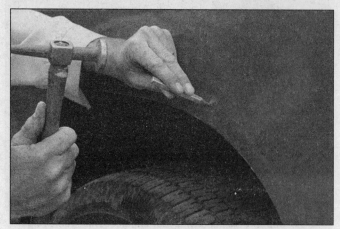

1 If you can't access the backside of the body panel to hammer out the dent, pull it out with a slide-hammer-type dent puller. In the deepest portion of the dent or along the crease line, drill or punch hole(s) at least one inch apart . . .

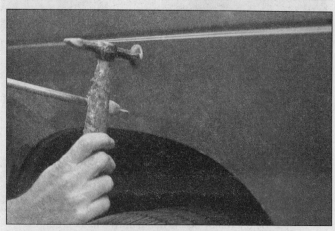

2 . . . then screw the slide-hammer into the hole and operate it. Tap with a hammer near the edge of the dent to help 'pop' the metal back to its original shape. When you're finished, the dent area should be close to its original contour and about 1/8-inch below the surface of the surrounding metal

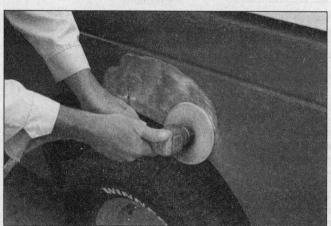

3 Using coarse-grit sandpaper, remove the paint down to the bare metal. Hand sanding works fine, but the disc sander shown here makes the job faster. Use finer (about 320-grit) sandpaper to feather-edge the paint at least one inch around the dent area

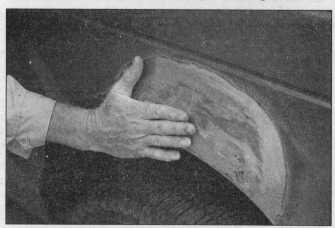

4 When the paint is removed, touch will probably be more helpful than sight for telling if the metal is straight. Hammer down the high spots or raise the low spots as necessary. Clean the repair area with wax/silicone remover

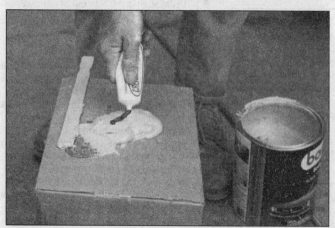

5 Following label instructions, mix up a batch of plastic filler and hardener. The ratio of filler to hardener is critical, and, if you mix it incorrectly, it will either not cure properly or cure too quickly (you won't have time to file and sand it into shape)

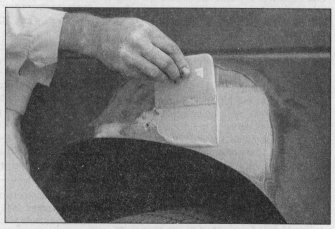

6 Working quickly so the filler doesn't harden, use a plastic applicator to press the body filler firmly into the metal, assuring it bonds completely. Work the filler until it matches the original contour and is slightly above the surrounding metal

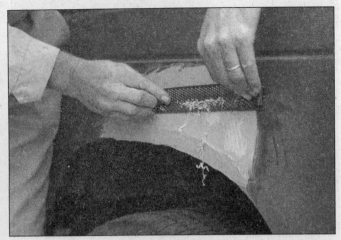

7 Let the filler harden until you can just dent it with your fingernail. Use a body file or Surform tool (shown here) to rough-shape the filler

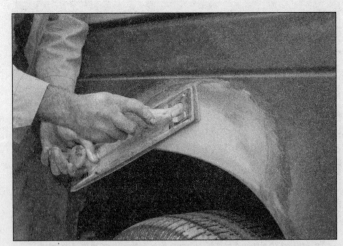

8 Use coarse-grit sandpaper and a sanding board or block to work the filler down until it's smooth and even. Work down to finer grits of sandpaper - always using a board or block - ending up with 360 or 400 grit

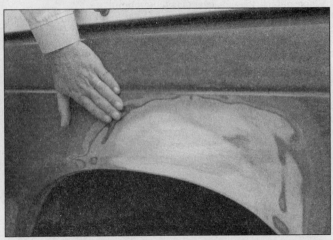

9 You shouldn't be able to feel any ridge at the transition from the filler to the bare metal or from the bare metal to the old paint. As soon as the repair is flat and uniform, remove the dust and mask off the adjacent panels or trim pieces

10 Apply several layers of primer to the area. Don't spray the primer on too heavy, so it sags or runs, and make sure each coat is dry before you spray on the next one. A professional-type spray gun is being used here, but aerosol spray primer is available inexpensively from auto parts stores

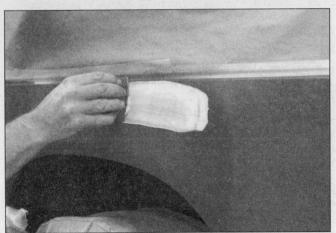

11 The primer will help reveal imperfections or scratches. Fill these with glazing compound. Follow the label instructions and sand it with 360 or 400-grit sandpaper until it's smooth. Repeat the glazing, sanding and respraying until the primer reveals a perfectly smooth surface

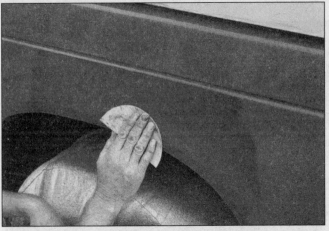

12 Finish sand the primer with very fine sandpaper (400 or 600-grit) to remove the primer overspray. Clean the area with water and allow it to dry. Use a tack rag to remove any dust, then apply the finish coat. Don't attempt to rub out or wax the repair area until the paint has dried completely (at least two weeks)

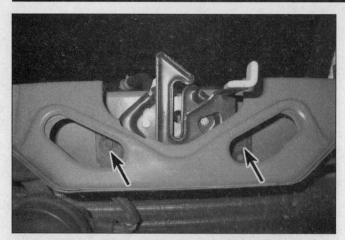

9.10 Scribe a line around the hinge to use as a reference point. To adjust the hood latch, loosen the retaining bolts (arrows), move the latch and retighten the bolts, then close the hood to check the fit

9.11 Adjust the hood closing height by turning the hood bumpers in or out

hood latch mounting bolts to provide a reference point, then loosen them and reposition the latch assembly, as necessary **(see illustration)**. Following adjustment, retighten the mounting bolts.

11   Finally, adjust the hood bumpers on the radiator support so the hood, when closed, is flush with the fenders **(see illustration)**.

12   The hood latch assembly, as well as the hinges, should be periodically lubricated with white, lithium-base grease to prevent binding and wear.

10   Hood release latch and cable - removal and installation

Latch

Refer to illustration 10.1

1   Remove the upper radiator cover **(see illustration)**.

2   Scribe a line around the latch to aid

alignment when installing, then detach the latch retaining bolts to the radiator support **(see illustration 9.10)** and remove the latch.

3   Disconnect the hood release cable by disengaging the cable from the latch assembly.

4   Installation is the reverse of the removal procedure. **Note:** *Adjust the latch so the hood engages securely when closed and the hood bumpers are slightly compressed.*

Cable

Refer to illustration 10.7

5   Disconnect the hood release cable from the latch assembly as described above.

6   Attach a piece of stiff wire to the end of the cable, then follow the cable back to the firewall and detach all the cable retaining clips.

7   Working in the passenger compartment, remove the driver's side kick panel (see Section 25). Then remove the two release lever mounting bolts and detach the hood release

lever **(see illustration)**.

8   Pull the cable and grommet rearward into the passenger compartment until you can see the wire. Ensure that the new cable has a grommet attached, then remove the old cable from the wire and replace it with the new cable.

9   Working from the engine compartment, pull the wire back through the firewall.

10   Installation is the reverse of removal. **Note:** *Push on the grommet with your fingers from the passenger compartment to seat the grommet in the firewall correctly.*

11   Bumper covers - removal and installation

Front bumper cover

Refer to illustrations 11.2, 11.3, 11.4 and 11.5

1   Apply the parking brake, raise the vehicle and support it securely on jackstands.

10.1 Remove the fasteners securing the upper radiator cover

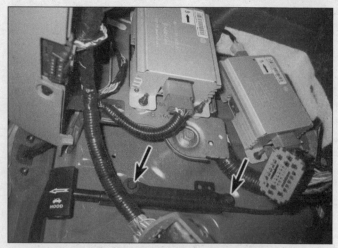

10.7 Remove the hood release lever mounting fasteners and pull the cable rearward into the passenger compartment

11.2 Bumper cover lower mounting fasteners

11.3 Inner fenderwell splash shield mounting fasteners

2   Working under the vehicle, remove the fasteners securing the lower edges of the bumper cover **(see illustration).**

3   Working in the front wheel opening, detach the retaining screws securing the bumper cover to the inner fenderwell splash shield **(see illustration).**

4   Pry out the lower edge of the splash shield, then reach up behind the bumper cover and remove the bumper cover to fender retaining nuts **(see illustration).**

5   Remove the fasteners securing the upper portion of the bumper cover and pull the bumper cover assembly out and away from the vehicle **(see illustration).**

6   Disconnect any electrical connections which would interfere with removal.

Rear bumper cover

Refer to illustrations 11.8, 11.9a, 11.9b and 11.10

7   Apply the parking brake, raise the vehicle and support it securely on jackstands.

8   Working under the vehicle, detach the

11.4 Peel back the splash shield and remove the bumper cover to fender retaining bolts

11.5 Remove the upper mounting fasteners from the bumper cover

fasteners securing the lower edges of the bumper cover **(see illustrations).**

9   Working in the trunk, pry out the fasteners securing the rear inside trunk finishing panel to allow access to the bumper cover

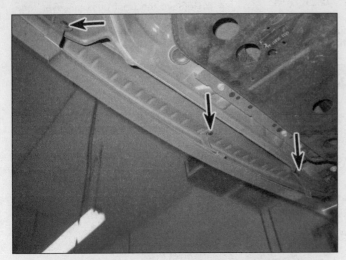

11.8 Remove the bumper cover lower mounting fasteners

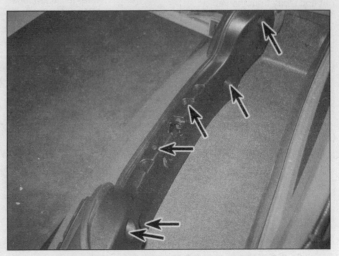

11.9a Remove the fasteners securing the rear inside trunk finishing panel . . .

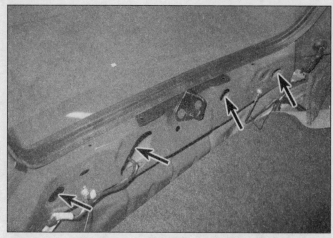

11.9b . . . to allow access to the bumper cover retaining bolts

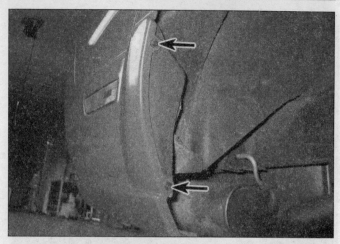

11.10 Remove the screws securing the bumper cover to the wheel opening

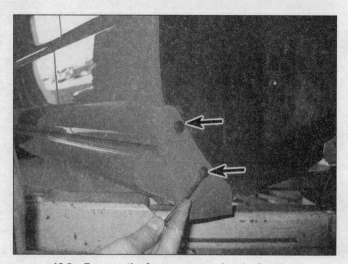

12.2a Remove the fasteners securing the front . . .

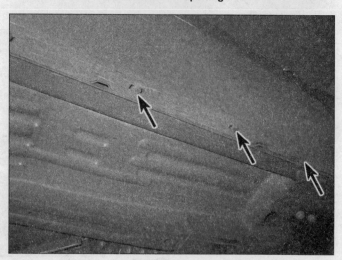

12.2b . . . and the bottom of the rocker panel

retaining bolts **(see illustrations).**
10   Detach the fasteners securing the bumper cover to the right and left quarter panels **(see illustration),** then pull the bumper cover out and away from the vehicle.
11   Installation is the reverse of removal.

12 Front fender - removal and installation

Refer to illustrations 12.2a, 12.2b, 12.5a, 12.5b and 12.5c

1   Raise the vehicle, support it securely on jackstands and remove the front wheel.

2   Remove the rocker panel **(see illustrations).**
3   Detach the inner fenderwell splash shield **(see illustration 11.3).**
4   Remove the front bumper cover (see Section 11).
5   Remove the fender mounting bolts **(see illustrations).**

12.5a Open the door to access and remove the fender-to-door pillar bolt

12.5b Remove the fender-to-body lower mounting bolts . . .

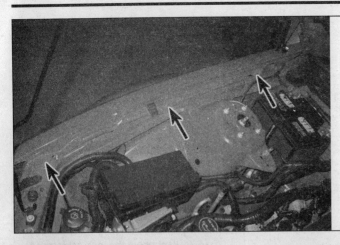

12.5c . . . and the fender-to-body upper mounting bolts

13.2 Use a small screwdriver to pry the clip out of its locking groove, then detach the end of the strut from the locating stud

6   Detach the fender. It's a good idea to have an assistant support the fender while it's being moved away from the vehicle to prevent damage to the surrounding body panels.

7   Installation is the reverse of removal.

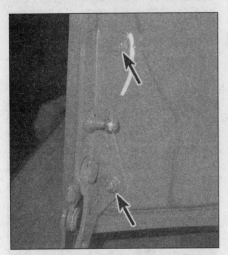

14.4 Scribe a mark around the hinge plate for realignment purposes - then remove the retaining nuts on each side of the trunk lid

13   Trunk lid support struts - removal and installation

Refer to illustration 13.2

1   Open the rear trunk lid and support it securely.

2   Using a small screwdriver, detach the retaining clips at both ends of the support strut. Then pry or pull sharply to detach it from the vehicle **(see illustration)**.

3   Installation is the reverse of removal.

14   Trunk lid - removal, installation and adjustment

Note: *The trunk lid is heavy and somewhat awkward to remove and install - at least two people should perform this procedure.*

Removal and installation

Refer to illustration 14.4

1   Open the trunk lid and cover the edges of the trunk compartment with pads or cloths to protect the painted surfaces when the lid is removed.

2   Disconnect any cables or wire harness connectors attached to the trunk lid that would interfere with removal.

3   Detach the trunk lid support struts from

the trunk lid (see Section 13).

4   Make alignment marks around the hinges with a marking pen, then while an assistant supports the trunk lid, remove the lid-to-hinge nuts **(see illustration)** on both sides of the trunk and lift it off.

5   Installation is the reverse of removal.
Note: *When reinstalling the trunk lid, align the marks made during removal.*

Adjustment

6   Fore-and-aft and side-to-side adjustment of the trunk lid is accomplished by moving the lid in relation to the hinge after loosening the bolts or nuts.

7   Scribe a line around the entire hinge plate so you can determine the amount of movement.

8   Loosen the nuts and move the trunk lid into correct alignment. Move it only a little at a time. Tighten the hinge nuts and carefully lower the trunk lid to check the alignment.

9   If necessary after installation, the entire trunk lid striker assembly can be adjusted up and down as well as from side to side on the trunk lid so the lid closes securely and is flush with the rear quarter panels. To do this, scribe a line around the trunk lid striker assembly to provide a reference point. Then loosen the nuts and reposition the striker as necessary. Following adjustment, retighten the mounting nuts.

10   The trunk lid latch assembly, as well as the hinges, should be periodically lubricated with white lithium-base grease to prevent sticking and wear.

15   Trunk lid latch and lock cylinder - removal and installation

Trunk lid latch

Refer to illustration 15.1 and 15.3

1   Open the trunk and remove the latch trim cover **(see illustration)**.

2   Scribe a line around the trunk lid latch assembly for a reference point to aid the installation procedure.

15.1 Carefully pull on the latch trim cover to release the mounting clips

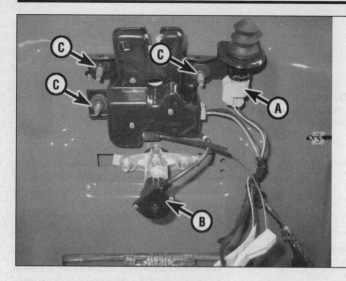

15.3 Trunk lid latch details

A   Electrical connector
B   Actuator cable
C   Latch mounting nuts

15.7 Disconnect the lock actuator (A), then remove the mounting fasteners (B)

3   Disconnect the electrical connectors and the trunk lock actuator from the lock cylinder **(see illustration)**.
4   Detach the retaining nuts and remove the latch.
5   Installation is the reverse of removal.

Trunk lock cylinder

Refer to illustration 15.7

6   Open the trunk and remove the latch trim cover **(see illustration 15.1)**.
7   Disconnect any electrical connectors and the trunk lock actuator from the lock cylinder **(see illustration)**.
8   Remove the lock cylinder mounting fas-

teners, then remove the lock cylinder from the vehicle.
9   Installation is the reverse of removal.

---

16   Door trim panel - removal and installation

Refer to illustrations 16.1, 16.2, 16.3a, 16.3b, 16.4a, 16.4b, 16.5 and 16.7

1   Remove the side view mirror trim cover **(see illustration)**.
2   Remove the inside door handle cover **(see illustration)**.
3   Pry out the armrest switch control plate and disconnect the electrical connector **(see illustrations)**.
4   Detach the retaining screw located behind the armrest trim panel **(see illustrations)**.
5   Remove the remaining door panel retaining fasteners **(see illustration)**.
6   Once all of the fasteners are removed, detach the trim panel, disconnect any electrical connectors and remove the trim panel from the vehicle by gently pulling it up and out.

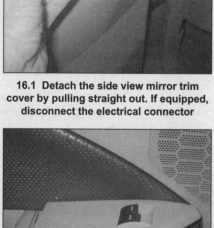

16.1 Detach the side view mirror trim cover by pulling straight out. If equipped, disconnect the electrical connector

16.2 Remove the inside door handle mounting fastener, then remove the cover

16.3a Using a trim stick, pry out the armrest switch control plate . . .

16.3b . . . and disconnect the electrical connector

16.4a Using a trim stick, remove the armrest trim panel . . .

16.4b ... then remove the retaining screw

7   Disconnect the door handle actuating cable and the electrical connector for the power door lock switch **(see illustration)**.

8   For access to the inner door, peel back the watershield, taking care not to tear it. To install the trim panel, first press the water-shield back into place. If necessary, add more sealant to hold it in place.

9   The remainder of the installation is the reverse of removal.

17   Door - removal, installation and adjustment

Note: *The door is heavy and somewhat awkward to remove and install - at least two people should perform this procedure.*

Removal and installation

Refer to illustrations 17.7 and 17.9

1   Lower the window completely in the door and then disconnect the negative cable from the battery

2   Open the door all the way and support it on jacks or blocks covered with rags to prevent damaging the paint.

3   Remove the door trim panel and water deflector as described in (Section 16).

4   Remove the door speaker (see Chapter 12).

5   Unplug all electrical connections, ground wires and harness retaining clips from the door. **Note:** *It is a good idea to label all connections to aid the reassembly process.*

6   Working through the door speaker hole and the door opening, detach the rubber conduit between the body and the door. Then pull wiring harness through conduit hole and remove from door.

7   Remove the door check strap mounting bolts **(see illustration)**.

8   Mark around the door hinges with a pen or a scribe to facilitate realignment during reassembly.

9   Have an assistant hold the door, then remove the hinge-to-door bolts **(see illustration)** and lift the door off.

10   Installation is the reverse of the removal.

Adjustment

Refer to illustration 17.13

11   Having proper door-to-body alignment is a critical part of a well functioning door assembly. First check the door hinge pins for excessive play. Fully open the door and lift up and down on the door without lifting the body. If

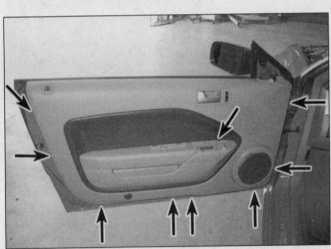

16.5 Door panel retaining fasteners

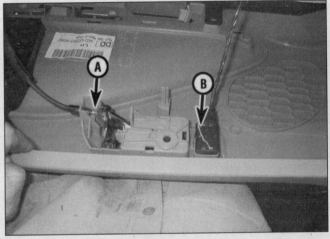

16.7 Disconnect the door handle actuating cable (A) and electrical connector (B)

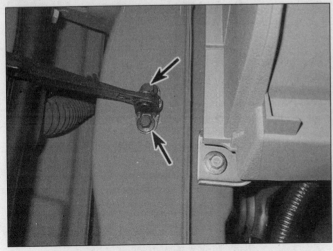

17.7 Remove the door check strap retaining bolts

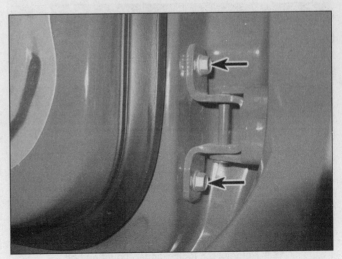

17.9 Remove the door hinge bolts

a door has 1/16-inch or more excessive play, the hinges should be replaced.

12   Door-to-body alignment adjustments are made by loosening the hinge-to-body or hinge-to-door bolts and moving the door. Proper body alignment is achieved when the top of door is aligned with the top of front fender and rear quarter panel and the bottom of the door is aligned with the lower rocker panel. If these goals can't be reached by adjusting the hinge-to-body or hinge-to-door bolts, body alignment shims may have to be purchased and inserted behind the hinges to achieve correct alignment.

13   To adjust the door closed position, first check that the door latch is contacting the center of the latch striker. If not, remove the striker and add or subtract shims to achieve correct alignment.

14   Finally adjust the latch striker as necessary (up-and-down or sideways) to provide positive engagement with the latch mechanism (see illustration) and so the door panel is flush with rear quarter panel.

18   Door latch, lock cylinder and handle - removal and installation

Door latch

Refer to illustration 18.3

1   Raise the window, then remove the door trim panel and watershield as described in Section 16.

2   Disconnect the door handle actuator rod and electrical connectors from the latch.

3   Remove the screws securing the latch to the door (see illustration).

4   Disconnect the manual lock and door lock cylinder actuating rods from the latch.

5   Disconnect the actuating cable for the inside door handle, then remove the latch assembly from the door.

6   Installation is the reverse of removal.

Door lock cylinder and outside handle

Refer to illustration 18.9

7   To remove the lock cylinder, raise the window and remove the door trim panel and watershield as described in Section 16.

8   Working through the large access hole, disengage the plastic clip that secures the lock cylinder-to-latch rod.

9   Remove the lock cylinder mounting bolt and remove the lock cylinder from the door (see illustration).

10   To remove the outside handle, work through the access hole and disengage the plastic clip that secures the outside handle-to-latch rod.

11   Remove the outside handle retaining nuts and pull the handle from the door.

12   Installation is the reverse of removal.

19   Door window glass - removal and installation

Refer to illustration 19.2

1   Remove the door trim panel and the plastic watershield (see Section 16).

2   Raise the window just enough to access the window retaining bolts through the hole in the door frame (see illustration), then remove the bolts.

3   Remove the glass by pulling it up and out.

4   Installation is the reverse of removal.

20   Door window glass regulator - removal and installation

Refer to illustration 20.3

1   Remove the door trim panel and the plastic watershield (see Section 16).

2   Remove the window glass (see Section 19).

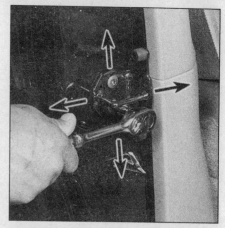

17.13 Adjust the door lock striker by loosening the mounting screws and gently tapping the striker in the desired direction

18.3 Remove the latch retaining screws from the end of the door

3   Disconnect the electrical connector, then remove the window regulator motor mounting bolts (see illustration).

4   Remove the window regulator mounting bolts (see illustration 20.3).

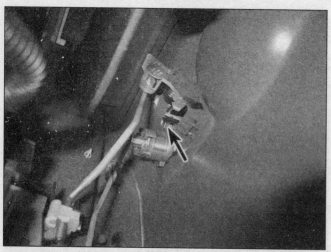

18.9 Remove the lock cylinder mounting bolt

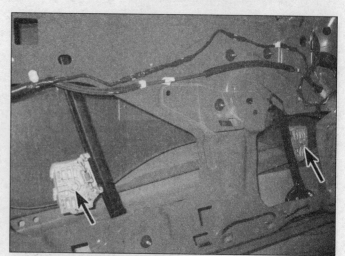

19.2 Window glass retainer bolts

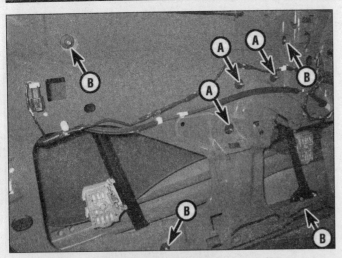

20.3 Window regulator motor (A) and regulator mounting bolts (B)

21.1 Detach the side view mirror trim cover by pulling straight out. If equipped, disconnect the electrical connector

5   Pull the regulator and motor assembly through the service hole in the door frame to remove it.
6   Installation is the reverse of removal.

21   Sideview mirrors - removal and installation

Refer to illustration 21.1 and 21.2

1   Remove the mirror trim cover from the door panel **(see illustration)**.
2   Remove the three mirror retaining nuts and detach the mirror from the vehicle **(see illustration)**.
3   Installation is the reverse of removal.

22   Center console - removal and installation

Refer to illustrations 22.1, 22.2, 22.3, 22.4, 22.5a and 22.5b

1   Pry out the gear selector trim bezel **(see**

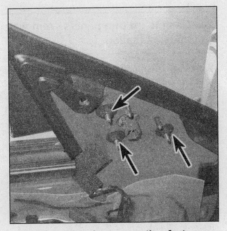

21.2 Outside mirror mounting fasteners

22.1 Using a trim stick, carefully pry out the gear selector trim bezel

illustration). Vehicles equipped with manual transmissions will require unscrewing the shift lever knob to remove the trim bezel.
2   Apply the parking brake, remove the two

center console trim panel mounting screws, then remove the trim panel **(see illustration)**.
3   Detach the retaining screws securing the front half of the console **(see illustration)**.

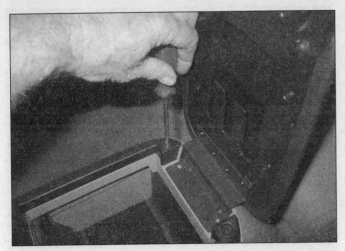

22.2 Remove the two center console trim panel mounting screws

22.3 Remove the retaining screws securing the front half of the console

22.4 Carefully detach the center console side trim panels

22.5a Remove the pad at the bottom of the center console glove box . . .

4   Detach the center console side trim panels **(see illustration)**.
5   Working in the console glove box, remove the center console rear retaining screws **(see illustrations)**.
6   Lift the console up and over the shift lever. Disconnect any electrical connections and remove the console from the vehicle.
7   Installation is the reverse of removal.

23 Steering column cover - removal and installation

Refer to illustration 23.2

1   Remove the driver side knee bolster (see Section 24).
2   Remove the screws from the lower steering column cover **(see illustration)**.
3   Separate the cover halves and detach them from the steering column.
4   Installation is the reverse of removal.

24 Dashboard trim panels - removal and installation

Warning: *The models covered by this manual are equipped with Supplemental Restraint Systems (SRS), more commonly known as airbags. Always disable the airbag system before working in the vicinity of any airbag system components to avoid the possibility of accidental deployment of the airbags, which could cause personal injury (see Chapter 12).*
1   Disconnect the cable from the negative battery terminal (see Chapter 5, Section 1).

Knee bolster

Refer to illustration 24.2

2   Working in the driver's side passenger compartment, detach the retaining screws along the lower edge of the knee bolster **(see illustration)**.
3   Pull outward on the lower edge of the knee bolster and detach it from the vehicle.

22.5b . . . then remove the glove box fasteners

4   Detach the retaining screws securing the knee bolster reinforcement panel.
5   Installation is the reverse of removal.

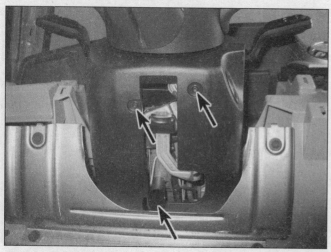

23.2 Steering column cover mounting fasteners

24.2 Remove the retaining screws along the lower edge of the knee bolster

24.7 Remove the retaining screws from the center trim panel

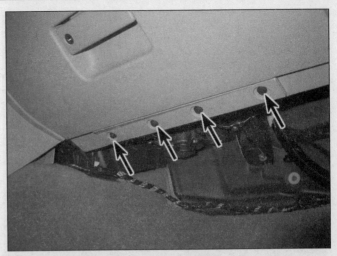

24.9 Remove the door hinge retaining screws along the lower edge of the glove box

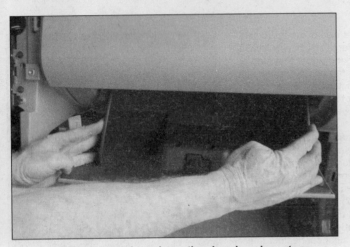

24.10 Push inward to release the glove box door stops

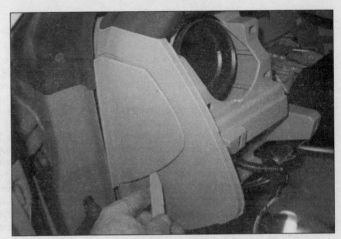

24.12 Using a trim stick, carefully pry around the panel and release the clips

Center trim panel

Refer to illustration 24.7

6   Remove the center console (see Section 22).

7   Detach the retaining screws from the center trim panel **(see illustration)**. Disconnect any electrical connections and remove the trim panel from the vehicle.

8   Installation is the reverse of removal.

Glove box

Refer to illustrations 24.9 and 24.10

9   Detach the door hinge retaining screws along the lower edge of the glove box **(see illustration)**.

10   Open the glove box door. Release the door stops by pressing in on the sides, then pull straight out to remove the glove box assembly **(see illustration)**.

11   Installation is the reverse of removal.

Instrument panel end caps

Refer to illustration 24.12

12   Using a trim stick, carefully pry around the panel and release the clips, then detach it from the dashboard **(see illustration)**.

13   Installation is the reverse of removal.

Instrument cluster bezel

Refer to illustration 24.14

14   Using a trim stick, carefully pry around the panel and release the clips, then detach it from the dashboard **(see illustration)**.

15   Installation is the reverse of removal.

25   Instrument panel - removal and installation

Refer to illustrations 25.2a, 25.2b, 25.6, 25.10a and 25.10b

Warning: *The models covered by this manual are equipped with Supplemental Restraint Systems (SRS), more commonly known as airbags. Always disable the airbag system before working in the vicinity of any airbag system components to avoid the possibility of accidental deployment of the airbags, which*

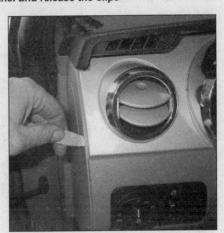

24.14 Using a trim stick, carefully pry around the instrument cluster bezel and release the clips

could cause personal injury (see Chapter 12).
Note: *This is a difficult procedure for the home mechanic. There are many hidden fasteners, difficult angles to work in and many electrical connectors to tag and disconnect/connect.*

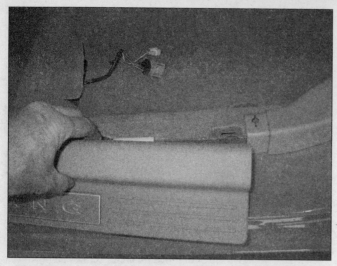

25.2a Pry up the front end of the door sill plates with a trim tool

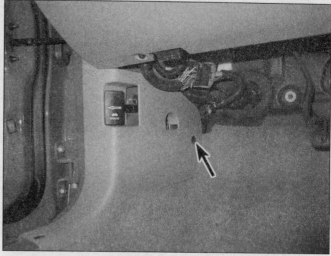

25.2b Remove the push-pin fastener, then remove the kick panel

1   Turn the front wheels to the straight-ahead position and lock the steering column, disconnect the negative battery cable and disable the airbag system (see Chapter 12).

2   Pry up the front end of the door sill plates with a trim tool, then remove the right and left side cowl trim ("kick") panels **(see illustrations)**.

3   Detach the bolt securing the steering shaft (see Chapter 10). **Note:** *There will be more working room if the front seats are moved back as far as possible, and even more room if the driver's seat is removed (see Section 28).*

4   Remove the center console (see Section 22).

5   Remove the instrument cluster (see Chapter 12).

6   Working in the instrument cluster opening, remove the fastener securing the cowl top bracket **(see illustration).**

7   Under the driver's side of the instrument panel, disconnect the large electrical bulkhead connector. While working under that side of the instrument panel, tag and disconnect all other electrical connectors connected to the instrument panel. **Note:** *Watch for ground straps bolted to the cowl or the area behind the kick panels.*

8   Working on the passenger's side of the instrument panel, disconnect the antenna connector (see Chapter 12). Disconnect any remaining electrical connectors.

9   Remove the instrument panel end caps **(see illustration 24.12)**.

10   Remove the instrument panel retaining fasteners **(see illustrations)**.

11   Detach any electrical connectors interfering with removal, then pull the instrument panel towards the rear of the vehicle to remove it.

12   Installation is the reverse of removal.

25.6 Remove the fastener securing the cowl top bracket

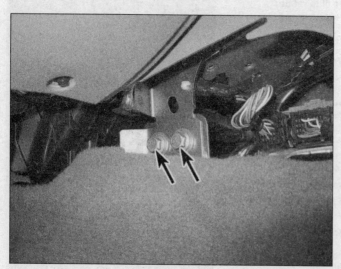

25.10a Remove the instrument panel retaining fasteners at the center support . . .

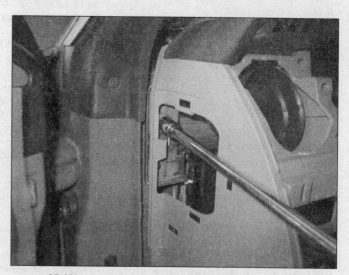

25.10b . . . and at the sides of the instrument panel

26.3 With the rear seat folded forward, remove the two remaining fasteners securing the quarter trim panel

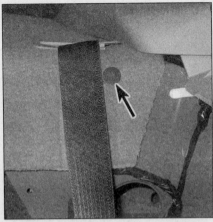

26.10 Remove the fasteners securing the parcel shelf

26   Rear trim panels - removal and installation

Rear quarter trim panel

Refer to illustration 26.3

1   Pry up the front end of the door sill plates with a trim tool.
2   Remove the lower pin-type fastener securing the quarter trim panel.
3   Fold the rear seat forward, then remove

the two remaining fasteners securing the quarter trim panel and remove the panel **(see illustration)**.
4   Installation is the reverse of removal.

Quarter window trim panel

5   Remove the rear quarter trim panel (see Steps 1 through 3).
6   Remove the cover from the coat hook, then remove the coat hook mounting bolt.
7   Remove the seat belt upper mounting bolt, then remove the trim panel.
8   Installation is the reverse of removal.

Parcel shelf

Refer to illustration 26.10

9   Remove the rear quarter window trim panel (see Steps 5 through 7).
10   Remove the four fasteners securing the parcel shelf, then remove the parcel shelf **(see illustration)**.
11   Installation is the reverse of removal.

27   Cowl cover - removal and installation

Refer to illustration 27.2

1   Remove the windshield wiper arms (see Chapter 12).
2   Remove the retaining fasteners securing the cowl cover **(see illustration)**.
3   Remove the right side cowl cover, then the left side.
4   Installation is the reverse of removal.

28   Seats - removal and installation

Front seat

Refer to illustration 28.2

1   Position the seat all the way forward or all the way to the rear to access the front seat retaining bolts.
2   Detach any bolt trim covers and remove the retaining bolts **(see illustration)**.
3   Tilt the seat upward to access the underneath, then disconnect any electrical connectors and lift the seat from the vehicle.
4   Installation is the reverse of removal.

Rear seat

Refer to illustration 28.5

5   Working at the front of each seat cushion, push the retaining tabs in, then lift up on the front edge of the cushion and release the seat cushion retaining tabs **(see illustration)**. Remove the cushion from the vehicle.
6   Detach the retaining bolts at the lower edge of the seat back.
7   Lift up on the lower edge of the seat back and remove it from the vehicle.
8   Installation is the reverse of removal.

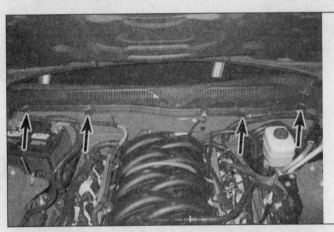

27.2 Remove the fasteners securing the cowl cover

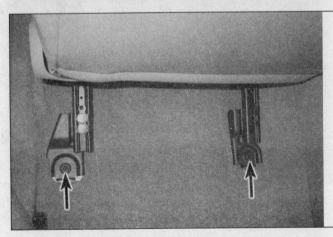

28.2 Detach the trim covers to access the front seat retaining bolts

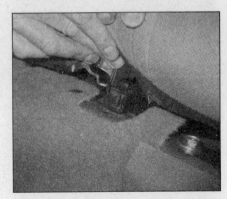

28.5 Press down on the retaining tabs to remove the rear seat cushion

Notes

Chapter 12
Chassis electrical system

Contents

1 General information

The electrical system is a 12-volt, negative ground type. Power for the lights and all electrical accessories is supplied by a lead/acid-type battery, which is charged by the alternator.

This Chapter covers repair and service procedures for the various electrical components not associated with the engine. Information on the battery, alternator, distributor and starter motor can be found in Chapter 5.

It should be noted that when portions of the electrical system are serviced, the cable should be disconnected from the negative battery terminal (see Chapter 5, Section 1) to prevent electrical shorts and/or fires.

2 Electrical troubleshooting - general information

Refer to illustrations 2.5a, 2.5b, 2.6 and 2.9

A typical electrical circuit consists of an electrical component, any switches, relays, motors, fuses, fusible links or circuit breakers related to that component and the wiring and connectors that link the component to both the battery and the chassis. To help you pinpoint an electrical circuit problem, wiring diagrams are included at the end of this Chapter.

Before tackling any troublesome electrical circuit, first study the appropriate wiring diagrams to get a complete understanding of what makes up that individual circuit. Not-ing if other components related to the circuit are operating properly, for instance, can often narrow down trouble spots. If several components or circuits fail at one time, chances are the problem is in a fuse or ground connection, because several circuits are often routed through the same fuse and ground connections.

Electrical problems usually stem from simple causes, such as loose or corroded connections, a blown fuse, a melted fusible link or a failed relay. Visually inspect the condition of all fuses, wires and connections in a problem circuit before troubleshooting the circuit.

If test equipment and instruments are going to be utilized, use the diagrams to plan ahead of time where you will make the nec-

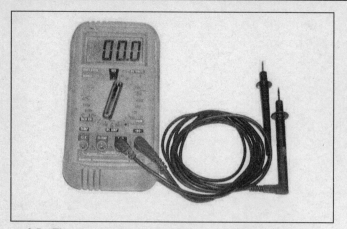

2.5a The most useful tool for electrical troubleshooting is a digital multimeter that can check volts, amps, and test continuity

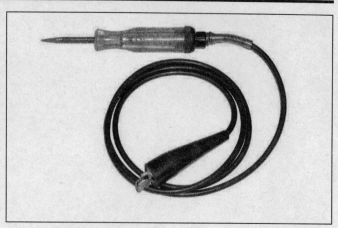

2.5b A simple test light is a very handy tool for testing voltage

essary connections in order to accurately pinpoint the trouble spot.

For electrical troubleshooting, you'll need a voltmeter, a circuit tester or a 12-volt bulb with a set of test leads; a continuity tester, which includes a bulb, battery and set of test leads; and a jumper wire, with a circuit breaker, which can be used to bypass electrical components **(see illustrations)**. Before attempting to locate a problem with test instruments, use the wiring diagram(s) to decide where to make the connections.

Voltage checks

Voltage checks should be performed if a circuit is not functioning properly. Connect one lead of a circuit tester to either the negative battery terminal or a known good ground. Connect the other lead to a connector in the circuit being tested, preferably nearest to the battery or fuse **(see illustration)**. If the bulb of the tester lights, voltage is present, which means that the part of the circuit between the

connector and the battery is problem free. Continue checking the rest of the circuit in the same fashion. When you reach a point at which no voltage is present, the problem lies between that point and the last test point with voltage. Most of the time the problem can be traced to a loose connection. **Note:** *Keep in mind that some circuits receive voltage only when the ignition key is in the ACC (accessory) or ON/RUN position.*

Finding a short

One method of finding shorts in a live circuit is to remove the fuse and connect a test light in place of the fuse terminals (fabricate two jumper wires with small spade terminals, plug the jumper wires into the fuse box and connect the test light). There should be voltage present in the circuit. Move the suspected wiring harness from side-to-side while watching the test light. If the bulb goes off, there is a short to ground somewhere in that area, probably where the insulation has rubbed through.

Ground check

Perform a ground test to check whether a component is properly grounded. Disconnect the battery and connect one lead of a continuity tester or multimeter (set to the ohm scale), to a known good ground. Connect the other lead to the wire or ground connection being tested. If the resistance is low (less than 5 ohms), the ground is good. If the bulb on a self-powered test light does not go on, the ground is not good.

Continuity check

A continuity check is done to determine if there are any breaks in a circuit - if it is passing electricity properly. With the circuit off (no power in the circuit), a self-powered continuity tester or multimeter can be used to check the circuit. Connect the test leads to both ends of the circuit (or to the power end and a good ground), and if the test light comes on the circuit is passing current properly **(see illus-**

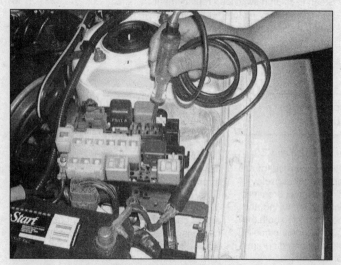

2.6 To use a test light, clip the lead to a known good ground, then test connectors, wires or electrical sockets with the pointed probe. If the bulb lights, the circuit that you're testing has battery voltage

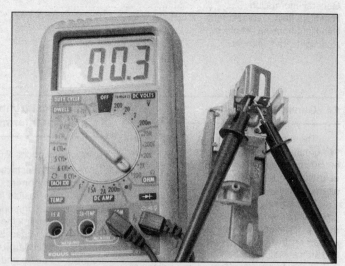

2.9 With a multimeter set to the ohm scale, you can check the resistance across two terminals. When checking for continuity, a low reading indicates continuity, a high reading or infinity indicates lack of continuity

3.1a One fuse and relay box is located in the right front corner of the engine compartment

3.1b The other fuse and relay box (the manufacturer calls it the Smart Junction Box) is located inside the vehicle, behind the right kick panel. There's an access door, but it's designed to provide access to the large electrical connector above the fuse box

tration). If the resistance is low (less than 5 ohms), there is continuity. If the reading is 10,000 ohms or higher, there is a break somewhere in the circuit. The same procedure can be used to test a switch, by connecting the continuity tester to the switch terminals. With the switch turned On, the test light should come on (or low resistance should be indicated on a meter).

Finding an open circuit

When diagnosing for possible open circuits, it is often difficult to locate them by sight because the connectors hide oxidation or terminal misalignment. Merely wiggling a connector on a sensor or in the wiring harness may correct the open circuit condition. Remember this when an open circuit is indicated when troubleshooting a circuit. Intermittent problems may also be caused by oxidized or loose connections.

Electrical troubleshooting is simple if you keep in mind that all electrical circuits are basically electricity running from the battery, through the wires, switches, relays, fuses and fusible links to each electrical component (light bulb, motor, etc.) and to ground, from which it is passed back to the battery. Any electrical problem is an interruption in the flow of electricity to and from the battery.

Connectors

Most electrical connections on these vehicles are made with multi-wire plastic connectors. The mating halves of many connectors are secured with locking clips molded into the plastic connector shells. The mating halves of large connectors, such as some of those under the instrument panel, are held together by a bolt through the center of the connector.

To separate a connector with locking clips, use a small screwdriver to pry the clips apart carefully, then separate the connector halves. Pull only on the shell; never pull on the wiring harness as you may damage the individual wires and terminals inside the connectors. Look at the connector closely before trying to separate the halves. Often the locking clips are engaged in a way that is not immediately clear. Additionally, many connectors have more than one set of clips.

Each pair of connector terminals has a male half and a female half. When you look at the end view of a connector in a diagram, be sure to understand whether the view shows the harness side or the component side of the connector. Connector halves are mirror images of each other, i.e. a terminal shown on the right side end-view of one half will be on the left side end view of the other half.

3   Fuses and fusible links - general information

Fuses

Refer to illustrations 3.1a, 3.1b, 3.1c and 3.2

The electrical circuits of the vehicle are protected by a combination of fuses, circuit breakers and fusible links. There are two fuse and relay boxes. One is located in the engine compartment **(see illustration)** and the other is located inside the vehicle, behind the right kick panel **(see illustrations)**. (For help with removing the right kick panel, refer to Chapter 11.) Each of the fuses is designed to protect a specific circuit, and the various circuits are identified on the fuse panel cover. You'll also find a guide to these fuses and relays in your owner's manual.

Miniaturized fuses are employed in the fuse block. These compact fuses, which use blade-type terminals, can be removed and installed without any special tools. If an electrical component fails, always check the fuse

3.1c To remove the cover from the interior fuse and relay box, you'll need to remove the right kick panel (see Chapter 11)

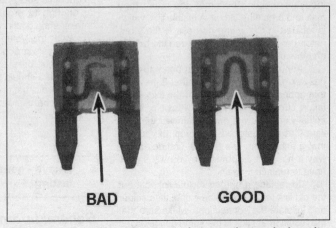

BAD   **GOOD**

3.2 When a fuse blows, the element between the terminals melts

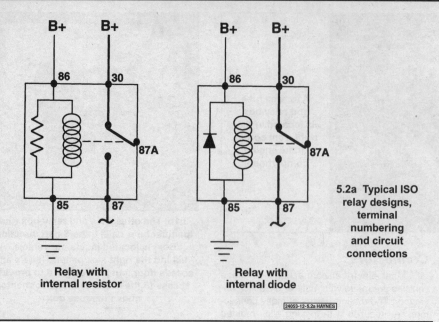

Relay with
internal resistor

Relay with
internal diode

5.2a Typical ISO
relay designs,
terminal
numbering
and circuit
connections

24053-12-5.2a HAYNES

5.2b Most relays are marked on the
outside to easily identify the control
circuits and the power circuits - four
terminal type shown

first. The best way to check a fuse is with a test light. Check for power at the exposed terminal tips of each fuse. If power is present on one side of the fuse but not the other, the fuse is blown. A blown fuse can also be confirmed by visually inspecting it **(see illustration)**.

Be sure to replace blown fuses with the correct type. Fuses of different ratings are physically interchangeable, but only fuses of the proper rating should be used. Replacing a fuse with one of a higher or lower value than specified is not recommended. Each electrical circuit needs a specific amount of protection. The amperage value of each fuse is molded into the fuse body.

If the replacement fuse immediately fails, don't replace it again until the cause of the problem is isolated and corrected. In most cases, this will be a short circuit in the wiring caused by a broken or deteriorated wire.

Fusible link

Fusible links work just like fuses, i.e. when the electrical load in the circuit that they're protecting becomes too great, they melt and break the circuit. A fusible link, which is located in the wiring harness to the alternator, protects the circuit between the battery and the alternator.

Although a fusible link appears to be a heavier gauge than the wire that it protects, that appearance is because of the thick insulation surrounding the link. A fusible link is actually several wire gauges smaller than the wire that it's designed to protect. If you find that a fusible link has melted, first determine why it happened. Otherwise you will simply have to replace it again.

To replace a burned fusible link, cut out the old link, splice in the new one, then solder and insulate the connections. Make sure that the replacement link is the same gauge as the old link. Fusible links should be available at your dealer service department and at auto

parts stores. If you're unable to obtain the correct fusible link to make the repair, replace the entire cable between the battery and the alternator.

4   Circuit breakers - general information

Circuit breakers protect certain circuits, such as the power windows or heated seats. Depending on the vehicle's accessories, there might be one or two circuit breakers, and they're usually located inside the vehicle, where they're scattered throughout the area under the instrument panel.

Because circuit breakers reset automatically, an electrical overload in a circuit-breaker-protected system will cause the circuit to fail momentarily, then come back on. If the circuit does not come back on, check it immediately.

For a basic check, pull the circuit breaker up out of its socket on the fuse panel, but just far enough to probe with a voltmeter. The breaker should still contact the sockets.

With the voltmeter negative lead on a good chassis ground, touch each end prong of the circuit breaker with the positive meter probe. There should be battery voltage at each end. If there is battery voltage only at one end, the circuit breaker must be replaced.

Some circuit breakers must be reset manually.

5   Relays - general information and testing

General information

1   Several electrical accessories in the vehicle, such as the fuel injection system, horns,

starter, and fog lamps use relays to transmit the electrical signal to the component. Relays use a low-current circuit (the control circuit) to open and close a high-current circuit (the power circuit). If the relay is defective, that component will not operate properly. Most of the relays are mounted in the engine compartment fuse and relay box **(see illustration 3.1)**. If you suspect a faulty relay, simply remove it and test it using the procedure below. Or have it tested by a dealer service department. Defective relays cannot be repaired; they must be replaced with a new unit.

Testing

Refer to illustrations 5.2a and 5.2b

2   Most of the relays used in these vehicles are of a type often called ISO relays, which refers to the International Standards Organization. The terminals of ISO relays are numbered to indicate their usual circuit connections and functions. There are two basic layouts of terminals on the relays used in these vehicles **(see illustrations)**.

3   Refer to the wiring diagram for the circuit to determine the proper connections for the relay you're testing. If you can't determine the correct connection from the wiring diagrams, however, you may be able to determine the test connections from the information that follows.

4   Two of the terminals are the relay control circuit and connect to the relay coil. The other relay terminals are the power circuit. When the relay is energized, the coil creates a magnetic field that closes the larger contacts of the power circuit to provide power to the circuit loads.

5   Terminals 85 and 86 are normally the control circuit. If the relay contains a diode, terminal 86 must be connected to battery positive (B+) voltage and terminal 85 to ground. If the relay contains a resistor, terminals 85 and 86 can be connected in either direction with respect to B+ and ground.

6   Terminal 30 is normally connected to the battery voltage (B+) source for the circuit loads. Terminal 87 is connected to the ground side of the circuit, either directly or through a load. If the relay has several alternate termi-

6.3 To detach the turn signal/headlight dimmer switch from the multi-function switch assembly, remove these two screws

nals for load or ground connections, they usually are numbered 87A, 87B, 87C, and so on.

7   Use an ohmmeter to check continuity through the relay control coil.

a) *Connect the meter according to the polarity shown in* **illustration 5.2a** *for one check; then reverse the ohmmeter leads and check continuity in the other direction.*

b) *If the relay contains a resistor, resistance will be indicated on the meter, and should be the same value with the ohmmeter in either direction.*

c) *If the relay contains a diode, resistance should be higher with the ohmmeter in the forward polarity direction than with the meter leads reversed.*

d) *If the ohmmeter shows infinite resistance in both directions, replace the relay.*

8   Remove the relay from the vehicle and use the ohmmeter to check for continuity between the relay power circuit terminals. There should be no continuity between terminal 30 and 87 with the relay de-energized.

9   Connect a fused jumper wire to terminal

86 and the positive battery terminal. Connect another jumper wire between terminal 85 and ground. When the connections are made, the relay should click.

10   With the jumper wires connected, check for continuity between the power circuit terminals. Now, there should be continuity between terminals 30 and 87.

11   If the relay fails any of the above tests, replace it.

---

6   Steering column switches - replacement

Warning: *The models covered by this manual are equipped with Supplemental Restraint Systems (SRS), more commonly known as airbags. Always disable the airbag system before working in the vicinity of any airbag system component to avoid the possibility of accidental deployment of the airbags, which could cause personal injury (see Section 24).*

Turn signal/headlight dimmer switch

Refer to illustrations 6.3 and 6.4

1   Disconnect the cable from the negative terminal of the battery (see Chapter 5, Section 1).

2   Remove the upper and lower steering column covers (see Chapter 11).

3   Remove the turn signal/headlight dimmer switch mounting screws **(see illustration)**.

4   Remove the turn signal/headlight dimmer switch and disconnect the electrical connector **(see illustration)**.

5   Installation is the reverse of removal.

Windshield wiper/washer switch

Refer to illustrations 6.8 and 6.9

6   Disconnect the cable from the negative terminal of the battery (see Chapter 5, Section 1).

7   Remove the upper and lower steering

6.4 Remove the turn signal/headlight dimmer switch, depress this release tab and unplug the electrical connector

column covers (see Chapter 11).

8   Remove the windshield wiper/washer switch mounting screws **(see illustration)**.

9   Remove the windshield wiper/washer switch and disconnect the electrical connector **(see illustration)**.

10   Installation is the reverse of removal.

Multi-function switch

Refer to illustrations 6.14a and 6.14b

Note: *The multi-function switch assembly is located on the steering column. It includes - and serves as the mounting base for - the turn signal/headlight dimmer switch and the windshield wiper/washer switch. Either of these switches can be replaced separately (see above), but if you're planning to remove or replace the steering column, remove the multi-function switch as a single assembly.*

11   Disconnect the cable from the negative terminal of the battery (see Chapter 5, Section 1).

12   Remove the steering wheel (see Chapter 10).

13   Remove the upper and lower steering column covers (see Chapter 11).

6.8 To detach the windshield wiper/washer switch from the multi-function switch assembly, remove these two screws

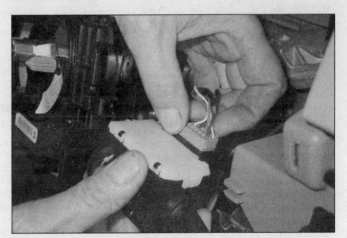

6.9 Remove the windshield wiper/washer switch, depress the release tab and disconnect the electrical connector

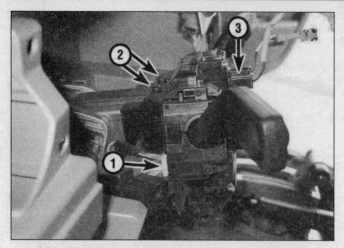

6.14a Multi-function switch assembly details (left side):

  1   *Turn signal/headlight dimmer switch electrical connector*
  2   *Upper mounting screws*
  3   *Electrical connector*

6.14b Multi-function switch assembly details (underside):

  1   *Electrical connector*
  2   *Lower mounting screws*

7.3 To disengage the ignition switch from the lock cylinder housing, depress these two release tabs

7.4 To disconnect the electrical connector from the ignition switch, depress this release tab and pull out the connector

7.6 To detach the PATS transceiver from the steering column assembly, disengage the retaining tabs from the lugs on the lock cylinder housing (lower lug not visible in this photo)

14   Disconnect the electrical connectors from the multi-function switch, the turn signal/headlight dimmer switch and the windshield wiper/washer switch **(see illustrations)**.

15   Remove the multi-function switch mounting screws and detach the switch from the steering column.

16   Installation is the reverse of removal.

---

7   Ignition switch and key lock cylinder - replacement

Warning: *The models covered by this manual are equipped with Supplemental Restraint Systems (SRS), more commonly known as airbags. Always disable the airbag system before working in the vicinity of any airbag system component to avoid the possibility of accidental deployment of the airbags, which could cause personal injury (see Section 24).*

1   Disconnect the cable from the negative terminal of the battery (see Chapter 5, Section 1).

2   Remove the lower steering column cover (see Chapter 11).

Ignition switch

Refer to illustrations 7.3 and 7.4

3   Remove the ignition switch **(see illustration)**.

4   Disconnect the electrical connector from the ignition switch **(see illustration)**.

5   Installation is the reverse of removal.

Key lock cylinder

Refer to illustrations 7.6 and 7.8

6   Remove the Passive Anti-Theft System (PATS) transceiver **(see illustration)**.

7   Insert the ignition key into the key lock cylinder and turn the key to the RUN position.

8   Insert an awl or small screwdriver into the this hole in the underside of the key lock cylinder, then depress the key lock cylinder retaining tab and remove the key lock cylinder **(see illustration)**.

9   Installation is the reverse of removal.

7.8 To remove the key lock cylinder, turn the ignition key to the RUN position, insert an awl or small screwdriver into this hole and depress the locking pin, then pull out the lock cylinder

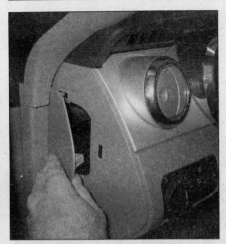

8.2 To access the headlight switch and instrument panel dimmer switch, remove this finish panel from the left end of the instrument panel

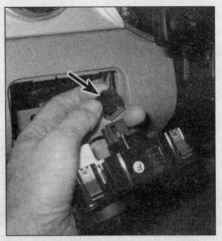

8.4a To disconnect the electrical connector from the instrument panel dimmer switch, depress this release tab and pull off the connector

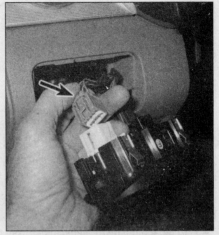

8.4b To disconnect the electrical connector from the headlight switch, depress this release tab and pull off the connector

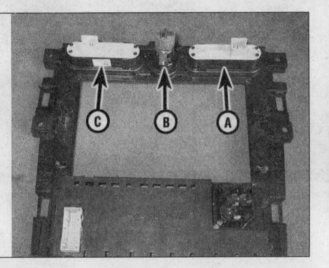

8.8a The center trim panel houses several electrical switches and the auxiliary power outlet:

A   Message center switch assembly
B   Auxiliary power outlet
C   Traction Control System (TCS) switch and hazard flasher switch assembly

8.8b To remove a switch from the center trim panel, simply squeeze the retaining tabs on the side of the switch and push it out from the backside

8   Instrument panel switches - replacement

Warning: *The models covered by this manual are equipped with Supplemental Restraint Systems (SRS), more commonly known as airbags. Always disable the airbag system before working in the vicinity of any airbag system components to avoid the possibility of accidental deployment of the airbag(s), which could cause personal injury (see Section 24).*

Headlight switch and instrument panel dimmer switch

Refer to illustrations 8.2, 8.4a and 8.4b

1   Disconnect the cable from the negative battery terminal (see Chapter 5, Section 1).
2   Remove the finish panel from the left end of the instrument panel **(see illustration)**.
3   Reach through the access hole in the left end of the instrument panel and push out the switch panel for the headlight switch and instrument panel dimmer switch.
4   Disconnect the electrical connectors from the headlight switch and rheostat **(see illustrations)** and remove the switch assembly.
5   When installing the switch assembly make sure that it snaps into place in the instrument panel. Installation is otherwise the reverse of removal.

Message center switches and Traction Control System (TCS) switch/hazard flasher switch

Refer to illustrations 8.8a and 8.8b

Note: *All of these switches are located in the center trim panel, right above the radio.*

6   Disconnect the cable from the negative battery terminal (see Chapter 5, Section 1).
7   Remove the center trim panel (see Chapter 11).
8   Remove the switch from the center trim panel **(see illustrations)**.
9   When installing one of these switch assemblies make sure that it snaps into place in the center trim panel.
10   Installation is otherwise the reverse of removal.

9   Instrument cluster - removal and installation

Refer to illustrations 9.3 and 9.4

Warning: *The models covered by this manual are equipped with Supplemental Restraint Systems (SRS), more commonly known as airbags. Always disable the airbag system before working in the vicinity of any airbag system components to avoid the possibility of accidental deployment of the airbag(s), which could cause personal injury (see Section 24).*

1   Disconnect the cable from the negative terminal of the battery (see Chapter 5, Section 1).
2   Remove the instrument cluster bezel (see Chapter 11).

9.3 To detach the instrument cluster from the instrument panel, remove these four screws

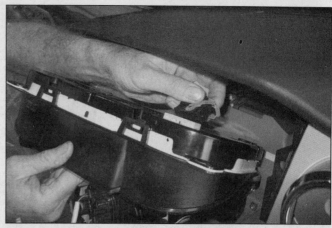

9.4 Lift up the locking lever to disconnect the electrical connector from the instrument cluster

10.5 To detach the windshield wiper motor and link assembly, remove these three mounting bolts

10.6 To disconnect the electrical connector from the windshield wiper motor, depress this release tab and pull off the connector

3   Remove the instrument cluster mounting screws **(see illustration).**
4   Pull the instrument cluster out of the instrument panel far enough to disconnect the electrical connector from the instrument cluster **(see illustration).**
5   Installation is the reverse of removal.

10 Windshield wiper motor - replacement

Refer to illustrations 10.5, 10.6 and 10.7

1   Pry off the trim caps from the windshield wiper arm retaining nuts.
2   Remove the windshield wiper arm nuts.
3   Before removing the wiper arms, put a piece of tape on the windshield to mark the position of the outer end of each windshield wiper arm, as a reference mark for reassembly. Then remove the windshield wiper arms.
4   Remove the left and right cowl covers (see Chapter 11).
5   Remove the windshield wiper motor and link assembly mounting bolts **(see illustration).**

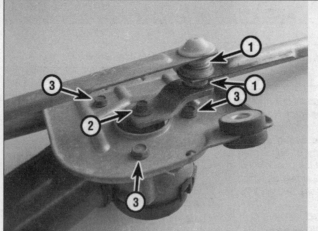

10.7 To separate the wiper motor from the link assembly:

1   *Pry off the linkage arms from the pin on the end of the crank arm*
2   *Remove the crank arm retaining nut and remove the crank arm*
3   *Remove the three wiper mounting bolts and remove the motor from its mounting bracket*

6   Turn the windshield wiper motor and link assembly over and disconnect the electrical connector from the wiper motor **(see illustration).**
7   Detach the windshield wiper linkage arms from the wiper motor crank arm **(see illustration).** The linkage arms are pressed onto a pin on the end of the motor crank arm. To detach each linkage arm, simply pry it off with a trim removal tool or a similar suitable tool.
8   Remove the wiper motor crank arm.
9   Remove the wiper motor from its mounting bracket.
10   Installation is the reverse of removal. Be sure to align the wiper blades with the pieces of tape applied in Step 3 before pushing them onto their shafts.

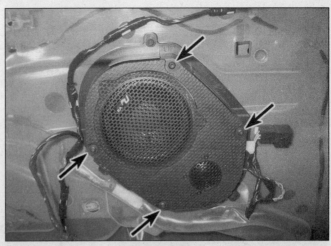

11.3 To detach the radio from the instrument panel, remove these four screws

11.7 To detach a front door speaker from the door, remove these four screws

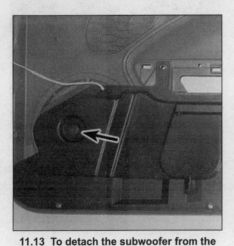

11.10 To remove the subwoofer grille trim panel, carefully pry it off with a trim removal tool or a screwdriver

11.13 To detach the subwoofer from the subwoofer enclosure, remove this screw

11.20 Disconnect the electrical connector from a parcel shelf speaker from inside the trunk area

11   Radio and speakers - removal and installation

Warning: *The models covered by this manual are equipped with Supplemental Restraint Systems (SRS), more commonly known as airbags. Always disable the airbag system before working in the vicinity of any airbag system component to avoid the possibility of accidental deployment of the airbag(s), which could cause personal injury (see Section 24).*
1   Disconnect the cable from the negative terminal of the battery (see Chapter 5, Section 1).

Radio

Refer to illustration 11.3
2   Remove the center trim panel (see Chapter 11).
3   Remove the radio mounting screws **(see illustration)**, then pull the radio out of the instrument panel just far enough to access the electrical connectors and the antenna cable on the backside of the unit.

4   Disconnect the electrical connectors and the antenna lead from the backside of the radio and remove the radio.
5   Installation is the reverse of removal.

Speakers

Door speakers

Refer to illustration 11.7
6   Remove the door trim panel (see Chapter 11).
7   Remove the speaker mounting screws **(see illustration)**.
8   Pull out the speaker, disconnect the electrical connector from the speaker and remove the speaker.
9   Installation is the reverse of removal.

Door subwoofers

Refer to illustrations 11.10 and 11.13
10   Remove the subwoofer grille trim panel **(see illustration)**.
11   Remove the door trim panel (see Chapter 11).
12   Remove the eight mounting screws from the front side of the subwoofer.

13   Remove the single subwoofer retaining screw from the backside of the subwoofer enclosure **(see illustration)**.
14   Pull out the subwoofer, disconnect the electrical connector and remove the subwoofer.
15   Installation is the reverse of removal.

Quarter panel speakers

16   Remove the quarter panel trim (see Chapter 11).
17   Remove the four speaker mounting screws.
18   Pull out the speaker, disconnect the electrical connector and remove the speaker.
19   Installation is the reverse of removal.

Parcel shelf speakers

Refer to illustrations 11.20 and 11.22
Note: *There are two speakers in the parcel shelf. This procedure applies to either speaker.*
20   Working from inside the trunk area, disconnect the speaker electrical connector **(see illustration)**.

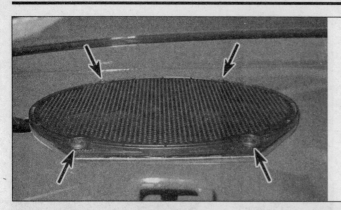

12.1 The antenna mast simply unscrews from the mounting base

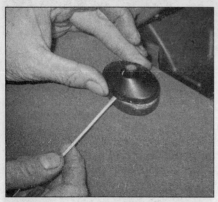

12.4 To remove the antenna mounting base cap, carefully pry it off

12.5 To detach the antenna mounting base from the fender, remove these three screws

Antenna mounting base and outer antenna cable

Refer to illustrations 12.4, 12.5, 12.7 and 12.11

3   Remove the antenna mast **(see illustration 12.1)**.
4   Remove the antenna mounting base cap **(see illustration)**.
5   Remove the antenna mounting base screws **(see illustration)**.
6   Remove the glove box (see Chapter 11).
7   Disconnect the outer antenna cable from the inner antenna cable **(see illustration)**.
8   Loosen the front right wheel lug nuts. Raise the front of the vehicle and place it securely on jackstands. Remove the right front wheel.
9   Remove the inner fender splash shield from the right front wheel well (see Chapter 11).
10   Trace the outer cable to the rubber grommet where the cable goes through the vehicle body. Note how the outer cable is routed. When you install the new cable, it must be routed exactly the same way, or it won't be long enough to reach the inner cable.
11   Working from inside the wheel well area, detach the outer antenna cable retainer pin,

21   Remove the parcel shelf (see Chapter 11).
22   Remove the speaker mounting screws **(see illustration)** and remove the speaker.
23   Installation is the reverse of removal.

---

12   Antenna and cables - replacement

Warning: *The models covered by this manual are equipped with Supplemental Restraint Systems (SRS), more commonly known as airbags. Always disable the airbag system*

before working in the vicinity of any airbag system components to avoid the possibility of accidental deployment of the airbag(s), which could cause personal injury (see Section 24).

Antenna mast, antenna base and antenna cable

Antenna mast

Refer to illustration 12.1

1   Unscrew the antenna mast from the base **(see illustration)**. Be careful not to scratch the paint.
2   Installation is the reverse of removal.

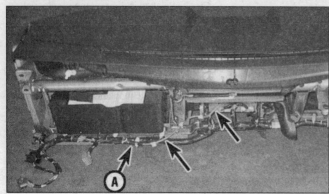

12.7 To disconnect the outer antenna cable from the inner antenna cable, reach through the glove box opening and disconnect the connector between the two cables (A) (instrument panel removed for clarity)

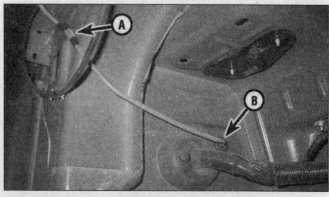

12.11 Working inside the wheel well, detach the antenna cable retainer pin (A), pry the cable grommet (B) out of its mounting hole, then pull out the cable through the grommet hole. (The grommet itself is part of the cable, so you can't remove it from the cable.)

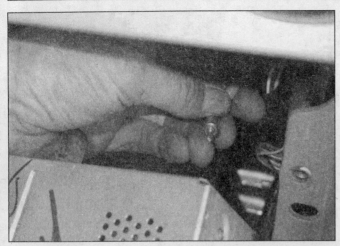

12.17 Disconnect the inner antenna cable from the backside of the radio

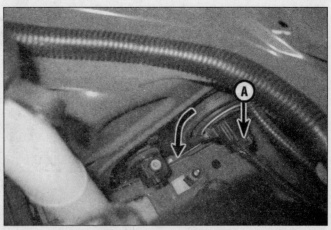

13.2 To replace a headlight bulb, depress the release tab (A) and pull off the electrical connector, then rotate the bulb holder counterclockwise . . .

then carefully pry out the cable grommet that seals the hole in the firewall **(see illustration)**.

12   Pull the cable through the grommet hole into the wheel well.

13   Remove the antenna mounting base and outer antenna cable from the wheel well area by pulling the base and the cable straight up through the base mounting hole.

14   Installation is the reverse of removal.

Inner antenna cable

Refer to illustration 12.17

Note: *The inner antenna cable connects the outer antenna cable to the radio.*

15   Remove the glove box (see Chapter 11).

16   Disconnect the inner antenna cable from the outer antenna cable **(see illustration 12.7)**.

17   Remove the radio (see Section 11) and disconnect the inner antenna cable from the backside of the radio **(see illustration)**.

18   Installation is the reverse of removal.

---

13   Headlight bulb - replacement

Refer to illustrations 13.2 and 13.3

Warning: *Halogen gas filled bulbs are under pressure and may shatter if the surface is scratched or the bulb is dropped. Wear eye protection and handle the bulbs carefully, grasping only the base whenever possible. Do not touch the surface of the bulb with your fingers because the oil from your skin could cause it to overheat and fail prematurely. If you do touch the bulb surface, clean it with rubbing alcohol.*

1   If you're going to replace the left headlight bulb, remove the air filter housing (see Chapter 4).

2   Disconnect the electrical connector **(see illustration)**. **Note:** *If you're simply removing the headlight bulb assembly to remove the headlight housing, it's not necessary to disconnect the electrical connector.*

3   Rotate the bulb holder counterclockwise and pull out the bulb and holder assembly **(see illustration)**.

4   The bulb and holder are a one-piece assembly. Do not try to pull the bulb out of the holder.

5   Installation is the reverse of removal. Be sure to rotate the bulb holder clockwise in the headlight housing until it stops. **Caution:** *Make sure that you don't touch the surface of the bulb during installation.*

---

14   Headlights - adjustment

Refer to illustrations 14.1 and 14.4

Note: *The headlights must be aimed correctly. If adjusted incorrectly they could blind the driver of an oncoming vehicle and cause a serious accident or seriously reduce your ability to see the road. The headlights should be checked for proper aim every 12 months and any time a new headlight housing is installed or front-end bodywork is performed. It should be emphasized that the following procedure is only an interim step, which will provide temporary adjustment until a properly equipped shop can adjust the headlights.*

1   These vehicles have a vertical adjustment screw located on the backside of the

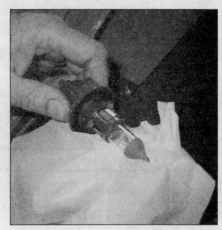

13.3 . . . and pull it out of the headlight housing. The bulb cannot be removed from the holder

headlight housing **(see illustration)**.

2   Adjustment should be made with the vehicle on a level surface, with a full gas tank and a normal load in the vehicle.

3   There are several methods of adjusting the headlights. The simplest method requires masking tape, a blank wall and a level floor.

4   Position masking tape vertically on the wall to indicate the vehicle centerline and the

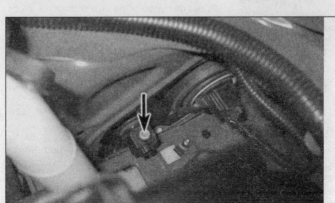

14.1 Headlight vertical adjustment screw location

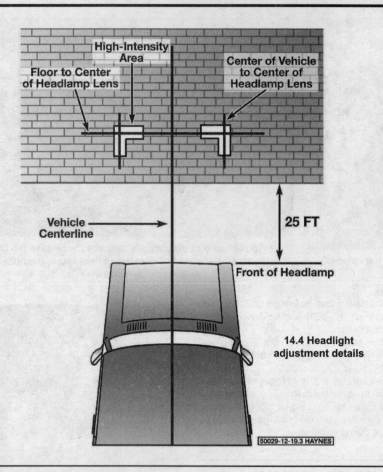

14.4 Headlight adjustment details

15.2 To detach the headlight housing from the radiator crossmember, remove the pushpin (A) and the upper mounting bolt (B)

When you're finished, be sure to adjust the headlights (see Section 14).

16 Horn - replacement

Refer to illustration 16.3
Note: *The horns are located between the grille and the radiator.*
1   To access the horn, remove the upper radiator cover (see Section 10 in Chapter 11).
2   Remove the four radiator support bracket bolts and remove the two radiator support brackets (see *Radiator - removal and installation* in Chapter 3). Then push the upper end of the radiator and the condenser toward the engine.
3   Remove the horn mounting bracket nuts **(see illustration)** and remove the horn assembly.
4   Disconnect the electrical connector from the horn assembly.
5   Installation is the reverse of removal.

centerline of each headlight bulb **(see illustration)**.
5   Position a horizontal tape line in reference to the centerline of all the headlights.
Note: *It may be easier to position the tape on the wall with the vehicle parked only a few inches away.*
6   Adjustment should be made with the vehicle parked 25 feet (7.6 meters) from the wall, sitting level, the gas tank half-full and no unusually heavy load in the vehicle.
7   Position the high intensity zone so it is two inches below the horizontal line and two inches to the side of the headlight vertical line,

away from oncoming traffic. Adjustment is made by turning the adjusting screw to move the beam up or down. The high beams on these headlights are automatically adjusted along with the low beam.
8   Have the headlights adjusted by a dealer service department or service station at the earliest opportunity.

15 Headlight housing - replacement

Refer to illustrations 15.2 and 15.3
Warning: *These vehicles are equipped with halogen gas-filled headlight bulbs, which are under pressure and may shatter if the surface is damaged or the bulb is dropped. Wear eye protection and handle the bulbs carefully, grasping only the base whenever possible. Do not touch the surface of the bulb with your fingers because the oil from your skin could cause it to overheat and fail prematurely. If you do touch the bulb surface, clean it with rubbing alcohol.*
1   Remove the front bumper cover (see Chapter 11).
2   Remove the upper headlight housing mounting bolt and pushpin **(see illustration)**.
3   Remove the lower headlight housing mounting bolts **(see illustration)**.
4   Pull out the headlight housing assembly and disconnect the electrical connector.
5   Installation is the reverse of removal.

15.3 To detach the lower side of the headlight housing, remove these two bolts

16.3 To detach the horn assembly, remove the two mounting bracket nuts (A) (other nut not visible in this photo), then remove the horn assembly and disconnect the electrical connector (B)

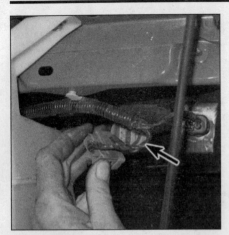

17.2a Disconnect the electrical connector from the front turn signal light

17.2b Disconnect the electrical connector from the front parking light

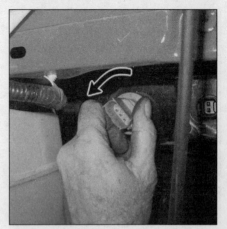

17.3 To remove the front parking light or front turn signal light from the housing, rotate it counterclockwise and pull it out of the housing

17   Bulb replacement

Exterior light bulbs

Front parking light/turn signal bulbs

Refer to illustrations 17.2a, 17.2b, 17.3 and 17.4

1   Raise the front of the vehicle and place it securely on jackstands.

2   Disconnect the electrical connector from the front parking light or turn signal bulb **(see illustrations)**.

3   To remove the front parking light or turn signal bulb holder from the headlight housing, turn it counterclockwise and pull it out of the housing **(see illustration)**.

4   Remove the bulb from the holder **(see illustration)**. Install the new bulb in the holder by pushing it straight into the holder until it stops.

5   Installation is the reverse of removal.

Front fog light bulbs

Refer to illustration 17.8

6   Remove the upper radiator cover (see Section 10 in Chapter 11).

7   Remove the upper bumper cover bolts

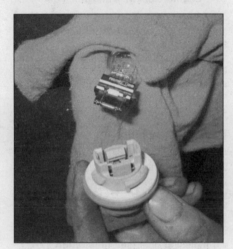

17.4 To remove the bulb, pull it straight out of the bulb holder

located on either side of the grille (see Section 11 in Chapter 11), then pull the upper edge of the grille forward to give yourself more room to work.

8   To remove the fog light bulb socket from the fog light housing, turn it counterclockwise and pull it out of the housing **(see illustration)**.

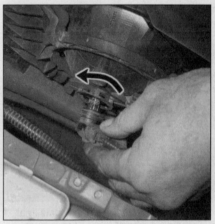

17.8 To remove the front fog light bulb from the fog light housing, turn the bulb holder counterclockwise and pull it out of the housing

9   Disconnect the electrical connector from the fog light bulb.

10   Installation is the reverse of removal.

High-mount brake light bulb

Refer to illustration 17.12

11   Remove the decklid latch trim cover (see *Trunk lid latch and lock cylinder - removal and installation* in Chapter 11).

12   Remove the bulb socket from the high-mount brake light housing **(see illustration)**.

13   Remove the old bulb from the socket by pulling it straight out.

14   To install a new bulb in the socket insert it straight into the socket until it stops.

15   Installation is the reverse of removal.

Brake/taillight/turn signal light bulbs

Refer to illustrations 17.17 and 17.18

16   Remove the rear inside trunk finishing panel (see *Bumper covers - removal and installation* in Chapter 11) to access the taillight housing mounting nuts.

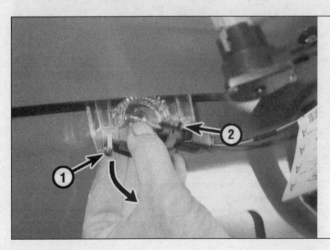

17.12 To remove the bulb socket from the high-mount brake light housing, carefully disengage the left locking lug (1), swing down the socket assembly, then disengage the right locking lug (2)

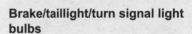

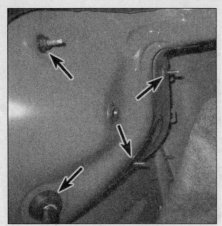

17.17 To detach the taillight housing from the vehicle, push the electrical harness grommet through the panel, then remove the three mounting nuts

17.18 To remove a bulb socket from the taillight housing, rotate it counterclockwise and pull it out

17.23 To remove the back-up light bulb socket from the taillight housing, rotate it counterclockwise and pull it out

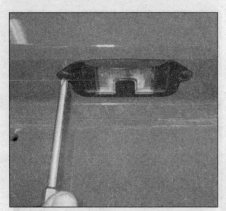

17.26 To remove the license plate light assembly, remove these two screws

17   Free the grommet for the taillight electrical harness and the three taillight housing mounting nuts **(see illustration)**, then pull off the taillight housing.

18   Rotate the bulb socket counterclockwise and pull it out of the taillight housing **(see illustration)**.

19   To remove the bulb from its socket, pull it straight out. To install a new bulb in the socket, push it straight into the socket until it stops.

20   Installation is the reverse of removal.

Back-up light bulbs

Refer to illustration 17.23

21   Remove the rear inside trunk finishing panel (see *Bumper covers - removal and installation* in Chapter 11) to access the taillight housing mounting nuts.

22   Remove the grommet for the taillight electrical harness and the three taillight housing mounting nuts **(see illustration 17.17)**, then pull off the taillight housing.

23   Rotate the bulb socket counterclockwise and pull it out of the taillight housing **(see illustration)**.

24   To remove the bulb from its socket, pull it straight out. To install a new bulb in the socket, push it straight into the socket until

it stops.

25   Installation is the reverse of removal.

License plate light bulbs

Refer to illustrations 17.26

26   Remove the license plate light assembly **(see illustration)**.

27   To remove the license plate light socket, rotate it counterclockwise from its mounting base.

28   Remove the bulb from the socket by pulling it straight out.

29   Install a new bulb in the socket by pushing it straight into the socket.

30   Installation is the reverse of removal.

Sidemarker light bulbs

Refer to illustrations 17.31, 17.32 and 17.34

31   Carefully pry the sidemarker light housing out of the rear bumper cover **(see illustration)**.

32   Disconnect the electrical connector from the sidemarker light bulb socket **(see illustration)**.

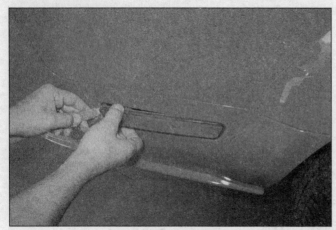

17.31 To remove the sidemarker light housing from the rear bumper cover, carefully pry it loose with a plastic trim tool

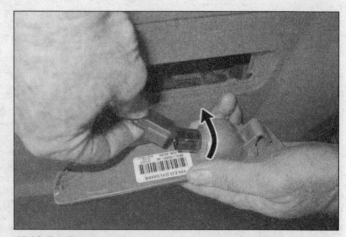

17.32 To disconnect the electrical connector from the sidemarker light bulb socket, depress this release tab and pull off the connector. To remove the bulb socket from the sidemarker light housing, rotate it counterclockwise and pull it out

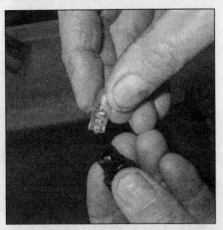

17.34 To remove a sidemarker bulb from is socket, simply pull it straight out

17.36 Carefully pop loose the overhead console (the trim cover for the map reading lights)

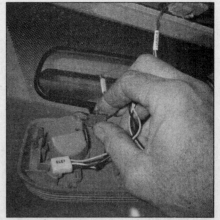

17.37 Remove the map reading light bulb socket from the trim cover

33   Remove the sidemarker light bulb socket from the sidemarker light housing.
34   To remove the sidemarker light bulb from the socket pull it straight out **(see illustration)**. To install a new bulb, push it straight into the socket.
35   Installation is the reverse of removal.

Interior light bulbs

Map reading light bulbs

Refer to illustrations 17.36, 17.37 and 17.38

36   Carefully pop loose the overhead console trim cover **(see illustration)**.
37   Remove the map reading light bulb socket from the trim cover **(see illustration)**.
38   Remove the bulb from its socket **(see illustrations)**.
39   Install a new reading light bulb in the socket.
40   Installation is the reverse of removal.

17.38 To remove a map reading light bulb, simply pull it straight out of the socket

18   Rear window defogger - check and repair

1   The rear window defogger consists of a number of horizontal elements baked onto the glass surface.
2   Small breaks in the element can be repaired without removing the rear window.

Check

Refer to illustrations 18.4, 18.5 and 18.7

3   Turn the ignition switch and defogger system switches to the ON position. Using a voltmeter, place the positive probe against the defogger grid positive terminal and the negative probe against the ground terminal. If battery voltage is not indicated, check the fuse, defogger switch and related wiring. If voltage is indicated, but all or part of the defogger doesn't heat, proceed with the following tests.
4   When measuring voltage during the next two tests, wrap a piece of aluminum foil around the tip of the voltmeter positive probe and press the foil against the heating element

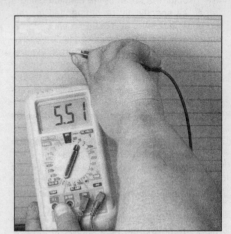

18.4 When measuring the voltage at the rear window defogger grid, wrap a piece of aluminum foil around the positive probe of the voltmeter and press the foil against the wire with your finger

with your finger **(see illustration)**. Place the negative probe on the defogger grid ground terminal.
5   Check the voltage at the center of each heating element **(see illustration)**. If the voltage is 5 or 6-volts, the element is okay (there is no break). If the voltage is zero, the element is broken between the center of the

18.5 To determine if a heating element has broken, check the voltage at the center of each element. If the voltage is 5 or 6 volts, the element is unbroken. If the voltage is 10 or 12-volts, the element is broken between the center and the ground side. If there is no voltage, the element is broken between the center and the positive side

element and the positive end. If the voltage is 10 to 12-volts the element is broken between the center of the element and ground. Check each heating element.

18.7 To find the break, place the voltmeter negative lead against the defogger ground terminal, place the voltmeter positive lead with the foil strip against the heating element at the positive terminal end and slide it toward the negative terminal end. The point at which the voltmeter reading changes abruptly is the point at which the element is broken

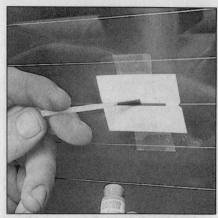

18.13 To use a defogger repair kit, apply masking tape to the inside of the window at the damaged area, then brush on the special conductive coating

6   Connect the negative lead to a good body ground. The reading should stay the same. If it doesn't, the ground connection is bad.

7   To find the break place the voltmeter negative probe against the defogger ground terminal. Place the voltmeter positive probe with the foil strip against the heating element at the positive terminal end and slide it toward the negative terminal end. The point at which the voltmeter deflects from several volts to zero is the point at which the heating element is broken **(see illustration)**.

Repair

Refer to illustration 18.13

8   Repair the break in the element using a repair kit specifically recommended for this purpose, available at most auto parts stores. Included in this kit is plastic conductive epoxy.

9   Prior to repairing a break turn off the system and allow it to cool off for a few minutes.

10   Lightly buff the element area with fine steel wool, then clean it thoroughly with rubbing alcohol.

11   Use masking tape to mask off the area being repaired.

12   Thoroughly mix the epoxy, following the instructions provided with the repair kit.

13   Apply the epoxy material to the slit in the masking tape, overlapping the undamaged area about 3/4-inch on either end **(see illustration)**.

14   Allow the repair to cure for 24 hours before removing the tape and using the system.

19   Electric side view mirrors - general information

1   Most electric rear view mirrors use two motors to move the glass; one for up and down adjustments and one for left-right adjustments.

2   The control switch has a selector portion that sends voltage to the left or right side mirror. With the ignition ON but the engine OFF,

roll down the windows and operate the mirror control switch through all functions (left-right and up-down) for both the left and right side mirrors.

3   Listen carefully for the sound of the electric motors running in the mirrors.

4   If the motors can be heard but the mirror glass doesn't move, there's a problem with the drive mechanism inside the mirror.

5   If the mirrors do not operate and no sound comes from the mirror motors, check the fuse (see Section 3).

6   If the fuse is OK, remove the mirror control switch. Have the switch continuity checked by a dealership service department or other qualified automobile repair facility.

7   Test the ground connections.

8   If the mirror still doesn't work, remove the mirror and check the wires at the mirror for voltage.

9   If there's not voltage in each switch position, check the circuit between the mirror and control switch for opens and shorts.

10   If there's voltage, remove the mirror and test it off the vehicle with jumper wires. Replace the mirror if it fails this test.

20   Cruise control system - general information

There are no conventional cruise control system components on these vehicles. The cruise control system is an integral subsystem of the Electronic Throttle Control (ETC) system. If the cruise control system isn't functioning correctly, take the vehicle to a dealer service department or other qualified repair shop for diagnosis.

21   Power window system - general information

1   The power window system operates electric motors, mounted in the doors, which lower and raise the windows. The system consists of the control switches, relays, the

motors, regulators, glass mechanisms and associated wiring.

2   The power windows can be lowered and raised from the master control switches by the driver or by remote switches located at the individual windows. Each window has a separate motor that is reversible. The position of the control switch determines the polarity and therefore the direction of operation.

3   The circuit is protected by a fuse and a circuit breaker. Each motor is also equipped with an internal circuit breaker; this prevents one stuck window from disabling the whole system.

4   The power window system will only operate when the ignition switch is ON. In addition, many models have a window lockout switch at the master control switch that, when activated, disables the switches at the rear windows and, sometimes, the switch at the passenger's window also. Always check these items before troubleshooting a window problem.

5   These procedures are general in nature, so if you can't find the problem using them, take the vehicle to a dealer service department or other properly equipped repair facility.

6   If the power windows won't operate, always check the fuse and circuit breaker first.

7   If only the rear windows are inoperative, or if the windows only operate from the master control switch, check the rear window lockout switch for continuity in the unlocked position. Replace it if it doesn't have continuity.

8   Check the wiring between the switches and fuse panel for continuity. Repair the wiring, if necessary.

9   If only one window is inoperative from the master control switch, try the other control switch at the window. **Note:** *This doesn't apply to the driver's door window.*

10   If the same window works from one switch, but not the other, check the switch for continuity. Have the switch checked at a dealer service department or other qualified automobile repair facility.

11   If the switch tests OK, check for a short or open in the circuit between the affected switch and the window motor.

12   If one window is inoperative from both switches, remove the trim panel from the affected door and check for voltage at the switch and at the motor while the switch is operated.

13   If voltage is reaching the motor, disconnect the glass from the regulator (see Chapter 11). Move the window up and down by hand while checking for binding and damage. Also check for binding and damage to the regulator. If the regulator is not damaged and the window moves up and down smoothly, replace the motor. If there's binding or damage, lubricate, repair or replace parts, as necessary.

14   If voltage isn't reaching the motor, check the wiring in the circuit for continuity between the switches and motors. You'll need to consult the wiring diagram for the vehicle. If the circuit is equipped with a relay, check that the relay is grounded properly and receiving voltage.

15   Test the windows after you are done to confirm proper repairs.

22   Power door lock system - general information

1   A power door lock system operates the door lock actuators mounted in each door. The system consists of the switches, actuators, a control unit and associated wiring. Diagnosis can usually be limited to simple checks of the wiring connections and actuators for minor faults that can be easily repaired.

2   Power door lock systems are operated by bi-directional solenoids located in the doors. The lock switches have two operating positions: Lock and Unlock. When activated, the switch sends a ground signal to the door lock control unit to lock or unlock the doors. Depending on which way the switch is activated, the control unit reverses polarity to the solenoids, allowing the two sides of the circuit to be used alternately as the feed (positive) and ground side.

3   Some vehicles may have an anti-theft system incorporated into the power locks. If you are unable to locate the trouble using the following general Steps, consult a dealer service department or other qualified repair shop.

4   Always check the circuit protection first. Some vehicles use a combination of circuit breakers and fuses.

5   Operate the door lock switches in both directions (LOCK and UNLOCK) with the engine off. Listen for the click of the solenoids operating.

6   Test the switches for continuity. Remove the switches and have them checked by a dealer service department or other qualified automobile repair facility.

7   Check the wiring between the switches, control unit and solenoids for continuity.

Repair the wiring if there's no continuity.

8   Check for a bad ground at the switches or at the control unit.

9   If all but one of the lock solenoids operates, remove the trim panel from the door with the non-operational solenoid (see Chapter 11), then check for voltage to the solenoid while operating the lock switch. One of the wires should have voltage in the LOCK position; the other should have voltage in the UNLOCK position.

10   If the inoperative solenoid is receiving voltage, replace the solenoid.

11   If the inoperative solenoid isn't receiving voltage, check the relay for an open or short in the wire between the lock solenoid and the control unit. **Note:** *It's not uncommon for wires to break in the portion of the harness between the body and door (opening and closing the door fatigues and eventually breaks the wires).*

23   Daytime Running Lights (DRL) - general information

The Daytime Running Lights (DRL) system, which is required on new Canadian models, illuminates the headlights when the engine is running. The DRL system supplies reduced power to the headlights so they won't be too bright for daytime use, which also prolongs headlight life.

24   Airbag system - general information and precautions

General information

1   All models are equipped with a frontal impact airbag system, which is referred to as the Supplemental Restraint System (SRS). The SRS is designed to protect the driver and the front seat passenger from serious injury in the event of a head-on or frontal collision. The SRS is controlled by the Restraints Control Module (RCM), which is mounted on the center tunnel, between the seats. The SRS uses a pair of airbags to protect the front-seat occupants: the driver's airbag in the steering wheel and the passenger airbag, which is located in the right end of the instrument panel, beneath the instrument panel top pad and above the glove box. Other important components in the SRS include the clockspring (a wind-up coil that delivers battery voltage to the steering wheel airbag), and the airbag readiness light on the instrument cluster.

2   The seatbelts, which are considered a critical part of the SRS, are equipped with pre-tensioners that remove excess slack from the seatbelt webbing. When the RCM detects an impact that exceeds the specified threshold, it activates the pre-tensioners.

Driver airbag

3   The airbag inflator module, which is

mounted in the center of the steering wheel, contains a housing incorporating the airbag and the inflator unit. The inflator assembly is mounted on the back of the housing over a hole through which gas is expelled, inflating the bag almost instantaneously when an electrical signal is sent from the system. The clockspring assembly on the steering column under the steering wheel carries this signal to the module. The clockspring assembly can transmit an electrical signal regardless of steering wheel position. The igniter in the airbag converts the electrical signal to heat and ignites the powder, which inflates the bag.

Passenger airbag

4   The airbag is mounted in the right end of the instrument panel, behind the aluminum trim panel above the glove box. It uses the same components as the driver's airbag, except that the passenger airbag is larger than the steering wheel-mounted unit. When the inflator module is activated, the trim panel splits open as the bag inflates.

Restraints Control Module (RCM)

5   In the event of a collision, the RCM supplies current to the SRS, *even if battery power is cut off*. Simultaneously, it also activates the seatbelt pre-tensioners to remove all slack from the seat belts.

6   The RCM checks the SRS every time the vehicle is started, and indicates that it is doing so by turning on the airbag readiness light, which is located on the instrument cluster. If the SRS is operating properly, the RCM turns off the airbag readiness light. If it detects a fault in the system, the airbag readiness light will remain on. If this condition occurs, take the vehicle to your dealer immediately for service.

Disarming the system and other precautions

Warning: *Failure to follow these precautions could result in accidental deployment of the airbag and personal injury.*

7   Whenever you are working in the vicinity of the driver airbag in the steering wheel or any of the other airbags on your vehicle, **DISARM THE SYSTEM**. To disarm the system:

a) Point the wheels straight ahead and turn the ignition key to the LOCK position.

b) Disconnect the cable from the negative battery terminal. Isolate the cable terminal so it won't accidentally contact the battery post.

c) Wait at least two minutes for the backup power supply to be depleted. (Backup power is supplied by a capacitor that takes about two minutes to fully discharge. During this two-minute interval the SRS is still capable of deploying.)

8   Whenever handling an airbag, always keep the airbag opening (the trim side) pointed away from your body. Never place the airbag on a bench or other surface with the airbag opening facing the surface. Always place the airbag module in a safe location with the air-

bag opening facing up.

9   Never measure the resistance of any SRS component. An ohmmeter has a built-in battery supply that could accidentally deploy the airbag.

10   Never dispose of a "live" airbag. Return it to a dealer service department or other qualified repair shop for safe deployment and disposal.

Component removal and installation

Driver airbag and clockspring

11   Refer to Chapter 10, *Steering wheel - removal and installation*, for the driver's side airbag module and clockspring removal and installation procedures.

Passenger airbag

12   We don't recommend removing the passenger airbag. If you ever have to remove the instrument panel (see Chapter 11), you will have to remove the airbag with it, but it's not necessary to actually detach the airbag from the instrument panel, and we don't recommend doing so. If the passenger airbag must be serviced, take the vehicle to a dealer service department or other qualified repair shop.

Restraints Control Module (RCM)

13   We don't recommend removing the RCM. If the RCM must be serviced, take the vehicle to a dealer service department or other qualified repair shop.

25   Wiring diagrams - general information

Since it isn't possible to include all wiring diagrams for every year covered by this manual, the following diagrams are those that are typical and most commonly needed.

Prior to troubleshooting any circuits, check the fuse and circuit breakers (if equipped) to make sure they're in good condition. Make sure the battery is properly charged and check the cable connections (see Chapters 1 and 5).

When checking a circuit, make sure that all connectors are clean, with no broken or loose terminals. When unplugging a connector, do not pull on the wires. Pull only on the connector housings.

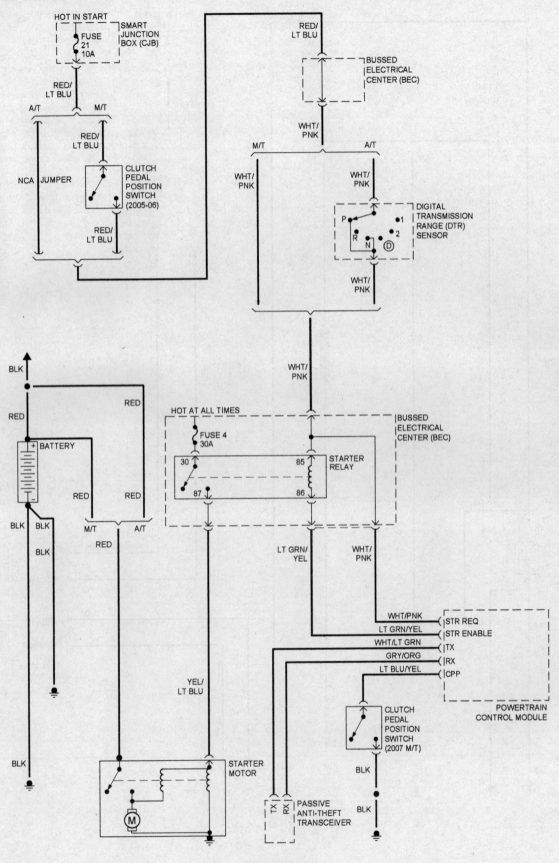

Starting system

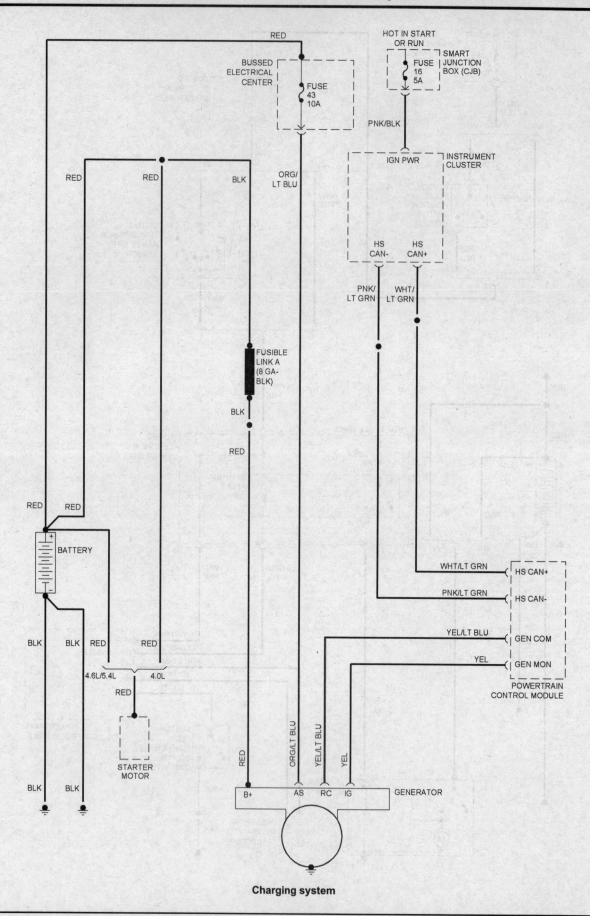

Charging system

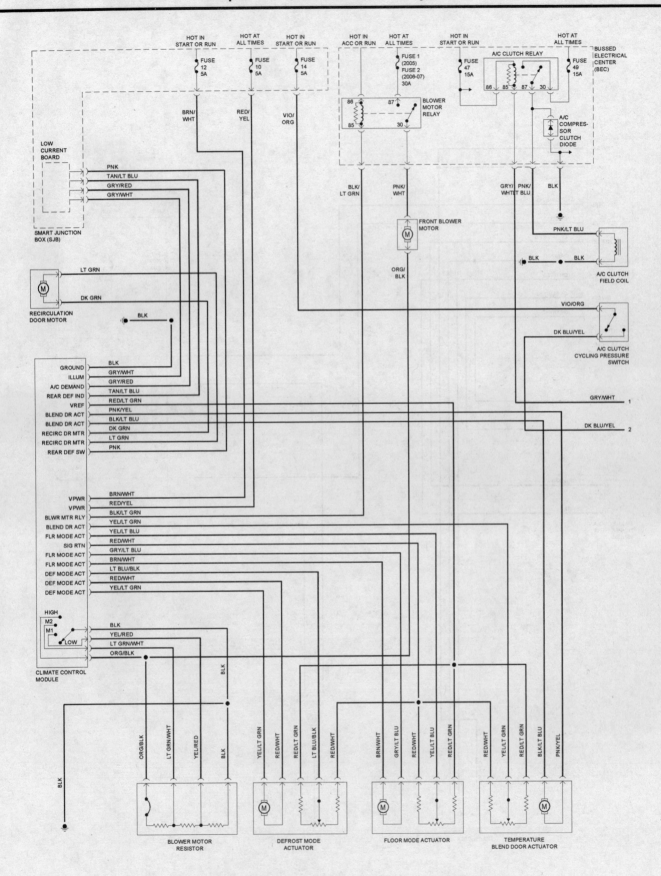

Air conditioning and engine cooling systems - V6 models (1 of 2)

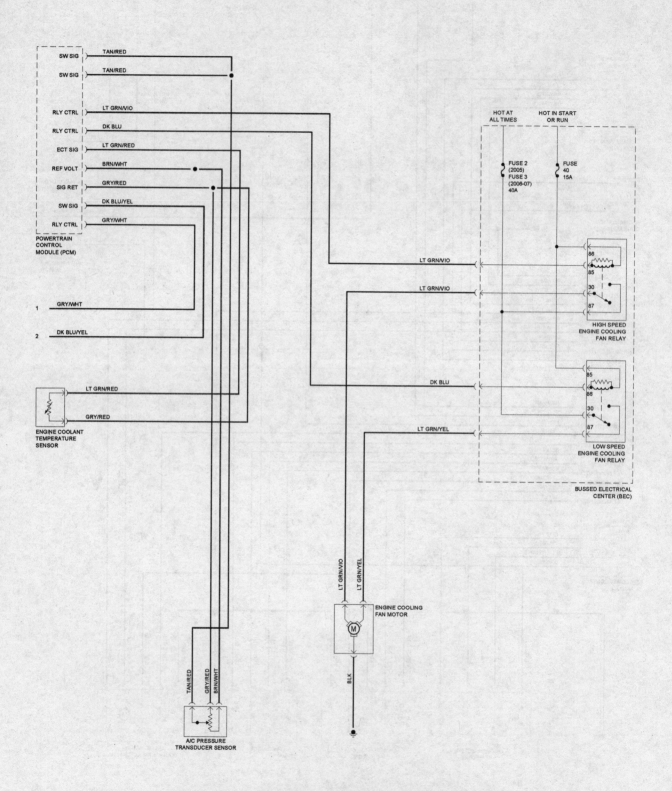

Air conditioning and engine cooling systems - V6 models (2 of 2)

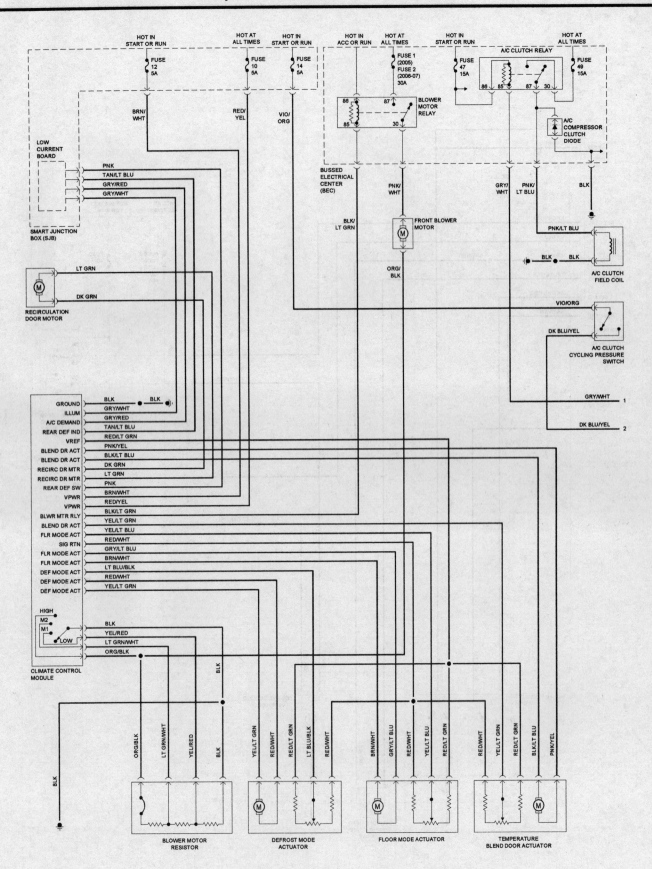

Air conditioning and engine cooling systems - V8 models (1 of 2)

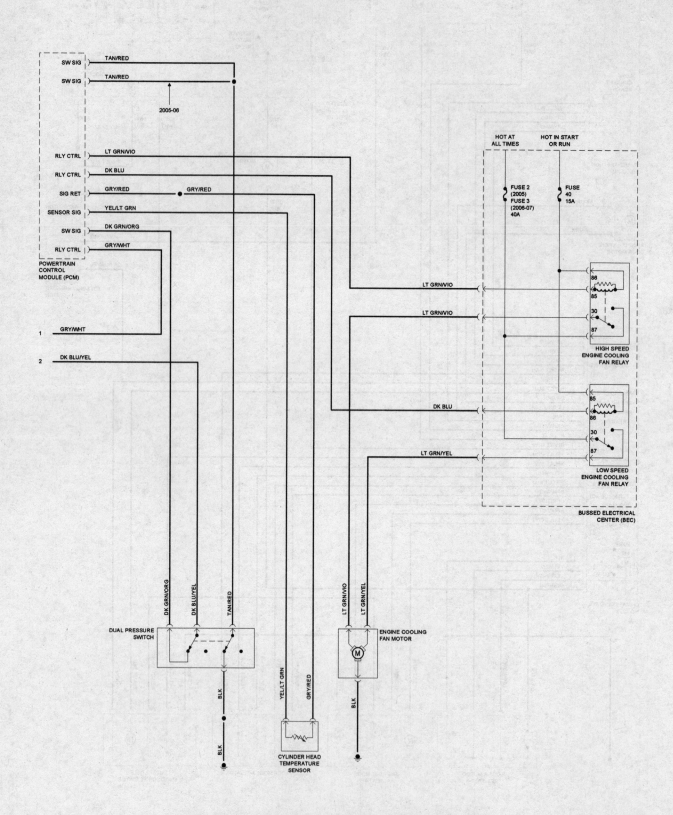

Air conditioning and engine cooling systems - V8 models (2 of 2)

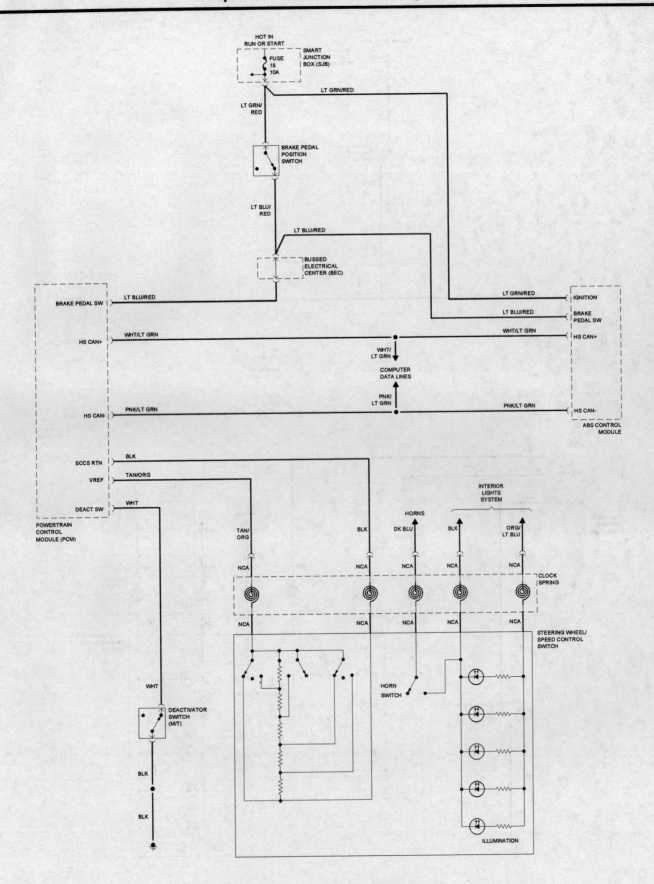

Cruise control system (2005 and 2006 models)

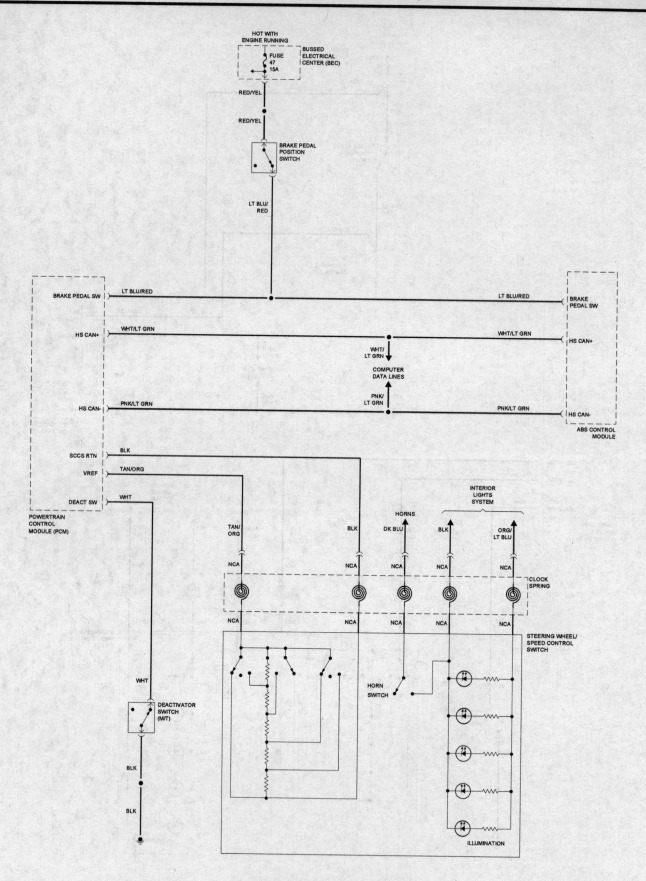

Cruise control system (2007 models)

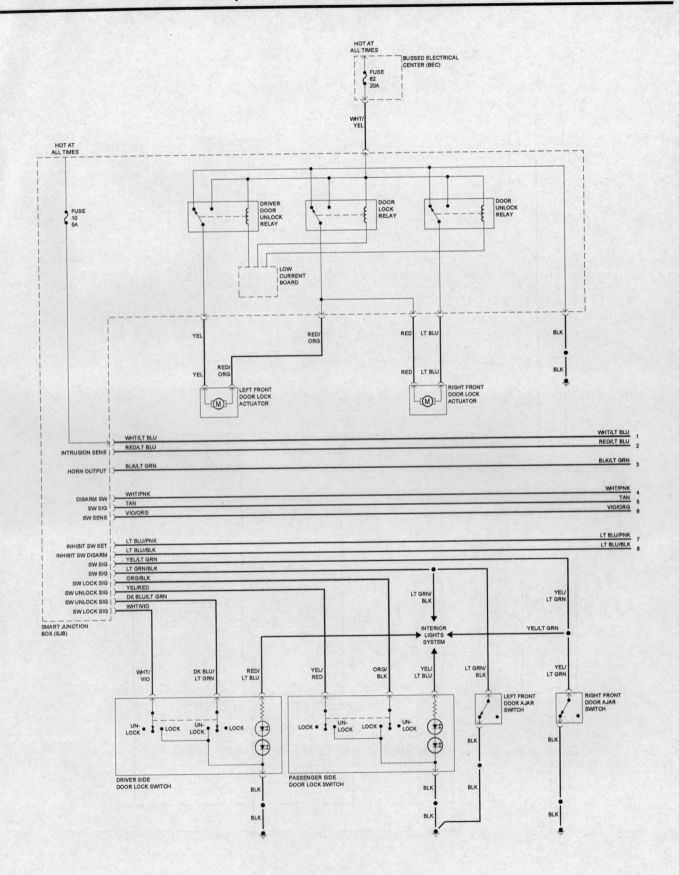

Power door lock system (1 of 2)

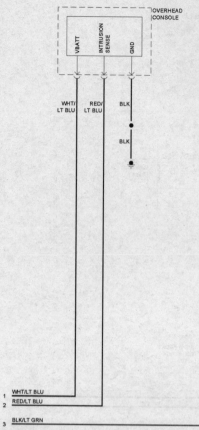

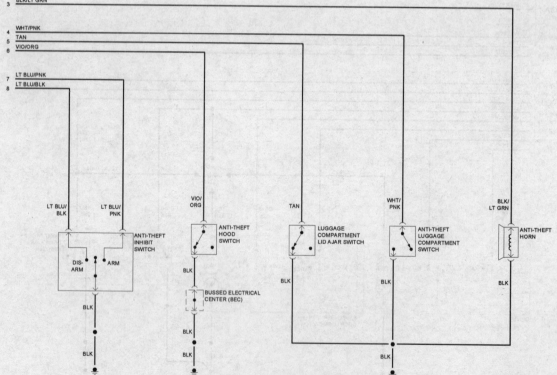

Power door lock system (2 of 2)

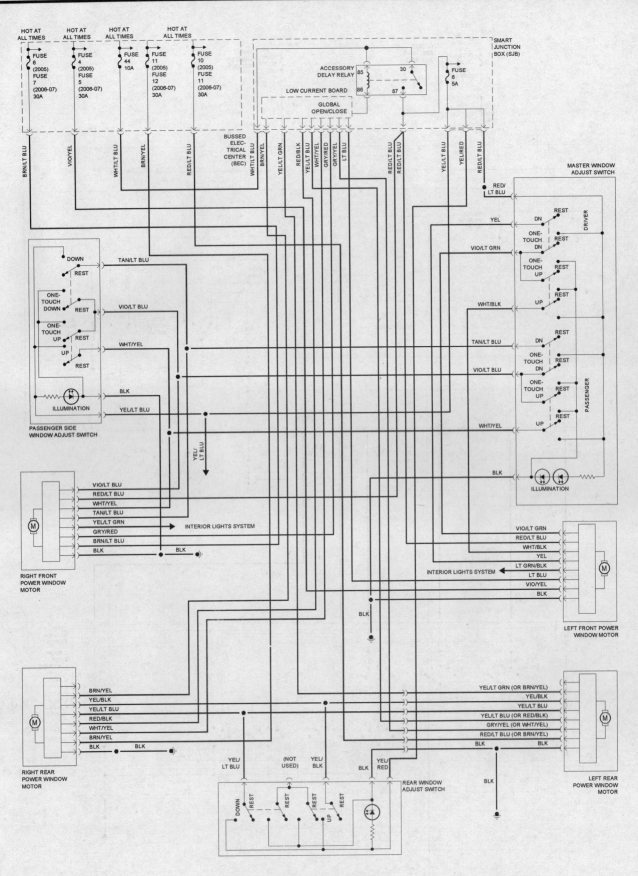

Power window system

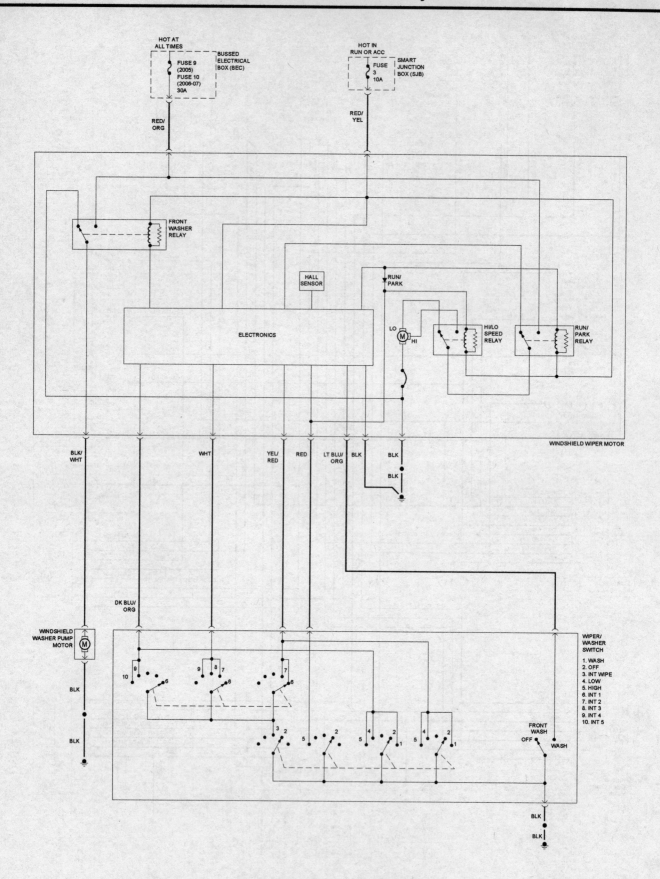

Windshield wiper/washer system

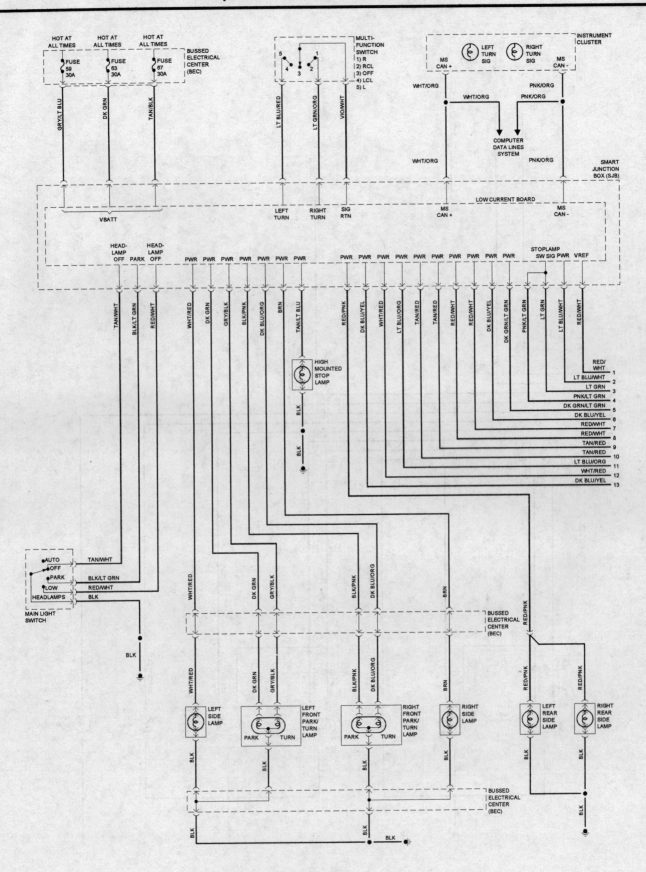

Park/brake/turn signal lighting system (1 of 2)

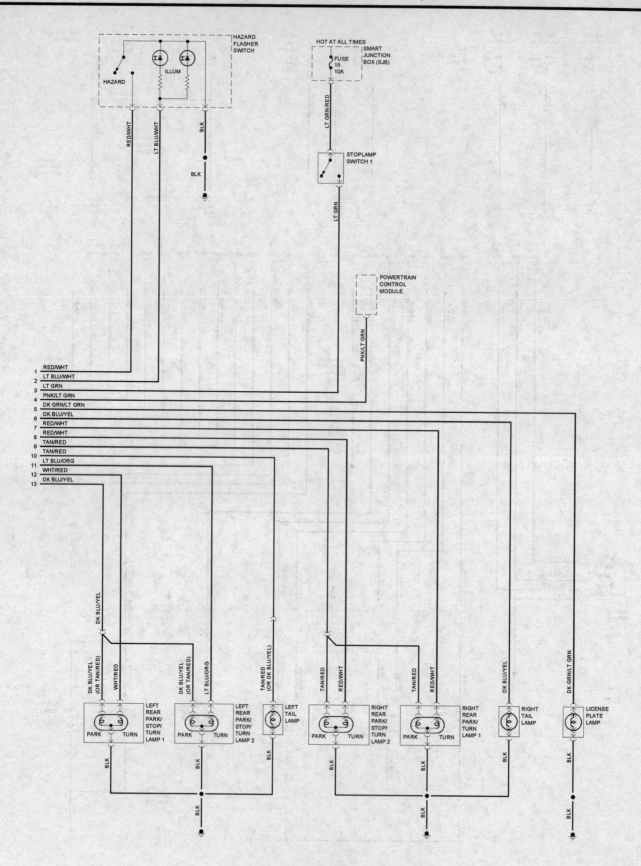

Park/brake/turn signal lighting system (2 of 2)

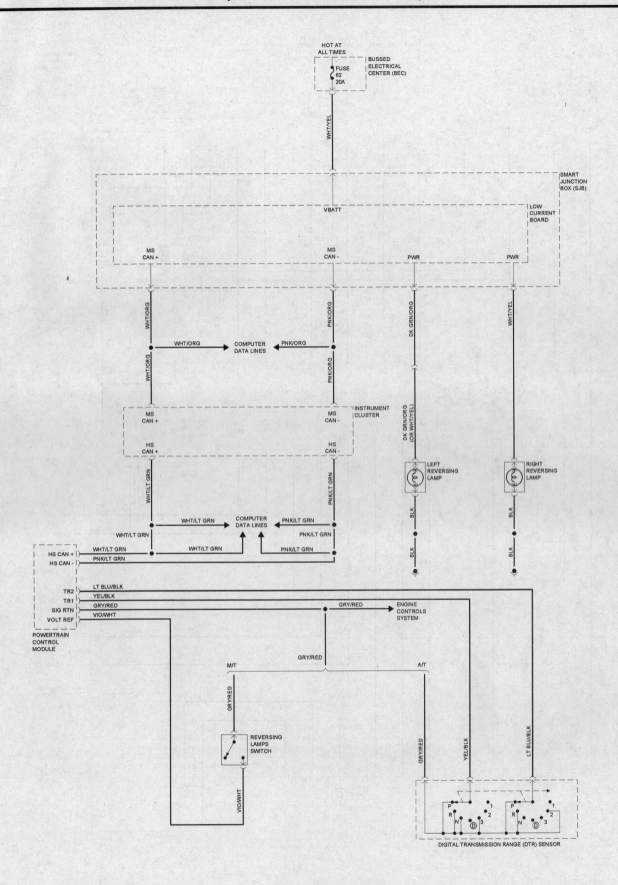

Back-up lighting system

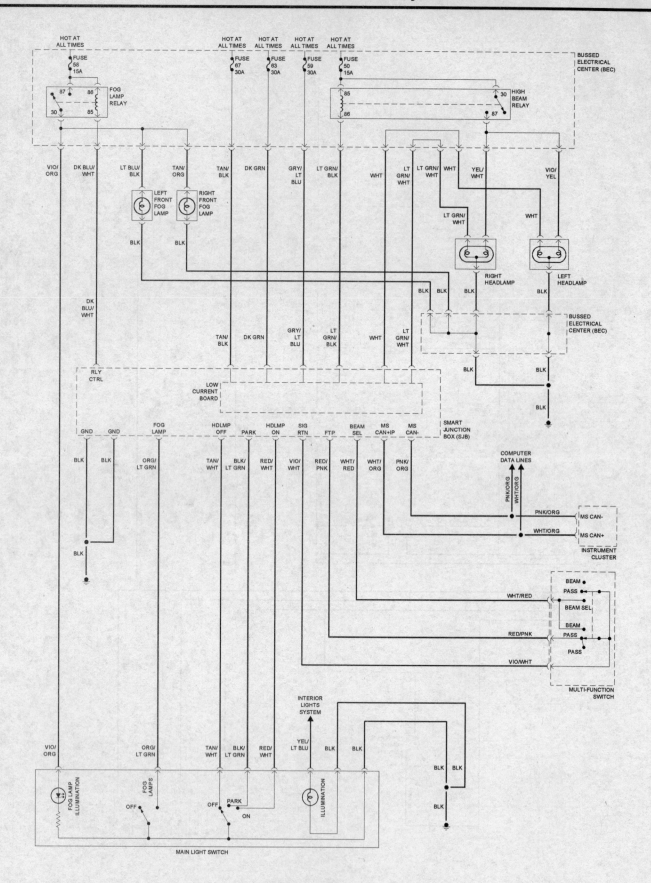

Headlight system

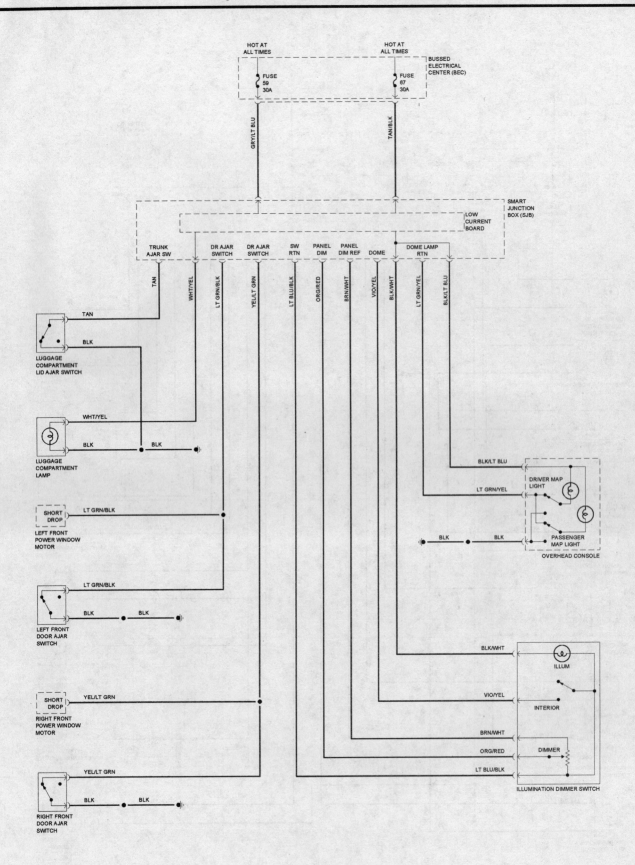

Interior lighting system (1 of 2)

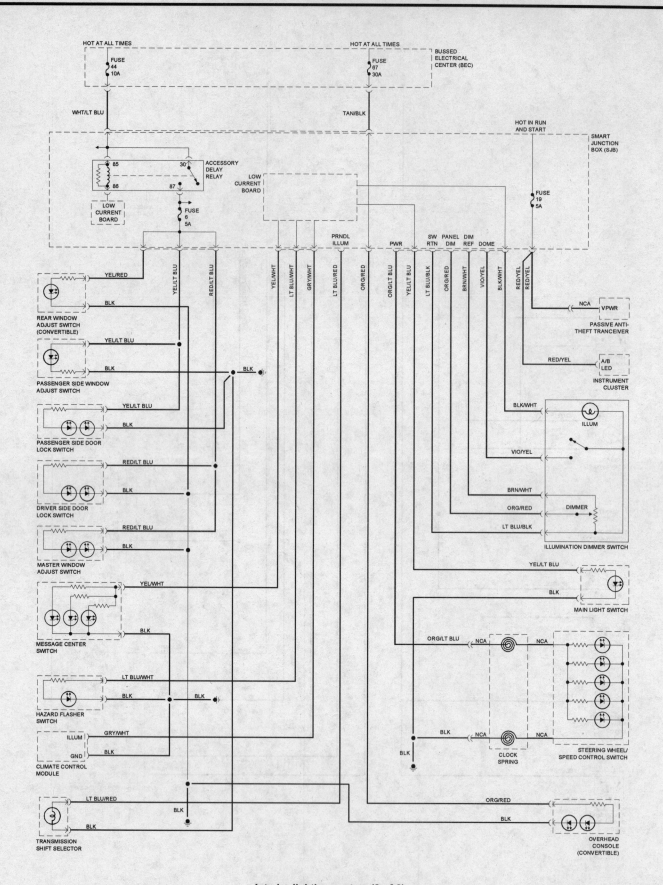

Interior lighting system (2 of 2)

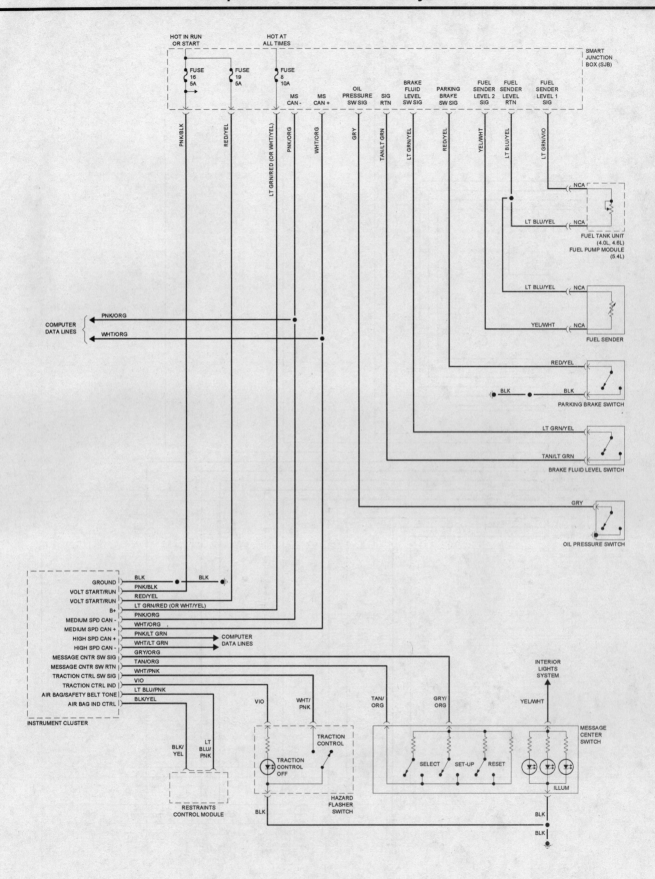

Warning light system

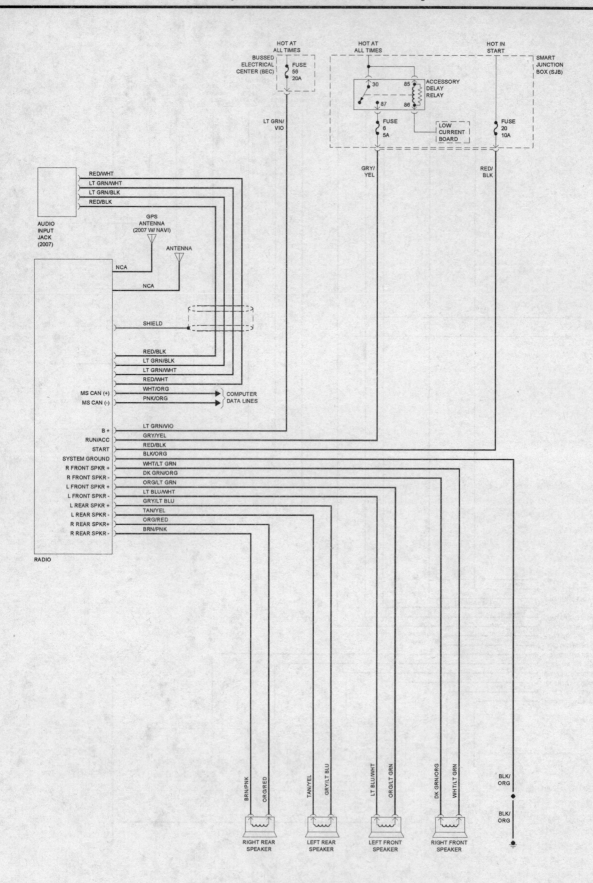

Radio and speaker system

Index

Notes